Quantity & Quiddity

Essays in U.S. Economic History

Quantity & Quiddity

Essays in U.S. Economic History

Peter Kilby, Editor

Jeremy Atack, Paul A. David, Lance E. Davis,
Robert E. Gallman, Claudia Goldin,
Teresa D. Hutchins, John A. James, William N. Parker,
Jonathan Skinner, Thomas Weiss, Gavin Wright

Wesleyan University Press
Middletown, Connecticut

Copyright © 1987 by Wesleyan University
All rights reserved
Frontispiece photograph of Stanley Lebergott by Richard Miller

LIBRARY OF CONGRESS CATALOGING-IN-PUBLICATION DATA

Quantity and quiddity.

Papers presented at a symposium in honor of Stanley Lebergott, Mar. 28th and 29th, 1985.
"Selected publications of Stanley Lebergott": p.
Bibliography: p.
Includes index.
1. United States—Economic conditions—To 1865—Congresses. 2. United States—Economic conditions—1865-1918—Congresses. 3. United States—Industries—History—Congresses. 4. Lebergott, Stanley. I. Kilby, Peter. II. Atack, Jeremy. III. Lebergott, Stanley. IV. Title: Quantity and quiddity.
HC105.Q36   1987   330.973   86-22448
ISBN 0-8195-5154-6

All inquiries and permissions requests should be addressed to the Publisher, Wesleyan University Press, 110 Mt. Vernon Street, Middletown, Connecticut 06457.

Distributed by Harper & Row Publishers, Keystone Industrial Park, Scranton, Pennsylvania 18512.

Manufactured in the United States of America

FIRST EDITION

Contents

Tables	vii
Figures	xi
Acknowledgments	xiii
Stanley Lebergott: An Appreciation Peter Kilby	xv
1. Historical Introduction William N. Parker	3
2. New England's Early Industrialization: A Sketch William N. Parker	17
3. Industrial Labor Market Adjustments in a Region of Recent Settlement: Chicago, 1848–1868 Paul A. David	47
4. Postbellum Southern Labor Markets Gavin Wright	98
5. The Gender Gap in Historical Perspective Claudia Goldin	135
6. Demographic Aspects of the Urban Population, 1800–1840 Thomas Weiss	171
7. Investment Flows and Capital Stocks: U.S. Experience in the Nineteenth Century Robert E. Gallman	214
8. Sources of Savings in the Nineteenth Century United States John A. James and Jonathan S. Skinner	255
9. Economies of Scale and Efficiency Gains in the Rise of the Factory in America, 1820–1900 Jeremy Atack	286
10. The Structure of the Capital Stock in Economic Growth and Decline: The New Bedford Whaling Fleet in the Nineteenth Century Lance E. Davis, Robert E. Gallman, and Teresa D. Hutchins	336
Selected Publications of Stanley Lebergott	399
Contributors	407
Index	409

Tables

2.1 Direct Labor Requirements in Subsistence Agriculture Over the Agricultural Year, New England, 1840–1860   20
2.2 Approximate Land and Labor Requirements for Grain and Meat, and Consumption Allowance, for Family of Six (at Uniform Allowances) on a New England farm, 1840–60   20
3.1 Percentage Average Skill Margins in Chicago, 1846–47 to 1898   88
3.2 Average Daily Money Wage Rates in Chicago for Selected Years (in dollars)   90
4.1 Labor Turnover on Four Plantations in Adam County, Mississippi, 1871–1874   102
4.2 Turnover on Leatherwood and Camp Branch Plantation, Henry County, Virginia, 1871–1877   102
4.3 Turnover on Waverly Plantation, Worth County, Georgia, 1871–1876   103
4.4 Term of Occupancy of Share Tenants, 1910   104
4.5 Percentage of Explained Variance in Southern Farm Wage Attributable to U.S. Industrial Wage, 1890–1915 and 1916–1930   108
4.6 Male Farm Laborers, 1890   113
4.7 Years on Present Farm by Tenure Class, South Atlantic Region, 1910   117
4.8 Years on Present Farm by Tenure Class, East South Central Region, 1910   117
4.9 Aggregate Statistics, Virginia Male Wage Farmers, 1907   127
4.10 Upward Occupational Mobility by Race, Birmingham, 1880–1914   128
5.1 Female Labor Force Participation Rates by Marital Status, Race, and Nativity, 1890–1980   142
5.2 Wage Ratios for Males and Females in Manufacturing Employment, 1815–1970 and Across All Occupations, 1955–1983   144
5.3 Full-time Earnings and Occupational Distributions of the Female and Male Labor Forces, 1890, 1930, and 1970, Entire United States   150
5.4 Ratios of Female to Male Earnings, 1890, 1930, and 1970   151
5.5 Partitioning Change in the Ratio of the Log of Female to Male Earnings, 1890 and 1970 Weights   152
6.1 Male and Female Participation Rates by Age, Residence, and Region, 1900   173

6.2	Urbanization by State and Region, 1800–1840	175
6.3	Comparison of Urban and State Populations, by State, Selected Dates	176
6.4	Slave Shares of City Populations, Southern Cities, 1800–1840	177
6.5	Comparison of All Southern Urban Slavery with Goldin's Selected City Data, 1800–1860	180
6.6	Male Shares of the Urban and Total Population, by Region, 1800–1840	181
6.7	Male Shares of the Free White Urban Population, by State, 1800–1840	183
6.8	Males 16 years and over as a Share of the Urban and Total Population, by Region, Free Whites, 1800–1840	186
6.9	Hypothetical Variations in the White Work Forces, by Region, in 1820	187
6.10	Urban Population by State, 1800–1840: Free White Males, All Ages and Ages 0–9	192
6.11	Urban Population by State, 1800–1840: Free White Males, Ages 10–15 and 16 and Over	193
6.12	Urban Population by State, 1800–1840: Free White Females, All Ages and Ages 0–9	194
6.13	Urban Population by State, 1800–1840: Free White Females, Ages 10–15 and 16 and Over	195
6.14	Urban Population by State, 1800–1840: Free Black Males, All Ages and Ages 0–9	196
6.15	Urban Population by State, 1800–1840: Free Black Males, Ages 10–15 and 16 and Over	197
6.16	Urban Population by State, 1800–1840: Free Black Females, All Ages and Ages 0–9	198
6.17	Urban Population by State, 1800–1840: Free Black Females, Ages 10–15 and 16 and Over	199
6.18	Urban Population by State, 1800–1840: Male Slaves, Ages 0–9 and 10 and Over	200
6.19	Urban Population by State, 1800–1840: Female Slaves, Ages 0–9 and 10 and Over	201
6.20	Estimated Share of Urban Population, 1800–1840	202
6.21	Discrepancies Between Present Estimates and Census Figures	205
6.22	Comparison of Population Mean and Sample Data, Selected States	207
7.1	Construction Price Indexes, 1789–1798 through 1889–1898 (base, 1860), and Estimates of the Ratio of Acquisition Cost to Reproduction Cost for "Other Construction" Capital Stock, 1839–1899, Decennial Intervals	221
7.2	Manufactured Producers' Durables Price Indexes, 1839–1848 through 1899–1908 (base, 1860), and Estimates of the Ratio of Acquisition Cost to Reproduction Cost, 1890, 1900, 1908	223

7.3	Ratios of Gross and Net Perpetual Inventory Capital Stock Estimates to Census-Style Capital Stock Estimates, 1840–1900 230
7.4	Ratios of Net Perpetual Inventory Capital Stock Estimates (adjusted for Civil War losses) to Census-Style Capital Stock Estimates, 1840–1900 235
7.5	Ratios of Net Perpetual Inventory Capital Stock Estimates (adjusted for Civil War losses) to Census-Style Capital Stock Estimates, 1840–1900, Land Improvements 238
7.6	Rates of Growth, Structure of the Capital Stock, and Capital Output Ratios, Two Versions, 1860 Prices, 1840–1900 241
7.7	Levels and Decennial Rates of Change of Stocks of "Other" Improvements and Producers' Durables, 1860 Prices, Perpetual Inventory Estimates, Various Versions, 1840–1900 243
7.8	Annual Perpetual Inventory Cumulations, Midyears of Years Specified (Millions of 1860 dollars) 246
8.1	Annual Growth Rates of Total Real Wealth and Income, by Decade, 1850–1890 259
8.2	U.S. Wealth Estimates Compiled by the Census, in Comparison with Alternative Measures, in Current Prices, 1850–1890 270
8.3	Nonlinear and Linear Wealth Regressions, 1850 and 1870 to 1890 273
8.4	Component Factors of the Total Change in Real Wealth per Head, 1850–1890 278
9.1	The Changing Use of Steam and Water Power by American Industry, 1850–1870 294
9.2	The Percentage of Industry Value-Added Originating from Different Methods of Production in Selected Industries, 1850–1870 296
9.3	Labor Productivity and the Switch from Hand to Machine Production, 1835–1897 300
9.4	Firms Producing Less than $1,000 Value-Added, by Industry, in 1850 and 1870, as a Percentage of All Firms in the Industry from the Samples and the Implied Number of Such Firms in the Population 321
9.5	Sample Sizes by Industry and Year 328
9.6	Statistically Significant and Economically Sensible Translog Production-Function Estimates, by Year and Industry, 1820–1870 332
9.7	Statistically Significant Variable Scale-Elasticity Production-Function Estimates Implying Well-Behaved Average Costs, 1820–1870 334
10.1	U.S. and New Bedford Whaling Fleets: Yearly Average Catch, 1816–1905 340
10.2	Annual Average Tonnage: U.S. and New Bedford Whaling Fleets, 1816–1905 343
10.3	Whaling Fleet Size 344
10.4	Vessel-Class Composition, New Bedford Fleet, 1816–1905 348

10.5	Technical Character of the New Bedford Whaling Fleet, 1816–1905	352
10.6	Tonnage Relatives by Hunting Ground, Ships and Barks, 1816–1905	354
10.7	Technical Characteristics Vessels in the New Bedford Whaling Fleet, 1816–1905 (Ships & Barks Only)	356
10.8	Crew and Crew per Ton, New Bedford Whalers, 1816–1905	361
10.9	Voyages by Vessel Class by Grounds, 3,369 Returning Voyages, 1816–1905	364
10.10	Specialization in Catch, New Bedford Whalers, All Voyages, By Ground, 1816–1905	365
10.11	Vessel Class by Grounds, New Bedford Whalers, 3,369 Returning Voyages, 1816–1905	368
10.12	Average Tonnage, by Class and Ground, New Bedford Whalers, 1816–1905	372
10.13	First and Last Voyages by Class and Ground, All Years and Vessel Weights, Time at Sea (interval), and Vessel Characteristics (tonnage, age, years in fleet), New Bedford Whalers, 1816–1905	373
10.14	Voyage Length, by Vessel Class and Ground, New Bedford Whalers, 1816–1905	375
10.15	Average Vessel Age, at Departure, by Year and Ground, New Bedford Whalers, 1816–1905	376
10.16	Years in Fleet, by Decade and Ground, New Bedford Whalers, 1816–1905	378
10.17	Men per Ton, by Decade and Ground, New Bedford Whalers, 1816–1905	379
10.18	Annual Loss Rate, by Ground and Class, New Bedford Whalers, 1816–1905	382
10.19	Average Yearly Earnings per Ton, New Bedford Whalers, by Ground and Class, 1816–1905	387
10.20	Average Catch per Vessel Ton per Year of Hunting, New Bedford Whalers, 1816–1905	390
10.21	Physical Index of Catch per Ton per Year of Hunt, by Product, Ground, and Decade, New Bedford Whalers, 1816–1905	391
10.22	Index of Profit Rates, by Rig and Ground, New Bedford Whalers, 1816–1905	392

Figures

2.1 Potential uses of farm-family labor in three colonial regions in agricultural crops   18
2.2 Direct labor requirements for each of two workers in subsistence agriculture   19
2.3 Interrelation of environment, culture, and economy in New England's industrialization, 1630–1840   22
4.1 Farm-Labor Wage Rates Per Day, Without Board, Deflated by WPI, 1866–1942, Selected States   105
4.2 Allocation of Plantation Acreage between Wage System and Tenant Plots   113
4.3 Aggregate Wage Distributions, Virginia, 1907   127
4.4 Gross Farm Income, South Atlantic States, 1900   129
4.5 Gross Farm Income, South Central States, 1900   129
5.1 The Gender Gap in Historical Perspective: The Manufacturing Sector and the Entire Labor Force, United States, 1815–1983   145
8.1 Savings and Investment: Perfectly Inelastic Savings Schedule   261
8.2 Savings and Investment: Positively Sloped Savings Schedule   261
9.1 Relative Efficiency by Industry, 1820   303
9.2 Relative Efficiency by Industry, 1850   304
9.3 Relative Efficiency by Industry, 1860   305
9.4 Relative Efficiency by Industry, 1870   306
9.5 Translog Production Functions: Flour   313
9.6 Flour-Milling Unit Costs, 1820–1870   316
9.7 The Spatial Distribution of Competition   324

Acknowledgments

Like most worthwhile ventures, this volume has been carried forward by a multitude of hands, those that are seen in the table of contents and those that labored invisibly and without reward. William Barber and Michael Lovell assisted in the planning of the enterprise. The papers were first presented at a two-day symposium, where they benefited from the criticism of discussants, especially those of Moses Abramovitz, Stanley Engerman and members of the Wesleyan Economics Department. Professor Engerman's public lecture on nineteenth century slavery in North and South America further added to the symposium's success. This joyous celebration of March 28th and 29th, 1985 was superbly orchestrated by Nancy Campbell, Joan Halberg, and Anne Crescimanno. Finally, Murray Foss provided valuable information for the editor's essay on Stanley Lebergott's early career.

In bringing the symposium papers to print we are much indebted to Karen Barrett of Wesleyan University Press and the fearless copy editor Jan Fitter.

Last but not least is the generous financial support provided by President Colin G. Campbell of Wesleyan University that made possible both the symposium and the publication of this volume. A small but welcome contribution was made by W. W. Norton & Co.

Stanley Lebergott: An Appreciation

PETER KILBY

For the past twenty-three years Stanley Lebergott has contributed his very special talents to the corporate life of the Wesleyan Economics Department. As measured in the quality of courses offered, new faculty attracted, and collective decisions made and avoided, the exercise of these talents has had a most happy effect. His contributions to quantitative economic history span an even longer period and have, in both their frequency and virtuosity, earned him a place with the top practitioners of that art. For one so appreciated and accomplished, a decision to convene a symposium in his honor should be no surprise. The fruits of that symposium, held in March 1985, are presented in this volume.

As is well known to his readers, Stanley Lebergott is a vociferous consumer of—among other forms of literature—biography and autobiography. Yet in his own case he effectively discourages the would-be Lytton Strachey. Thus the reader must be satisfied with the barest chronology of our subject's earthly progress. He was born in the closing months of World War I to Isaac and Lillian Lebergott of Detroit, Michigan. Isaac, a dentist, had immigrated to the United States from the Ukraine in his early twenties. Stanley attended Detroit public schools and at age sixteen entered Wayne State University. An improvement in finances permitted a transfer in his sophomore year to the University of Michigan, where he pursued a Bachelor's degree in social studies. Following graduation in 1938, he stayed on at Michigan to do advanced work in economics.

With no job openings in the academic world, Lebergott terminated his formal studies with a Master's degree and considered himself lucky to find a position as Junior Economist with the Bureau of Labor Statistics in Washington in 1940. It was in this same year that Stanley married Ruth Wellington, who had also been at Michigan. Five years in Washington was followed by a two-year stint with the International

Labour Office. It was while in Montreal that the first of two children was born.

In 1948 the Lebergotts returned to Washington and for the ensuing thirteen years Stanley held the positions of Economist and then Assistant Division Chief of the Office of Statistical Standards at the Bureau of the Budget. At that time the OSS played an important role not only in overseeing the budgets of Federal statistical agencies but also in coordinating the activities of different agencies so that outputs of some could usefully serve as inputs to others. For instance, Lebergott was instrumental in making the statistical activities of agencies like the Census Bureau and Internal Revenue Service more useful and timely for the construction of the national income and product accounts. This is a process that has undergone a vast growth and continues to this day. Lebergott took a special interest in sectors where basic source data were commonly acknowledged to be weak, such as income of nonfarm proprietors.

Our subject also served as the Budget Bureau representative on the National Accounts Review Committee, whose report in 1957 gave considerable impetus to the subsequent publication of quarterly estimates of real GNP and the integration of the income and product accounts with input-output, flow of funds and balance of payments accounts. In recognition of his many contributions to the improvement of Federal statistics Lebergott was given a Rockefeller Public Service award, which enabled him to do graduate studies at Harvard for the year 1957–58. Before coming to Wesleyan he held a Visiting Professorship at Stanford in 1961. And on top of all this, as evidenced in the bibliography in the back of this volume, he managed to publish one monograph and twenty-one articles during this period.

The work for which Stanley Lebergott is best known was published in 1964. An immense scholarly undertaking that drew on more than two decades of research, *Manpower in Economic Growth: The American Record Since 1800* provided both an analysis of nineteenth century American growth and time-series statistical estimates of various components of labor employment over the preceding 160 years. The later consists of two segments. In the first segment the investigator takes up various official series—consumer prices, labor force, employment, unemployment, employees by industry, full-time equivalent earnings—that go back no further than 1940 and extends them in a consistent and comprehensive manner to 1900. In the second segment he turns to the nineteenth century and, beginning with 1800, estimates at approximately decennial intervals total labor force, employment by industry, unemployment, the cost of living, and wage rates by industry, occupation and region.

Published at the flood tide of the "new economic history" and filling a major gap in nineteenth century economic statistics, it is perhaps not surprising that scholarly attention fixed upon the 300 pages of statistical derivation. Perhaps the idiosyncratic format also played a part. And yet, for many, a richer fare is to be found in the book's first half. There the same scholarly breadth and ingenuity of the statistical estimates is combined with an original analysis of the interacting institutional and quantitative factors that shaped the American economic experience. Given that Lebergott's interpretive analysis has had less exposure than his statistics, a brief outline of his thesis may not be inappropriate.

In the early chapters the author considers and rejects three traditional explanations of the wellsprings of U.S. economic growth: the resource endowment, the size of the market, and entrepreneurial ability. Rather he characterizes the distinctive ingredients of American economic success as immigrants and immigrant ideas within the matrix of space, hope, and ignorance. Hope of economic and social betterment peopled the continent with wave upon wave of immigrants to whose energy and initiative no bar was placed in a unique open society. Although not as skilled as European counterparts, the labor force showed a reckless adaptiveness that was one component of "the permanence of innovation": "Here a wine merchant who had been a railway director, a watch merchant who now sold shoes, there a farmer selling butter but formerly the proprietor of a factory making calicoes." As Leo Lesquereux wrote back to France, "The Yankees believe themselves suited to anything, there is no status to which they do not feel predestined, not a situation which they do not feel they understand to its depths. . . . Everything here speaks in dollars, everything is measured in dollars, everything is done for dollars."

The lack of specialization fit the state of knowledge, or rather, the state of ignorance—the second feature of Lebergott's matrix. American attempts to reproduce English and continental patterns of production failed. Given the vast differences in resource pattern and relative costs, new approaches had to be tried. And tried they were.

> For a given problem, the Sicilian procedure tried by one, the North Wales technique applied by another, the Baden-Baden system proposed by a third, all might work very badly. But one procedure, one dubious compromise did in fact work. And out of the

multitude of experiments conducted by astute men energetically seeking success, it was inevitable that the percentages favored technical advance.

The third distinctive element was space. The unlimited availability of fertile farmland kept wages and labor turnover high. The response to both, abetted by falling interest rates, was to move toward labor-saving mechanization, contributing to the phenomenal rise in labor productivity over the century.

This process took place within the context of markets. Lebergott analyzes the results of this process in terms of wage differentials and state migration rates, occupational mobility, the relation of immigration to unemployment, and the trend in real wages.

Professor Lebergott published his second book in 1976. Like the earlier volume, it consisted of two parts, a collection of thirteen short essays and five original data sets. The essays deal with a variety of questions bearing on historical change in the U.S. distributions of income and wealth, on the meaning and measurement of poverty, on upward and downward mobility, on the philosophical-ethical premises that underlie key public policies.

In the third decade of Lebergott's academic career, a third volume appeared. Published in 1984, *The Americans: An Economic Record* is a textbook, but of a rather unusual kind. Its 500-plus pages are divided into forty chapters, each an original essay based on new research. The standard institutional material is at a minimum, quantified relationships are abundant, and the focus is typically on a single "key" issue. Like the reviewer in the *Journal of Economic History,* students, not being prepared by an initial presentation of the ruling interpretations of the subject at hand, are unlikely to realize that the empirical estimates being presented to them have just altered that conventional wisdom. Here we have an introductory text in which, in many instances, the first word is also the last word.

The third decade also gives promise of a fourth volume. Its subject will be changing patterns of household consumption since 1900. Conforming to its predecessors, it will consist of original data sets and the methods by which they were derived, on the one hand, and a series of essays exploring their various economic and philosophical implications, on the other.

How might one characterize the corpus of Stanley Lebergott's scholarly work, his books, reports and sixty-odd articles? The answer is found in the title of this volume. First, Lebergott is an empiricist of the most thoroughgoing sort. Economic theory always informs the conceptualization of the problem and isolates the key relationship; but the number of links in the chain of reasoning is held to a minimum. It is quantification, systematic evidence, that alone is capable of selecting the "correct" explanation from the many equally valid competing theories. As for the process of drawing the evidence together, Lebergott has a breadth of sources at his command—census materials, state data sources, House and Senate reports, contemporary financial and commercial publications, the direct reports of travelers, novelists, and the participants themselves—that qualify him as an economic *historian* to a degree that few of his colleagues can match.

The second hallmark of Stanley Lebergott's research is its quiddity. Both the theoretical constructs and the econometric techniques that he applies, while wholly adequate to the task, are relatively modest. And yet, whether by unusual clarity of vision or intensity of concentration, Lebergott is able to pierce through both the obscuring plaque of previous analyses and the surface complexity of the problem itself to discern "that which makes a thing what it is." Having isolated the key ingredients, he proceeds with shattering simplicity to conceive a single measure that captures the direction and force of their effect. The economic quiddity of U.S. imperialism is revealed by a comparison of two profit-rate calculations, that on foreign investment and that on the U.S. capital stock; the fate of the Indian is disclosed by a calculation of the land requirements to sustain one hunter versus one agriculturalist; the efficiency of slavery is made clear by an estimate of the maintenance cost of a slave.

Quiddity in its secondary sense, a subtle nicety of argument or an elegant refinement, is also evident in every Lebergott presentation. While his own prose is fresh and terse, sometimes elliptic, it is the wealth of apt but unfamiliar quotations that fill the headnotes, footnotes, and text that has become a Lebergott trademark. Whether it be from Seneca, Learned Hand, a medieval cleric, Oliver Wendell Holmes, W. B. Yeats, or Ibn Khaldun, the quoted statement connects the problem at hand to a broader category of human endeavor, an unexpected perspective.

Given the virtues that have been claimed here and the volume and breadth of the contributions that are recorded in the bibliography at the back of this volume, why have the writings of Stanley Lebergott not attracted more attention? In part, his virtues are responsible: his writing is free of all self-advertisement, no straw men are erected, no methodological razzle-dazzle, no dramatic conclusion. But the vices of idiosyncrasy may have also played a role. Those readers whose learning is aided by an introduction—placing the subject in context, sketching in past and current opinion—will find cold comfort in the typical Lebergott paper. Nor will a Lebergott work find its way into the syllabus of those who demand that extra instructional dividend of a step-by-step development of the pertinent economic theory. Extreme parsimony in exposition, the avoidance of jargon (familiar terminology), the absence of recapitulation or closing discussion of the wider implications of what has been presented, all combine to require a high level of concentration. It is only when the reader, with self-supplied motivation, has successfully completed the course that he or she experiences the excitement of genuine discovery. The following section presents a sampling of Lebergott's work, with the aim of persuading the reader that the cost in concentration, the effort to pry open the oyster shell, is a small price to pay for the treasure found therein.

The first selection is a ten-page essay from the 1976 collection that would seem to be a fitting candidate for quiddity in its secondary meaning. Here Lebergott identifies and explores a hitherto unnoticed flaw in our basic measure of material welfare. In "Per Capita Income and the Angel of the Lord," he observes that it is a consequence of long division that the death of every older person raises society's measured welfare just as the birth of every child lowers it. A brief but learned review of ancient and modern views on the utility of increased numbers, ranging from the Jews to the Hindus, from Sir William Petty to Rachel de Queiroz, suggests that Bishop Jeremy Taylor's dictum of two hundred years ago, albeit perhaps in a somewhat more secular dress, still holds for most societies of the world: "Every child we feed is a new revenue, a new title of God's care and providence; so that many children are great wealth." If it is in the having of children that we come closest to the prospect of infinite utility, certainly per capita income is a perverse measure of our welfare.

To get at the core of the conundrum, our author focuses on a family of two, since welfare is not experienced by nations but by individuals. In stage I, family income is spent on final consumption goods. In stage II, the couple decides to adopt a child. The national income accountant notes a shift in family expenditures from airline tickets and restaurant meals to diapers and pediatric care, but the level of aggregate output and the size of the population that consumes it remain unchanged. In stage III, the couple is able to beget a child of its own; the national denominator now increases by one. We know the family's welfare has increased yet further, but their contribution to GNP per capita is reduced by a third.

By examining the problem at its most basic level and by postulating a fixed level of consumption expenditure, Lebergott is able to uncover the source of the difficulty. Per capita income comparisons require constancy of tastes, but every demographic event in human existence—birth, marriage, death—reflects a decisive change in tastes. Given that such shifts in individual indifference curves are part of the normal human cycle, the use of per capita GNP as an index of change in levels of welfare is, quite simply, invalid.

And there is much more in these ten pages—the relevance of expected satisfaction over realized satisfaction, statistics on wanted versus unwanted pregnancies, expenditures on foster children versus "own" children, choice quotes from Theognis of the fifth century, Richard Cantillon of the eighteenth, and others. What is not done, in the characteristic Lebergottian mode, is to hint to the reader what the far-reaching implications of this insight are. To wit, if Lebergott's conclusion is considered adequate by itself and if the aim of economic development is a higher level of welfare, a minor revolution is effected in how we assess economic performance. Growth in GNP and population are in; growth in per capita income is out. What at the outset seemed a quiddity of the second kind is now converted into one of the first kind.

A more successful candidate for "mere quiddity" might be found in *The Americans,* in the second chapter, which deals with the displacement of the Indians. In virtually all countries of the world the initial hunting and fishing societies have been pushed aside by agriculturalists. So it was in the United States between 1700 and 1840. But what was the motive force: religious imperative, greater savagery of the European, or differences in technology? No, Lebergott argues, it was diet. A way of life requiring deer and buffalo steaks was at an overwhelming disadvantage in competition with a way of life based on pig and cattle meat. An ingenious set of estimates, derived from over a hundred eighteenth and nineteenth century sources, shows that the land required to support twelve major Indian tribes ranged from 1,600 to 4,000 acres per capita as against 2 acres for the nonhunter. Given this thousandfold efficiency differential in land use, the European immigrant could buy up land at prices well above its value in the Indian way of life and still retain a wide margin for prospective capital gain. As is exhaustively documented in his 1985 "Land Sales" article, most of the value in all virgin farms (i.e., those that pushed the Indians off the land) was in that three-quarters of the acreage that was not cultivated, but rather held as a speculation. Lebergott's insertion of this mediating economic mechanism is an elegant refinement of the conventional explanation of a numerically unequal conflict between two incompatible cultures.

Much of our subject's non-historical work has focused on poverty. A 1970 note in the *Journal of Political Economy* set forth a proposal by which twice the number of public housing units as then were being constructed (50,000 units at $1.3 billion) could be obtained and at no cost. Because of the existing incentive structure for tenants, public housing depreciates at three times the rate of comparable private rentals. A reduction in the depreciation rate for low rent housing from 5 percent to 3 percent would add 100,000 units to the stock of available housing. HUD-aided housing projects allocate approximately 50 percent of rental proceeds to maintenance and repair. By rewarding those tenants who generate less than average repair costs with a bonus equal to the savings, the public landlord benefits from less destruction and higher asset values. And more housing is available for the poor at no additional cost.

The final paragraph of the paper reveals some of our author's more general views as to how social progress is best pursued and is best not pursued:

> The proposal pays for outputs, not inputs. It pays if those expenditures that now benefit plumbers and electricians, American Standard and General Electric, are reduced. It does not pay for inputs of exhortation, teaching, advertising. It offers no opportunities for the idealistic social worker to reform the poor, for great teachers of social analysis to lead them by the hand, or for militant doers who want them to march in step until the walls of Jericho are blown down, one way or the other. Neither Groupthink nor Groupaction is required. No savings, no rewards.

In a somewhat similar vein, Lebergott's work on poverty and the distribution of income (inter-generational mobility out of both the bottom and top decile of the income distribution is high) leads him to the conclusion that much intervention in the name of the poor is aimed more at reducing the rich, and satisfying the psychic needs of those that make and implement public policy, than at providing effective help for a lasting escape from low income.

We end our sampling of the Lebergott cornucopia with a 1983 paper that offers an answer, a shocking answer, to a perennial question of nineteenth century American history: Why did the South lose the war? For Lebergott it was not, as in the Indian case and as most writers contend, owing to some form of unequal conflict. Rather the contest was lost because in the Confederacy motives of personal gain among cotton planters were given priority over winning the war.

The argument is advanced in the typical Lebergott three-prong mode: assembling the known facts in an original way; citing previously unnoted qualitative evidence from contemporary letters, diaries and speeches; then clinching the argument with a single ingeniously constructed quantitative estimate of a phenomenon that no one had heretofore thought to measure.

Although a ban on cotton exports had been proposed for decades as the policy that would bring all Europe into opposition against the North, the planters in the event resisted the ban, and despite appeals to expand food production for the increasingly underfed Southern armies, bent all efforts to maintain cotton production. Just as the Confederate Congress urged but did not require the planting of food, so too did it refuse to tax lucrative cotton exports to finance war expenditures; the resulting catastrophic deficit financing and impressment were major forces in eroding the Confederacy's will to fight.

Maintaining cotton production meant that the slave labor force who cultivated it could not be diverted to other uses. So all efforts to mobilize any portion of that half of the southern labor force to repair roads and railways, construct fortifications, dig niter, and above all, grow critically needed food was resisted to the last. (The key calculation reveals that the manpower absorbed in cotton production until August 1864—2.3 million man-years—far exceeded the manpower of all Confederate armies combined.) As a Confederate Congressman observed in 1864,

> It is the strange conduct of our people during this war that they give up sons, husbands, brothers and friends to the army; but let one of their negroes be taken and what a howl you will hear. The love of money had been the greatest difficulty on our way to independence—it is now our chiefest obstacle.

And so it was that the planters' pursuit of short-run private profit insured the defeat of their armies.

The monographs that follow celebrate the work and person of Stanley Lebergott in a rich time-hallowed tradition. Their subject is his subject, that golden age of American economic history which witnessed the traverse from a rustic agrarian society of the early republic to the boisterous industrial economy at the turn of this century. Their method is his method. And their work is manifestly animated with that spirit which he has always sought: in the words of Oliver Wendell Holmes Jr., "To see as far as one may, and to feel the great forces that are behind every detail . . . to hammer out as compact a piece of work as one can, to try to make it first rate, and to leave it unadvertized."

Quantity & Quiddity

Essays in U.S. Economic History

I
Historical Introduction

WILLIAM N. PARKER

For thirty years there has existed within the body of American scholarship a group of economic historians who have called themselves new. But newness does not last for thirty years, so some of them have become old, and some middle-aged. They might have died out and carried with them all their values, their data, their techniques, their harsh criticisms of the "old," had they not generated a *Nachwuchs,* an aftermath, a group of younger economic historians for whom the movement had an appeal. This volume and the conference from which it originated represent one point partway on this transition from the old "new" group to their new "new" successors. If the presiding genius of the conference, Professor Stanley Lebergott, is included in the group, the score is nearly even: five of the original group (Parker, Gallman, David, Davis, Lebergott) and seven (Wright, Goldin, Hutchins, Weiss, James, Skinner, Atack) who were either their students or students of their contemporaries. Before going further into their essays collected in this book, the reader may wish to consider briefly the thirty years of intellectual history from which it all derives, the better to assess its significance and accomplishment and to project the directions of scholarship toward which it—along with much other contemporary quantitative history—seems to point.[1] Just what, in the last analysis, has been added to our understanding of the American economy's past by all of this work and in particular by the group of essays collected here?

For such an exercise, one must of course go to the primary contem-

---

[1] Of the many other examples of a new appreciation of a historical perspective in economic work, one might cite work in the fields of trade theory and finance (Eichengreen), macroeconomics (Bernstein and Weir), labor (Piore), and finance (Sylla, Rockoff). The major journals—*Journal of Economic History, Explorations in Economic History, Economic History Review*—contain numerous other examples. A discussion at the 1984 meetings of the American Economic Association appeared in *The American Economic Review*, Vol. 75, No. 2, (May, 1985), pp. 320–37, reproduced with significant extensions in William N. Parker ed. *Economic History and the Modern Economist* (Oxford: Basil Blackwell, 1986) defending the validity of economic history in the graduate education of the economist.

poraneous sources, which, however, are not quantitative, but literary, in nature.[2] The movement all began, I think it would be widely agreed, in 1955 when Solomon Fabricant of the National Bureau of Economic Research approached the Economic History Association to propose a joint conference for 1957. His attitude—perhaps understandably—was a bit like that of an Egyptologist visiting a peasant village along the Nile. "You people," he seemed to say to us, "have a great deal of data, bits of old prices, scraps of wages and output series. You really don't know what to do with them; you use them to hold open doors, give them to your children to play with. Pile them all up in a heap and bring them to us, and we will show you how they fit into the great pattern of statistical history." What this pattern might be was not clear to the historians, nor indeed to most young economists. The historians were continuing to pursue business histories, to update the texts of Shannon, Kirkland, Faulkner in rather faded versions, or to argue over essentially political issues—the Bank War, Populism, the morality of Big Business, the New Deal. Only in the work of some students of Paul Gates—notably Allan Bogue and Harry Scheiber—did the effort to bring statistics and some understanding of social and economic theory to bear on the analysis and revision of Turner bear fruit. It was indeed not by economic historians trained in history eager to bring their presents of data to the altar, but, rather, by ambitious young economists looking for a research topic that Fabricant's call was answered. Considering Fabricant's vision and the efforts of the planners and editors of the two Income and Wealth conferences into which much of the work was poured, it is at first surprising that its pieces were really much more interesting than the whole.

The concern of the economists who entered the field was a rather surprisingly philosophical one: the problem of historical explanation. The session of the Economic History Association planned by Alexander Gerschenkron for Fabricant's joint conference centered on questions of methodology. A formidable pair, Alfred Conrad and John Meyer, prepared two papers—one on slavery and one on method. The economists were indeed slaves to a method—or at least vigorous apostles of an approach. They took as their starting point the uneasiness they felt

---

[2] In preparing this introduction, I have drawn freely and sometimes literally, on my own rather obscure writings of the period 1960–1975, in which this history is reviewed. See especially, "Through Growth and Beyond," Louis P. Cain and Paul Uselding, eds., *Business Enterprise and Economic Change* (Kent, OH: Kent State University Press, 1973), 15–48, and "Work in Progress," Report to Ernst Söderlund, *Scandinavian Economic History Review* 10 (1962), pp. 233–244, from which portions of the text have been extracted.

at the work of historians in the field. Historians, they said, were either asking no questions of history at all—merely endeavoring to produce verbal syntheses of total experience, or were asking questions about political, rather than economic, phenomena. The economists wanted to know how important certain elements were in producing economic change, and whether the adjustment of the system to its opportunities was optimal. To ask these questions involved them in stating hypotheses and interrelating them through a model. They objected that historians had not made this explicit and had not used data to test such models in any strict fashion. The economists objected to the grandeur of the historians' hypotheses, the use of such vague, untestable, sociological and psychological constructs as "national character," "spirit of the times," "intellectual climate" in the explanation of economic change.

From the studies of the 1960s and early 1970s, one immensely reassuring conclusion was derived: it turned out that, after all, the market in the nineteenth century had really worked very well. Should the Southerners have invested in slaves? Yes; slaves yielded a normal rate of return, and their price covered their reproduction costs. Should the railroads have been built? Yes, though by a narrow margin. Should the railroads have been financed by generous land grants? Yes; the returns were just sufficient to cover the risks of the necessary investment. Should the Pennsylvania iron industry have clung to charcoal as long as it did? Yes; there was a good market justification for its shifting to anthracite, then to coke just when it did. Should the reaper have been introduced in the 1850s? Yes; on farms of Midwest scale and organization, its introduction any earlier would not have been profitable.

But, in all fairness, the concern in many of the leading studies in the new vein was not whether the market worked; that was simply one hypothesis that was tested. The analysis was directed beyond test of the market, to an inquiry into precisely which elements in the environment were the active and which the passive ones, whether and in what proportion the influences on the market came from the side of supply or of demand. Here results were in a sense both more complex and more understandable than historians had thought them to be. Was the reaper important in raising productivity in American agriculture? Was the railroad important in the growth of the iron industry? The historian, averse to quantification and model-building, had no technique for answering such questions. The economists, however, asserted that importance had a definite meaning. It could be measured by what would have happened if in a situation everything had been constant except the one factor whose importance was under consideration. If this factor were given a different value, or were wholly removed, how would the result have been affected? Such a bald unvarnished use of imagination and analysis in historical work made many traditional historians gasp. But some of the work and, I would contend, some of the most interesting of the "new" economic history work has been just such efforts, done within a very carefully specified set of assumptions. Such work did not deny the interrelatedness of all historical phenomena. It only analyzed them to try to find out what some of those lines of interrelation may have been. In this sense, it showed economic history to have been even more complex than the historian would have imagined.

In their actual studies, rather than as claimed in their programmatic statements, the "new" economic historians employed economics to test the static efficiency of the market mechanism for the allocation of resources and new investment, and to solve the identification problem in specific situations. How else, indeed, could economics have been used for historical explanation unless really dynamic models of economic growth could be elaborated and made sufficiently complex to be used in econometric studies of income growth over long periods?

Scholars are not such special animals, so different from the rest of mankind, as they sometimes like to believe. Intellectual movements or networks like the new economic history share many common features with political or religious sects. Their organization possesses a sociology, however inchoate; an ideology, or common aim and values; and a vocabulary, expressing a certain common method. In the years between two Income and Wealth conferences (1957, 1963)[3] and the early 1970s, when the publication of two major syntheses of all this work appeared, an institutionalization of the efforts occurred around the annual Purdue seminar, which experienced a continuing life at other locations until the formal organization, through the initiative of Donald McCloskey, as the Cliometric Society in 1985. At the same time, an outlet was found for publication of the studies in the *Journal of Economic History,* which opened up to the new work, and the *Explorations in Economic History,* which grew into the castoff shell of the earlier *Explorations in Entrepreneurial History.* No new association, like the recently formed Social Science History Association, was organized, and room was left within this flimsy network for a mingling of new with old, al-

---

[3] National Bureau of Economic Research, Conference on Research on Income and Wealth, *Studies in Income and Wealth,* Vol. 24 (Princeton: Princeton University Press, 1960); Vol. 30 (New York: National Bureau of Economic Research, 1966).

though that did not mean that much real mingling actually occurred.

Instead, a collection of quantifiers grew up, with the strength of a pack, able to repel attacks from without by applying the time-honored maxim that "offense is the best defense." The tangle of intricately interrelated statistics springing up junglelike around it gave the most diligent or most hostile critic pause. Profound resentment or sheer misanthropy was needed to push the uninitiated into the tangle, and once entered, the heavy perfumes, the charm of the foliage, and the sense of communion with the esoteric often overpowered enmity. Constructive historians, chary of their time and vulnerable to criticism, came gladly to exchange confidence and congratulation for the right to pursue their own concerns. They accepted income estimates, quoted them on occasion, and sometimes imagined that they existed as facts, rather than as the frail structures of hypotheses their authors had intended. And along with the impress made by a like-minded group of scholars upon the world at large, there was felt the reinforcement of inner strength that comes from family life. When the number of scholars grew to a certain point, they produced their own conferences, became one anothers' reviewers and critics, established a private language and tradition. Jokes and allusions sprang up, and the corporate life grew through students and through friendly attachments. They formed an example of the social equivalent of what in atomic physics is known as a critical—or in this case, some would say, uncritical—mass.

It is an odd vestige of the individualist ethic in American institutionalist thinkers following Veblen to confuse sociology with satire. To spot a social group is—Adam Smith to the contrary notwithstanding—not necessarily to sniff conspiracy. Every exchange of information requires a certain sympathy and mutual understanding, combined with individual differences, among the participants. As sympathy and insight grew, exchange became freer and more meaningful and discriminating. Beyond a pooling of sources and techniques, there was the purely statistical reinforcement that the checks within an interdependent system of estimates could provide. The famous advantages deriving from the division of labor also appeared. Each of a dozen or so new economic historians, without plan or prearrangement, took one subject or one sector unto himself—national income, labor force, technology, capital, manufactures, agriculture, transportation, population, demand—an allocation of topics by taste and comparative advantage, a miraculous instance of wonder-working providence, of the Invisible Hand!

The effectiveness of such a division of labor, however, depends intimately on the structure of markets by which the individual labors are interrelated. It was at this point that the national-income framework for the nineteenth century, combined with a rough industrial breakdown of interindustry transfers and factor utilization, came into its own. Around such a structure the 1963 Conference on Research in Income and Wealth was organized, with special emphasis on the "supply side." Gallman's massive reconstruction of the national product from 1834, Lebergott's basic series on the movement of the labor force and employment, the several series on the output of certain final and intermediate products, productivity measures, and partitionings for agriculture and transportation made up the volume. The dream was to build toward a full display of the basic outputs and inputs, factor payments, and the Keynesian categories of demand, and to show these as the surrounding elements of a Leontief input-output matrix, in which the interindustry structure would be exactly filmed as it operated through the decades of the nineteenth and twentieth centuries.

When, in 1972, twelve apostles of new economic history whose work had been displayed in the two Income and Wealth conference volumes came to codify their findings in what purported to be textbook form, this schema was amplified and formed a central place in the Introduction. It was recognized that as presented in the two conference volumes, it was still missing many of its limbs. Account needed to be taken, on the supply side of capital, not just of machinery, machine tools, and equipment, but of capital in the financial sense; the problem of financial capital as a factor, or at least a facilitator of production, had to be directly faced. Lance Davis's work in that volume, and the magisterial chapter by Davis and Gallman in the *Cambridge Economic History,* Vol. VII, went a long way to filling that gap. The evolution of government as a productive force and an organizational form was directly attacked by D. C. North in a confrontation between scholar and subject that is still continuing. Dorothy Brady, at the 1963 conference, had looked at shifting demand, but much work obviously remained to be done to follow up the ancient insights of Veblen into the tastes of a consuming society. At one point, even within the inner recesses of their citadel, the new band of quantifiers was compelled to beat a retreat: the concoction of a Leontief input-output matrix even for any year of the nineteenth century proved, despite several valiant and not wholly unsuccessful attempts, to be an impossible job.

Most serious of all, the classic set of problems of market and industrial organization, which had been the breath of life to the old economic historians, had yet to be submitted to the inquisitorial torture of quantification. The schema, as presented in the 1972 textbook, made a sharp separation between what it called "elements of Opportunity"—Resources, Labor Force, Technology, and a growing Demand—together with the operation of the markets for productive factors and intermediate products that translated opportunities into profitable production, and "elements of Response," the private and public structures within which the economic agents acted, and the psychological and cultural influences that made for an "achieving society." But even then, the massive researches of Alfred D. Chandler were beginning to show that, by collecting enough business histories, something might eventually be done in specifying and measuring the development of the corporate structure.

It is truly astonishing to observe how a society's cultural artifacts reflect in finest detail the general movement of its political and social history, how the social body in every organ and cell gives the prognosis of the general state of health. The publication of the two texts, in 1971 and 1972, represents the climax of the work of the first generation of new economic historians.[4] But it came at a time when the foundations of the national-income approach to economic growth and development, which helped to structure research and policy during the two decades we now call "the American Century," were crumbling. The structure was not torn down; indeed, many scholars continued to inhabit it and to construct new chambers, new extensions, and more elaborate ornamentation. But, as happens in intellectual styles, it was increasingly by-passed. Vines grew up the sides; children no longer played on its porch, but ran off with their games to sunnier and more dangerous precincts. Still, for better or for worse, the obsolescence of the structure did not implicate the quantitative methods and scientific instincts that had caused its original builders to work there. Instead, a continuity with the original inspiration was preserved, not through the filling out of the national-income structure, but through the use of quantitative methods and statistical materials to attack an ever-widening range of narrower problems. The way was shown, not by the synthesis made in the text produced by twelve apostles of the field, *American Economic Growth: An Economist's History of the United States,* but through a collection of an entirely different character: *The Reinterpretation of*

[4] Robert W. Fogel and Stanley L. Engerman, eds., *The Reinterpretation of American Economic History* (New York: Harper & Row, 1971). Lance E. Davis, Richard A. Easterlin and William N. Parker, eds., *American Economic Growth: An Economist's History of the United States* (New York: Harper & Row, 1972).

American Economic History, edited with generous (occasionally even overgenerous) introductions and commentary by Robert W. Fogel and Stanley L. Engerman.

This second of these two syntheses did not seek substantive unity by relating the parts to a form of *histoire totale*. Its aim was, rather, to display the "new" methods in a varied succession of problems, of which economic growth or industrial expansion was only one. Technological change, education, capital markets, slavery, immigration and urbanization, economic policies—all served as fields of encounter between the new methods and emphases and old problems. In all, *The Reinterpretation* offered thirty-five such studies by twenty-seven different authors, including pieces by most of the twelve authors of *American Economic Growth*. Through many windows it shone the light of social science in many directions over the scene of American economic history. Yet it was not so much a "reinterpretation" of the whole landscape as a picture gallery of selected pieces of the artists of a single school. It was as if Fogel and Engerman had arranged a monster exhibition of works by these artists—not surely Impressionists, but perhaps Abstract Expressionists—who had been toiling in their studios all over the country in the 1950s and 1960s.

Reinterpretation, however, was afoot and it came out not in gigantic new syntheses in the manner of Turner and Beard, but in monographs, and collections of studies on many of the topics foreshadowed in the Fogel and Engerman collection. Already Fogel and Engerman had turned their colossal research energies in admirable complementarity to the original subject of the Conrad and Meyer 1958 article: American Negro slavery. *Time on the Cross,*[5] despite or perhaps by virtue of its well-publicized technical flaws, represented a new departure in the quantitative history done by economists in that it endeavored not simply to measure quantities and values, but also to construct from objective evidence an interpretation of the psychic state of the enslaved. Indeed, the great danger of the book, in my opinion, was that for the unwary reader a kind of halo effect would be felt, carrying over from the mounds of evidence on diets, prices, and the like to the suggestions about slaves' welfare and achievement under the plantation system, for which the hard evidence was not at hand. As a result, the picture of blacks' sufferings and indignities under slavery that Kenneth Stampp built up by the historian's traditional methods—description based on anecdote, documentary records, contemporary comment—had, in the

---

[5] Robert W. Fogel and Stanley L. Engerman, *Time on the Cross: The Economics of American Negro Slavery* (Boston: Little, Brown and Company, 1974).

opinion of many and despite criticisms by Guttman and others, a strong claim to credence.

Fogel and Engerman, however, had let the genie out of the bottle. Historians attacked their conclusions and their methods furiously, and in the latter they were joined by some of the new economic historians as well. But in the vast, dignified, stately ranks of the historical profession, the younger members—graduate students with minds still unformed enough to absorb statistical methodology—began to exploit what were undeniably the tangible evidences of past life encased not in letters, diaries, and newspapers, but in numbers. The nineteenth-century censuses, for example, are, after all, not simply, or even principally, bodies of economic data. They are vast storage tanks of quantitative data on social history—on education, religion, crime, insanity, urbanity, and much else.[6] And in a democracy, far more than in a kingdom, the numerical records of votes and elections which exist in abundance count for more, at one level of history, than the speeches of politicians or the correspondence of diplomats. So it was that the techniques and attitudes of the new economic history were diffused into political and social history. It was a general intellectual movement, and one should not overemphasize the leadership Fogel and Engerman, or the twelve apostles of *American Economic Growth,* gave to the profession as a whole. But several 'new' economic historians, notably Allan Bogue and Robert Fogel, had important roles in the establishment of the Social Science History Association, and promoted the diffusion of the techniques in many directions.

What, then, has all this history to do with the volume at hand? It is no part of the job of the writer of an introduction to explain what the pieces say. They stand as independent works of art, and represent the state of art of quantitative economic history as it stands today. They should be judged, in my opinion, in the same way as the essays in Fogel and Engerman's *Reinterpretation,* as examples of skilled work, offering new evidence and often new light on their respective topics. Their unity is first of all one of methodology. The book is not, to use Isaiah Berlin's well-worn image, the work of "hedgehogs," wrapping the parts into a single self-contained ball, and hurtling toward some single point. It is conceived, perhaps under the inspiration of him to whom it is dedicated, as the work of foxes, a whole den of foxes, demonstrating their statistical and rhetorical tricks, each in a different part of the field.

[6] Stanley Engerman points out that the most frequent early use of the censuses was in studies of geographic and social mobility.

Nevertheless, as a transitional work, this book carries with it, on its pages as in the minds and files of its authors, the outlines of the old national-income schema. That it contains no new measures of aggregate income or of the labor force is testimony to the exhaustive thoroughness with which Gallman and Lebergott, respectively, did their original work. But uneasiness continues to be felt, about the measure of the capital stock—an uneasiness well justified in view of the profound theoretical, even philosophical, conundrums involved, and the variety of definitions which the nineteenth-century wealth censuses had at their disposal. To this both Gallman's paper and—to a degree—James and Skinner's are addressed. Not all the puzzles of the "capital controversy" are solved, even by Gallman's lucid exposition. The concept Capital has seemed to wrap its mantle tighter around itself under the strong winds that blow from the caves of the theorists. But here, under the open sunlight of measurement, it sheds some of its flapping outer garments. In Gallman's paper, some of the problems of measurement, definition, and intertemporal comparison presented for capital measurement by the phenomena of falling prices of capital goods and a secular fall in the rate of interest are opened up and skillfully dissected.

The paper by James and Skinner goes at the problem from the side of the supply of savings, and mines the census data of wealth and of the age distribution of the population to find a bit of confirmation for the now popular "life cycle" model of savings behavior. And it goes beyond this, to estimate the effect of the changing occupational mix, getting in this way a statistical accounting for 44% of the increase in the value of capital per head in 1850–1890; it then partitions the remaining growth among nondemographic developments; the rise in real income per head, the growth of financial intermediation, and the decline in the prices of capital goods, relating these, of course, through their effects on household decisions during American economic growth. Such an approach to capital formation preserves much of the spirit of the earlier quantitative work. If one believes that savings come from individuals' expressing their preferences through the financial and corporate organization of the economy, then the James and Skinner essay will give a comfortably satisfying intellectual experience.

Thomas Weiss's essay is the first step in an effort to develop further the Lebergott labor-force estimates. It is based on regional demographic data, broken down by urban or rural residence, and, as is customary, by age and sex, so as to show clear variations in labor-force participation rates arising from age and sex distributions, alone in different parts of the country. The treatment continues into three essays, by David, Wright, and Goldin, looking at variations by region, race, and sex in the conditions of labor. Under the heading of labor, the contribution of these essays is directed even more specifically to dissected, decomposed elements in the experience of the labor force.

Paul David, in an essay of real scope and great significance, examines the varied questions of "labor shortage" and the wage differential of skilled over unskilled labor. To the classic discussions of the New England case, including the somewhat novel, recently published findings and interpretation of Goldin and Sokoloff, David's paper adds a new dimension by examining Chicago, an urban industrialization on the labor-scarce frontier. Those familiar with a Paul David essay will know how prolix a richness of allusion, reference, and insight is to be found in the illuminating and original discussion. The paper includes, among other things, a short economic history of Chicago, and opens up vistas of history of that great *terra incognita* of American economic history: the industrial Middle West.

Gavin Wright's essay on the Southern labor force has similar qualities of scope and novelty; indeed, the novelty is in a sense produced by the scope. The South, Wright says, offers a case of what other authors have called "a segmented labor force" at two separate points. On commercial farms, one segment was composed of restless or shiftless men who became wage laborers on plantations, moving from place to place and equalizing wages across regions, and across the color line to a surprising degree. The other segment was those farmers, black and white, who, though without land or capital, used their labor to establish credit-worthiness as tenants on shares, and even to climb to a modest degree up the 'agricultural ladder'. In the industrial labor force, segmentation appeared in racial discrimination in job assignment. In both cases the market 'worked' in solving the immediate allocation problem; equal pay was given for equal work, and workers were taken on in each category to yield a marginal product equal to his wage. But blacks were kept down by the failure to provide rewards for experience or ambition in industry, and in farming by initial poverty and the need to prove credit-worthiness by "good" behavior, a part of which was simply to refrain from rising too high.

The male/female earnings differential does not yield in Claudia Goldin's study to such a direct explanation. She finds the existence over a very wide spectrum of time and occupational category of a "gender gap," narrowing within farming and manufacturing, when women left farms for factory labor, and varying some over long periods, but remaining relatively stable during the past thirty-five years. Striking is

Goldin's search for "real" factors as the basis of changes in the differential: changing job requirements relative to differences in physique, changing patterns in education and in expected lifetime participation, and changing relative supply. Here again, the market appears to have made the best use at each moment, as it were, of an underlying structure of what she would call "gender distinction". Yet in learning how the market does its work we learn a great deal. In racial discrimination, the obstacle is the refusal of whites to work indoors alongside blacks: the "solution" was to confine blacks to selected classes of occupation—a caste system. In "gender" distinction, the source lay deeper: basically in the social definition of women's domestic role. The effect was to establish separate male and female markets within occupations, and to pay women wages in relation to their relative supply, which reflected their disadvantage in having failed to obtain the status of full-time, trained, and experienced workers.

The six essays just discussed open up new windows on phenomena related to some of the large topics of the national-income framework: capital, savings, population, and labor. The essay of Jeremy Atack derives its importance from taking up a side of income growth untouched in the traditional accounting (except through Denison's "residual"): the capture of economies of scale through changes in plant, firm, and industrial structure. The great lacuna in the efforts of the 1957 and 1963 conferences, the Purdue conferences, and the earlier texts was just this. Everyone knew that scale economies shared with technological change and capital accumulation the major responsibility for productivity growth. The whole subject was simply treated like a wayward uncle; one did not talk about it. Alfred Chandler's work drew these changes and adaptations into the limelight—not only the grand economies of specialization and market scale, but the narrower ones of firm structure, of plant and throughput organization. Yet the great deficiency in Chandler's treatment has been the lack both of a sophisticated theoretical basis and of any sort of comprehensive measurement. Atack's paper is only a beginning in this direction. It tries to measure, from the census data, the relation of productivity to scale of establishment, with the shift to the power-driven factory, and the persistence nonetheless of small workshops, protected by high transport costs in many industries as late as 1870. More remains to be done to move the measurement into the more intangible economics of system management, regularized throughput, and distribution and (even) transactions costs suggested by Chandler's studies and the recent theorizing of organizational economists like Douglass North.

We are left with two essays on Yankeedom. I realize that I must resist the temptation to co-opt the massive researches of Davis, Gallman, and Hutchins on whaling, into my own interpretive effort on New England's industrialization. They treat whaling as an industry in its own right, not simply as the source of spin-offs to land-based commerce and manufacture, or as an item in the Yankee balance of payments. My own essay I view as a sidelong attempt, by largely literary methods and rhetoric, to look at Yankees themselves, to deduce the springs of their Industrial Revolution from their culture and its environment. In this way I have had the ill-concealed hope to restore the word *puritan* to the economic historian's vocabulary. I suspect the attempt will fail, largely because, being at least 51% puritan in our instincts and values as economists, we find it hard to recognize anything peculiar in those traits when we find them in others. In their essay, Davis, Gallman, and Hutchins take it as perfectly natural that men would be willing to get in small sailing ships and travel halfway around the world to chase those monstrous fishy mammals, even in frozen arctic seas, where, they record, "twice a large fraction of the fleet was caught by a sudden freeze and lost in the ice." Since whaling, like the railroad, is one of those activities about which humans are endlessly curious, the reader may find that their long essay, exhaustive as it is of their data sources, will be hard to pick up, but equally hard to put down, even though it treats only the economist's subjects—technology, costs, productivity, and market decline.

It would be foolish to impose a forced unity on the group of essays published here; to do so would distort the authors' intentions and lop off much of the subtlety and special points of interest in the papers. Yet, they all move us one step deeper into the study of American economic growth, into details and peculiarities of market behavior that accompanied the general upward movement. The image of a growing competitive economy is retained, and we see in many of the essays some impediments and some implementing factors.

We look at all this detail until slowly the once-exciting task of explaining the end result begins to seem remote and even a bit hollow. But that means that, without losing our contact with the general and the universal, we are finding the data and developing the many insights that bring the history, even the history of magnitudes, of aggregates, averages, and distributions, closer to the actual experience of the men and women who lived in past times—to their behavior, their problems, their working and consuming lives, how and why they saved, what they earned, under what conditions they worked, what equipment they had—even for catching whales. Certainly the meticulous scholar and fastidious mind whom this book honors, whose own work spans the entire transit described in this introduction, from the quantitative study of manpower in American history to sensitive and intriguing general history, *The Americans,* and who knows how to treat a general theme by a polished use of a particular statistic and apposite literary quotation, must, and we hope will—along with the general public—approve the result.

2

New England's Early Industrialization: A Sketch

WILLIAM N. PARKER

"Grau, teurer Freund, ist aller Theorie"—Goethe, *Faust*

Omnis est America in partes tres divisa. At least since Henry Adams detected three "intellects" in the three regions,[1] the trifurcation of the North American colonies into New England, the Middle colonies, and the South has been standard historical practice. But an economic historian cannot begin with "intellect," however willing he may be to slip it in later as an "independent factor," "overdetermining," as Althusserians say, an "economic and social formation." Even under the implausible assumption of uniformity in the social psychology of the original colonists, a simple diagram relating potential income to hours of labor time can show the basis for a three-fold division.

Figure 2.1 is such a diagram. With respect to productive factors and markets, it is based on the following assumptions: (1) that for relevant ranges of population growth, the supplies of land at constant marginal physical productivity are unlimited; (2) that production is organized in family units, and that these units are joined to a limited degree in rural neighborhoods, allowing for local trade in some specialized artisan industrial products and in some agricultural products as well; (3) that external markets for agricultural and resource products differ sharply among the regions; (4) that productivity differences—due to soil, climate, and terrain—in crops grown for home consumption have little significance for the economic outcome, compared to differences in export opportunities.

The diagram then defines the usual impressions of the three regions' differing abilities to provide subsistence and, beyond that, to provide

---

The author is indebted to discussants at the Wesleyan Symposium, at the American History Seminar at the University of Cambridge under Professor Charlotte Erickson, and in the Yale Economic History Workshop. He thanks especially Jeremy Atack, for valuable written comments; Ann Kibbey for guidance into the literature on Puritanism; the Yale Program in the History of Science; and particularly, Carolyn Cooper for references on technology and Laura Owen for research assistance. Financial assistance from the Yale Department of Economics (AMAX fund) is also gratefully acknowledged.

[1] Henry Adams, *History of the United States during the First Administration of Thomas Jefferson* (New York: Scribner's, 1889). Chs. 3, 4, 5 reprinted as *The United States in 1800* (Ithaca: Cornell University Press, 1955). Today, Adams's good English word *intellect* would be replaced by the Franglais *mentalités*.

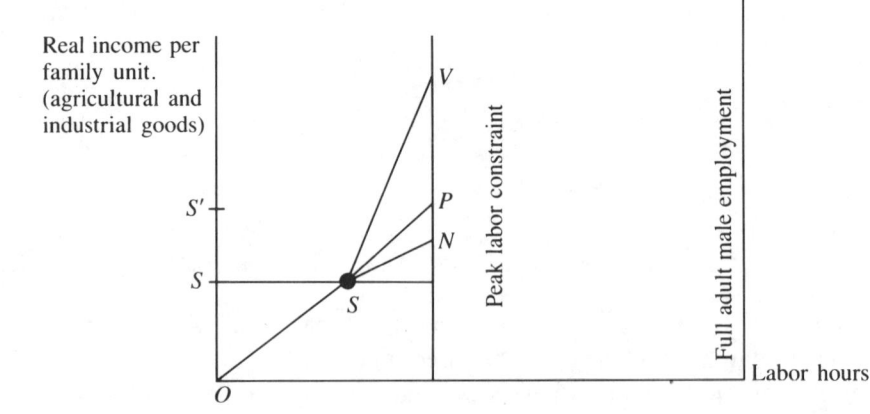

[Graph with axes: vertical "Real income per family unit. (agricultural and industrial goods)"; horizontal "Labor hours". Vertical lines labeled "Peak labor constraint" and "Full adult male employment". Points/rays labeled V, P, N, S, S' from origin O, with horizontal lines from S and S'.]

- *OS* Subsistence output.
- *SS'* Surplus above subsistence, food crops producible by family labor under two-person peak labor constraint. (See Figure 2.2.)
- *SN* Subsistence crops salable on local markets only in rural neighborhood against local craft and household manufactures (New England case).
- *SP* Subsistence crops salable on export market, permitting manufactured imports at favorable terms of trade ("Pennsylvania" case).
- *SV* Surplus peak labor convertible to export specialty crop ("Virginia" case).

Note: Figure 2.1 represents only a very primitive stage in the growth. Beyond this, various paths to higher income emerge:

(1) Employ off-peak and family labor on specialty crops and industrial goods for local trade.
(2) Replace subsistence farming with specialized local producers in farm and industrial products, improving terms of local trade for surpluses of subsistence food crops.
(3) Specialize in export specialty crop, importing some subsistence products.
(4) Hire, at rates below value of marginal product, or impress by force, additional labor at peak planting or harvesting times.
(5) Develop nonagricultural exports, particularly in a specialized nonfarm sector, which then trades with farm sector for surplus subsistence goods.
(6) Improve productivity in farming, especially in peak load operations.

In developing rural neighborhoods, all three colonial regions used methods (1) and (2) in varying degrees as opportunity offered. The regions differentiated as determined by agricultural export opportunities:

Virginia/Maryland, Carolinas, Georgia—(4) and to a degree (3).
Pennsylvania, New York, New Jersey—intensive local development—
   (1) and (2), plus some compulsion of indentured and family labor—(4).
New England—(5).

None of the regions shows much technological improvement in farming—(6)–before the early nineteenth century.

FIGURE 2.1 Potential uses of farm-family labor in three colonial regions in agricultural crops (subsistence and surplus production), with differences in resources and external markets, and with peak labor constraints, but no land constraint.

New England's Early Industrialization

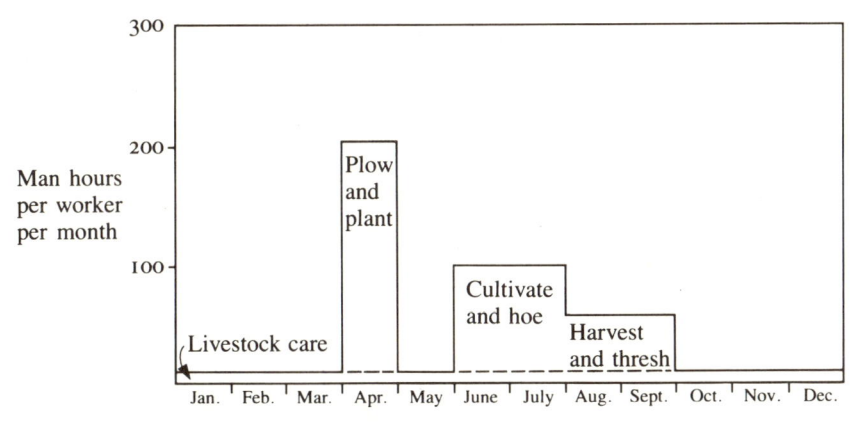

FIGURE 2.2 Direct labor requirements for each of two workers in subsistence agriculture (livestock care and field operations) over the agricultural year, pre-1840 (based on 1840–1860 data for New England). (See Tables 2.1 and 2.2 for basis of estimates.)

higher standards of physical comfort from farming, traditional crafts, local trade, exports of resource products and agricultural staples. The unit in which income is earned is considered to be not the isolated individual, but a mixture of family units within the rural neighborhood. Within such a neighborhood, a moderate degree of division of labor occurs. Some special craftsmen—millers, smiths, weavers—and professionals—a teacher, a public official, a minister—subsist through limited local trade or barter of their services and products with the farm households roundabout. The unit of labor is an individual farmer plus one family member. Conceptually, this work is considered to consist not only of livestock care throughout the year and work in the subsistence crops, but also of the maintenance of farm capital. It would include the heavy and constant labor of farm wives in food preparation, homemaking, and domestic industry.

Figure 2.2 shows a rough estimate of the labor hours required, in minimal operations in field and barn, to grow a family's subsistence in New England in the mid-nineteenth century. In this work, New England's conditions do not differ drastically from those there or in the other colonies in the eighteenth century. Two important conclusions derive from such an examination: (1) large amounts of time remain even in summer, and particularly in the winter months, for farm capital formation and for growing crops for the market, if a market is present; (2) as a result of (1), productivity differences in the subsistence activities are not very important from one region to another in determining

TABLE 2.1
Direct Labor Requirements in Subsistence Agriculture Over the Agricultural Year, New England, 1840–60

	Corn	Wheat	Hay	Total
(a) Acres	6	2	10	18
(b) Labor requirement per acre				
Plow (m-h/acre)	18	18	18	
Plant (m-h/acre)	14	2	—	
Cultivate (m-h/acre)	15	—	—	
Hoe (m-h/acre)	52	—	—	
Harvest (m-h/acre)	14	15	10	
Thresh (m-h/15 bu)		15		
TOTAL	113	50	28	n.a.
(c) Total man-hours (a × b)	678	100	280	1050

(d) Per-worker total = 520, or 52 10-hour days, spread over 180-day growing and harvesting season

(e) Livestock care for two animals estimated at 15 minutes a day each, milking labor excluded

SOURCE: Table 2.2.
NOTE: Basis of estimates for Figure 2.2.

TABLE 2.2
Approximate Land and Labor Requirements for Grain and Meat, and Consumption/Allowance, for Family of 6 (at Uniform Allowances) on a New England Farm, 1840–60
(Source Data for Figure 2.2)

Land:	wheat	1600 lb + 50 lb/bu = 32 bu	=	2 acres at 16 bu/acre
	corn	8900 lb + 50 lb/bu = 178 bu	=	6 acres at 30 bu/acre
	hay (est.)			10 acres
Labor:	wheat	2 acres at 49 m-h/acre	=	98
	corn	6 acres at 112 m-h/acre	=	672
	hay	10 acres at 28 m-h/acre	=	280
				1050 m-h

= 105 10-hour days = 18 6-day weeks = 4½ months

Consumption allowances for family of 6, with one cow and one work animal:

			Family
Daily	1½ lb meal:	.75 lb flour	4.50 lb
		.75 lb corn	4.50 lb
	⅓ lb meat		2.00 lb
Annual	wheat flour	1600 lb	
	corn flour	1600 lb	
	meat	730 lb × 10[a] = 7300 lb corn	

SOURCE: Land and labor requirements from Parker and Klein, "Productivity Growth in Grain Production in the United States, 1840/60 and 1900/10," in NBER, *Studies in Income and Wealth*, vol. 30, (New York: 1966). Among omissions in these calculations, the most notable are wild game and fish, milk, wild forage, straw, cornstalks and tops, orchard crops, poultry, etc. Account is also not taken of labor used in home manufactures or in farm capital formation.

[a] Rough estimates: hay acreage for two animals, 10 acres; corn-meat ratio for cattle or hogs, 10:1. Under pioneer conditions, wild forage replaces ⅔ or ¾ of the corn acreage.

relative income levels. The crucial question is the marketability of the surplus within the local community and in external markets.

With these elemental principles in mind, we may now make the comparison among the three regions: the South, the Middle colonies, and New England. Line *OS* in Diagram 1 shows the similar position of all the regions with respect to the minimum hours of labor and acreage required for a family's subsistence, allowing for traditional exchanges of produce for industrial goods within the rural area. Line *SV* shows the superior value of the uses of labor time beyond subsistence in the South (Virginia), once the tobacco culture was established. Even at the low price at which tobacco settled, farm and planter families could still have imported enough commodities to move well beyond any assumed "satiety" level of a peasant standard of living. Line *SP* shows the "satisficing" life of the Middle colonial (Pennsylvania) family, made possible to a degree by good land, but even more by the possibility of export shipments of food crops to the West Indies and by trade with the growing seacoast cities. Line *SN* shows the plight of New England families who, without rich resources or an exportable agricultural crop, were unable by traditional means to get beyond a moderate margin over subsistence. Their inferior position depended not on harsh winters, thin soils, or a growing season possibly twenty days shorter than that in the lower states. Even if New Englanders were only seeking to perpetuate a peasant economy not much above subsistence, their problem was markets.

It is obviously not very significant whether a "satisficing" level of living sometimes credited to pre-industrial societies existed in the mentality of villagers in the colonial economies. In New England, even "satisficing villagers" would have needed to strive and to innovate. In Virginia, despite ample supplies of family labor, the most strenuous efforts were made to obtain an auxiliary labor force so as to increase the number of acres from which a landowner could obtain a surplus income. At this level, Virginia-Maryland tobacco farming appears as a capitalistic operation from the outset. In Pennsylvania, where perhaps the best case for a peasant-like ideal of "limited good" could be made, the surplus even in the eighteenth century was beginning to be reinvested in urban construction, industrial plant and tools, and mercantile trade.

Arnold Toynbee, in his *Study of History,* cites New England in a list of such classic adverse environments as Central America, the "parched plains of Ceylon," the North Arabian Desert, Easter Island, and the Roman compagna whose "challenge" produced the response of an ag-

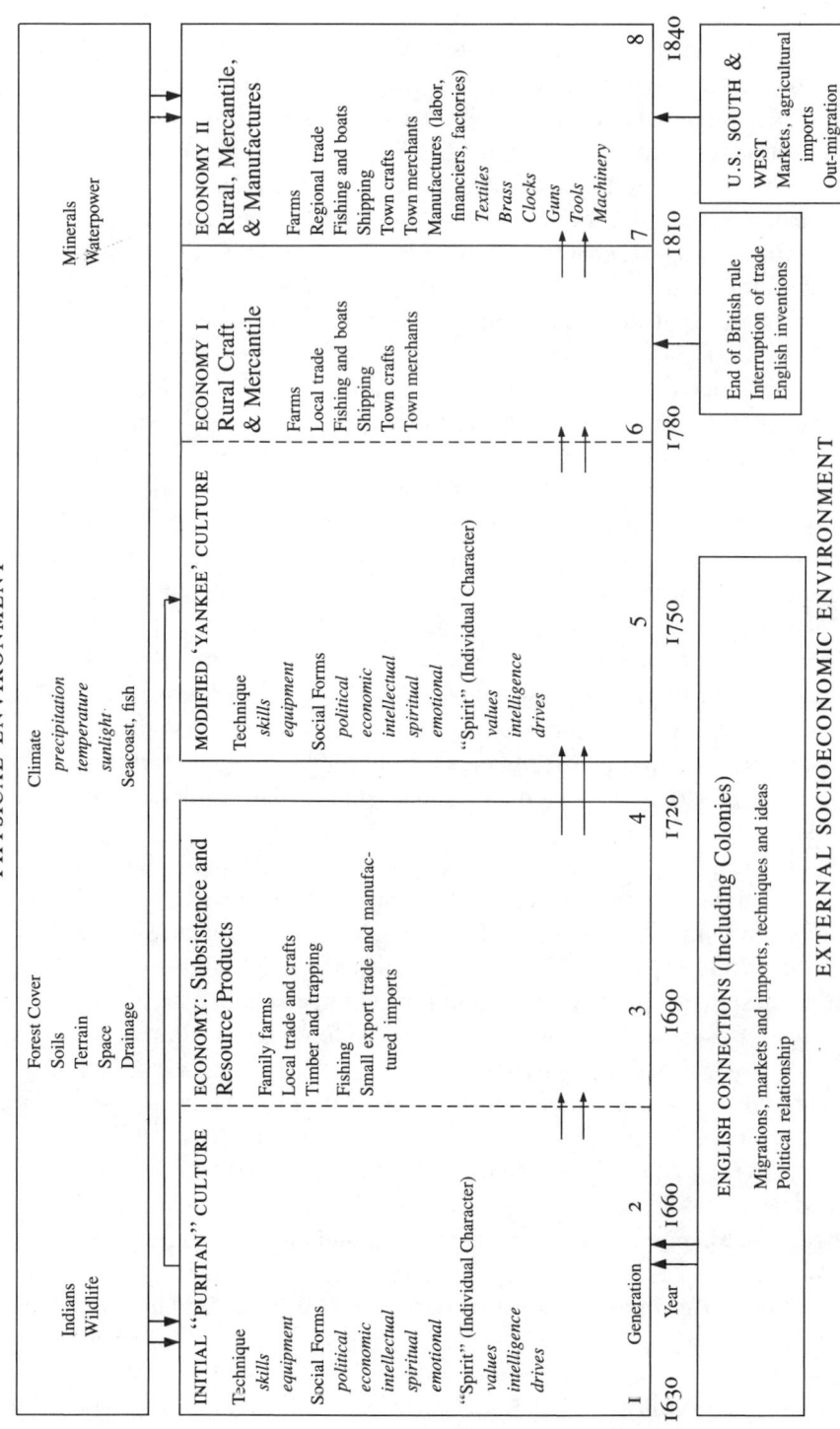

FIGURE 2.3 Interrelation of environment, culture, and economy in New England's industrialization, 1630–1840.

gressive civilization.[2] New Englanders found an economy in their region which, by traditional agricultural or industrial techniques adapted to New World soils and climate, was able to yield a surplus above subsistence. This surplus was adequate to give some room for trade and maneuver; but without the product markets enjoyed by the Middle and Southern colonies, the surplus forced New Englanders into novel endeavors outside the villages and farms, in trade and industrial occupations. It is at this point that the infamous "Puritan mentality" may have come into play, even as the economic social experience in the New World was changing the Puritan into the Yankee. This aspect of the matter is touched on only timidly and sketchily in the discussion below.

The historical analysis is developed here on a framework shown in skeletal form in Figure 2.3. If economic history is to be thought of as walking the line where nature—soils, climate, terrain, resources—meets society—techniques, organizational forms, and human individuals—then that line was formed in the seventeenth century by the western coastline of the Atlantic Ocean. To the white man, the Indians were a part of the natural environment, and the settlers brought into this setting the skills, knowledge, values, and social relationships that had been acquired in England. This constituted for the young colonial societies an original endowment, made clearer and more unequivocal in New England's case by the fact that the thirty thousand or so "Puritans" came in boatloads within twenty years. After 1660, or even 1650, the society they created did not open itself readily to new arrivals. At least 90 percent of the 1790 population of nearly one million must have been descendants of the original stock. Locked in this box, the Puritans created their farms, seaports, and domestic and artisan industries. The relations first with England, and the other colonies, then after 1810 with the American South and West form the third side of the box in which the successive economies of New England grew and in which the cultural transition from Puritan to Yankee was achieved. Within this space (the central boxes in Figure 2.3), New England's economic activities—farming, fishing, trade, and the industrial arts—had room to develop. With the Revolution, the interruption of foreign trade, and the freeing of industrial life from colonial restrictions, that environment changed and drew an ardent response from a population already trained to market commerce and industry and eagerly seeking outlets in ways that form the subject of section II of this sketch. The history leads thus into comparisons—not made here—with the other colonial areas, es-

---

[2] Arnold J. Toynbee, *A Study of History,* abridged by D. C. Somervell, 2 vols. (New York and London: Oxford University Press, 1947). See vol. I, Ch. VII(1) "The Stimulus of Hard Countries," esp. pp. 96–99.

pecially with Pennsylvania and New York, as their economies developed. It then leads further into a history of how the northern streams of westward migration, with some upper South admixture, blended to form the gigantic light and heavy industries of American Middle West after 1850.

II. A Bit of History[3]
"Und grün ist des Lebens goldner Baum."—Goethe, *Faust*

A Preparation for the Factories

The repeated action of the seventeenth century Puritan settlers over the harsh New England terrain, like water running over rock, created a landscape.[4] Their actions toward one another created a society. At Plymouth, Salem, and Massachusetts Bay, around the inlets along the southern shore, and up the Connecticut River, families formed and reproduced abundantly; towns were founded and lands subdivided; property was defined and crimes against persons, property, and public morality were invented and punished. Town meetings were held; selectmen, deputies, and local officers were elected. A few taxes were collected; money was printed; British regulations of trade and industry were received and sometimes obeyed. The Puritan churches maintained a privileged status; heresies appeared in Massachusetts Bay and were put down, to reestablish elsewhere in the region. Puritan morality, Puritan energy, Puritan intellectuality moved ceaselessly in new directions, and the ministry engaged in its tortured and vehement effort to reconcile the doctrines of Calvin to a world that was growing more optimistic, more materialistic, more rational and pragmatic, more American.

It is hardly fair to call such a stage of society—so little traditionalistic, so dynamic, and of so short a life—"pre-industrial." It was agricultural, with a significant mercantile enclave, and it engaged perforce in extensive fabrication of industrial items for domestic or local consumption. It contained relatively little manufacture for commercial purposes and none, except boats, for distant trade. Yet it was a society that stood in 1775 on the edge of industrialization. In retrospect one can see that the prerequisites for an Industrial Revolution, part native,

---

[3] I regret that time was not available to prepare detailed footnoting of the impressions of the history given throughout sections II and III of this essay. Relevant portions of the attached bibliography give some of the principal sources used.

[4] The interaction of Indians and settlers with New England's physical environment is examined in an interesting recent book, William Cronon, *Changes in the Land* (New York: Hill and Wang, 1983).

part an echo of England's, were in place. They appear in the three sectors—agriculture, trade, and artisan industry—and in three respects—(1) the physical flows of supplies and markets; (2) the skilled and motivated population, stable families, literate and competent craftsmen, and some spirited and original entrepreneurs; and (3) the "institutional" structure encased in political constitutions, in laws and law courts, and in the practiced behavior of merchants, farmers, workers, wives, and businessmen.

From farms, grazing meadows, and woodland pastures came that modest surplus of grain, meat, cider, and milk which, processed in the farmhouse or locally, could be exchanged at small trading points for a few imports—notably molasses, salt, tea, and rum—or for the services of local craftsmen and professionals. From woodlots and forested hillsides came firewood and the logs which, processed in a local sawmill, became the materials of construction for furniture and clapboard houses and, along the coast and the banks of the Connecticut, for fishing boats and merchant ships. Farms also harbored supplies of labor, seasonally idled by the weather and the agricultural calendar—the labor of adult males, of young men and women, of children beyond their few seasons of schooling, even on occasion, the labor of overworked wives. All this found outlets in farm capital formation—clearing, stone fencing, construction, and farm industrial tasks such as woodworking, spinning, leather working, even work at a forge. Farming and much of the domestic industry also required the services of specialized workers and equipment: plows, harrows and scythes required iron tips and edges; grain quickly encouraged the development of local mills, and water-powered sawmills succeeded the sawyer's pit. Even fishing had its linkages backward to boats, ships, nets, sails, ropes, hooks, and harpoons, to barrels and salt; forward to a growing commerce in which, as with lumber, a New England resource could enter as a principal staple. Boat building, stimulated by the fisheries, readily became an important industrial occupation in its own right, and merchant voyagers could depend not only on boats built in Maine or local shipyards, but also on a cadre of fishermen-sailors ready to go out on the high seas.

But linkages and spill-over effects demonstrable in physical flows on an input-output table were perhaps the least important of those interconnections among resource industries, trade, and artisan industry that led into the factories of the nineteenth century. The agricultural calendar, although slack in the winter, did put stress on time at peak seasons. A New England farmer often had to hurry to finish plowing, to get in the wheat, to prepare firewood for winter. The changeability of the weather was a continuous threat, and good days had to be utilized. Work outside in winter was done rapidly, both to keep the body warm and to reduce the time of exposure. The briskness of the Yankee, even the short speech, tight vowels, and nasalizations have been related—half in jest—to the cold air.

Clearer was the effect of the rather complex land system, with its classes of membership in the community—first the proprietors of a town's original grant, and then the freeholders who were grantees or purchasers, the inhabitants, the church members. A legalistic quality appeared early in New England Puritanism: the arrangement with God—although intensely emotional, even mystical in the personal experience of conversion—was contained in a law-like covenant with God, and that ruler, like the English King, was bound by his promises and previous acts, inscrutable but ultimately legal and righteous. Quarrels over boundaries, rights, and contracts are endemic in peasant villages, but New England towns were notably litigious; and if fee-simple ownership and partible inheritance made for a cleaner definition of rights in real property than had obtained under feudal law, it also made the disputes so much the more bitter and significant. Moreover, in a poor agriculture, the capitalization of land, with the hopes and dreams it compounds, appears as the only escape from the prison of a narrow income margin; with the accession of some urban landed and mercantile wealth, property had become, even by 1650, a notable object of speculation. A class of wealthier landholders, speculators, and merchants had arisen by the early eighteenth century, and the yeoman farmer himself, often with more land than he could farm, surely had an eye out for sharp deals by which a neighbor or a foreigner could be skinned.

In mercantile trade, among industrial New England's original activities, these intangible preparations for the later industrialization most patently occurred. The story is the familiar one that appears over and over again in the history of the trading world. A few of the merchants whose enterprise gave the colonial Atlantic ports one of the world's largest merchant marines by 1775 had been attracted there from England a century after first settlement. The urban atmosphere, even of Boston, developed a strong Anglican and royalist component. Yet new English families, too, became acculturated to the not-so-very-different provincial society, and the native impulse to trade had appeared at the very first. Over the countryside, trade occurred among farmers, and between them and villagers, perhaps to the maximum extent that resources permitted. Fishing and flourishing shipyards, utilizing the abundant timber, gave their stimulus. By 1750, most of the organizational devices of mercantile capitalism were in place—partnerships, contracts, bills and notes, accounts, wages to clerks and sailors, intangible forms of property, commercial law and lawyers, bonds and notaries. And intertwined with the commercial activities came information—the knowledge of prices, markets, sea lanes, and distant people, customs and technologies. From an information industry of colleges, churches, newspapers, and pamphleteers arose the intellectual currents that so disturbed the ministers of Calvinism, penetrating so far into the countryside as to produce in reaction, perhaps not the witchcraft episode, but surely the "Great Awakening." On the back of trade, all the modernizing ideas of the eighteenth century were borne in on the isolated religious society. More tangibly, the wealth of towns attracted artisans from England, as well as from the farms, and created a class with skills and equipment able to rival those of European craft producers—a body of high craftsmanship without which the industry of the nineteenth century, had it derived only from the simple crafts of rural industry, could hardly have so readily reproduced the English industrial revolution.

In New England in the decades from 1790 to 1840, influences from the Puritan countryside mixed with the mercantile culture of the somewhat more worldly towns. From this conjunction derived a powerful and complex stream of factory industry in the first four decades of the nineteenth century. But peasant arts have existed in many cultures, and high mercantile capitalism with craft shops had flourished in Greece, in Rome, in late medieval Europe, and in Italy, Spain, France, and Holland of the seventeenth and eighteenth centuries. Far from stimulating further industrialization, an excess of mercantile activity in those cases almost stifled it. Why did industry blossom in Massachusetts Bay and in the remote valleys of the Naugatuck and Connecticut rivers, 150 years after settlement, just as on the European continent, the Rhine, the Meuse, and the Rhone Valleys were coming into modern industrial life? A historian must answer such deep questions not at first with general speculation but with an examination of the specific episodes in question. Industrialization came into New England in two clusters: (1) the textile mills in Rhode Island and eastern Massachusetts and their smaller copies along all the tributary streams where water power could be captured; (2) the metals-smelting and working industry, unfolding in a series of steps to make the Connecticut and Naugatuck valleys a pair of long machinery workshops. The two histories are quite different and distinct in their origins, in their forms of organization, in their markets, and in their implications for later industrial development.

The Textile Development

In textile history, the putting-out system that preceded the textile factories is a notorious part of England's and Europe's industrialization, but it has only a slight echo in the tale of the North American colonies. Here the market was not concentrated; the population was too dispersed; British restrictions, especially on wool, were firm; and British imports could undersell anything except the products of home spinning for domestic use. When the opportunity for putting out came in the 1790s, that phase in spinning had been eliminated by the new factories, and this mode of merchant-organized industry was left to a few products—cut nails, straw hats, and longest and most notably, boots and shoes. On summary view, the familiar history of the New England textile mills presents instead several distinct features—circumstances whose existence seems to have been indispensable for the development as it occurred.

First, there were the English inventions held in the mind of the migratory Samuel Slater,[5] and the other information that Francis Cabot Lowell got on his spying trips to England. The level of craftsmanship was good in Providence and Boston—good enough to construct the machinery that the Browns had set up at Pawtucket and that Slater found inadequate. But the denser net of artisans, the larger scale of effort present in the English Midlands, was needed to invent machinery. Without this initial borrowing (or theft) in the textile industry, as later in the brass industry, New England's "revolution" would probably have been long delayed. Moreover, stolen machinery was not patented, and once stolen, its diffusion could be rapid. Only a small start, a small lift over a threshold of knowledge, was required.

A second striking element was the availability in a new country of enough craft workers in construction, in woodworking, in metal fabrication, and stone masonry to make the machinery and design the buildings, dams, and structures to bring the mills into quick operation. The carpenter whom Slater and Brown hired to work, in a barn with shades down, to construct the machinery as Slater remembered it, was indispensable, but his equals were to be found at a hundred locations inland and along the coast. And as the mills were built, the machine shops came into being and began their semi-independent history. Machinists trained in textile shops were an important component of machine-tool manufacture when that spectacular development climaxed New England's industrial achievement.

[5] Slater's story is told somewhat dramatically in Massena Goodrich, *Historical Sketch of the Town of Pawtucket* (Pawtucket, 1876).

Once the plants were erected, the availability of a domestic labor force, of women and children and some adult males, testifies to the existence of a "labor surplus" and to the real, but slight margin of income on which the farms subsisted. Partially skilled through home practice on the spinning wheel but still in need of some further training, the cardinal feature of this labor force was its diligence, its readiness for a temporary emancipation from home and family, its willing independence—in short its inculcated Puritan values. Later, it was succeeded by a labor force of very different ethnic origin and culture—French Canadians and Irish, who worked equally well. But the presence of the "ladies of the loom" in the first mills solved a problem for the manufacturers that in many peasant societies would have posed a difficult if not insurmountable obstacle.

Nor must one, in an effort to avoid the "great man" interpretation, downplay the boldness, cleverness, and entrepreneurial energies and ingenuity of the Browns, the Lowells, the Cabots, and all the rest. At heart, they were not manufacturers, and their factories were truly an extension of mercantile capital enterprise and its mentality. The raw material was imported, the product largely carried off in trade, as in Hong Kong or Taiwan today. The textile enclave in New England was simply a processing point between remote materials and remote markets, a collection of labor, power-driven machines, and cunning capitalists. It was really not textiles, but machinery that the native industry made and with which it transformed imported raw material. And there is evidence indeed that the textile entrepreneurs looked at matters in just that way—and considered their fixed assets as a collection of values to be formed, manipulated, increased, transformed, and sold as the market might dictate. There was perhaps as little feudalism, as little patriarchy in New England textile firms as has ever been exhibited in the early stages of a manufacturing operation. The Boston merchants confined their philanthropies to the distribution of their wealth after—not before—they had made up their balance sheets.[6]

As an extension of such considerations, one cannot help but observe the thoroughness with which textile opportunity was pursued up New England's streams away from the shore. In upstate Massachusetts into the Berkshire hills, in west-central Connecticut, in southern Vermont and New Hampshire, one may say with only slight exaggeration that no possible site for a waterwheel was left untouched. When sheep

---

[6] Some of the points are made in a recent reexamination of the records of the Boston Associates by Jeffrey Oxley in work in progress on his doctoral dissertation, *"The Textile Manufacturing Corporation and Boston Capital, the Era of the Lowell System, 1813–1860"* (see Bibliography I, A).

moved into Vermont, woolen mills followed them; in north-central Connecticut, even a silk mill was opened in midcentury, with Jacquard looms and initially using local mulberry trees and caterpillars.[7] Mills in Holyoke and Chicopee rivaled those on the Merrimack. Unlike the mills of the early southern Piedmont industry that proliferated later in the nineteenth century, many of these establishments were not collections of a few hundred spindles, but substantial operations. This development forms an effective antidote to excessive emphasis on the industry's great men. Enterprise and enough small capital lay ready across the area to come to life at the first demonstrated reduction of risk; the opportunity was hunted up the streams as the trappers had hunted beaver two hundred years earlier, with the grim fanaticism of a Javert stalking his criminal prey. Is it too fanciful to call this thoroughness, this almost vengeful obsessiveness in the hunt for profit, a Puritan trait?

Metals and Machinery

The story of the Connecticut Valley's industrialization does not start with Eli Whitney or even with guns and machinery. Almost wholly neglected because wholly obscure are the eighteenth century activities connected with smelting the few and rather poor deposits of copper, zinc, and iron in the region, and with the forging, beating, and casting of the so-called hollow-wares—pots, pans, kettles, and small metal objects; nails, ax heads, buttons, buckles, and harness fixtures, pewter tableware and kitchen utensils. In New Haven and Hartford, these craft objects showed some of the skill and ingenuity common in the more elegant work of the sophisticated shops in Boston and Philadelphia. In smaller villages, from dozens of small shops, came a clattering stream of such wares in the late eighteenth century.[8] Distribution was probably not so much downriver, where competition would come from the seacoast urban artisans, but by peddlers overland—first on foot into the Hudson Valley, and soon by pack horse, fanning out by 1820 into Pennsylvania, lower Canada, and the upper South. Very little but legend is known of this unique commerce, but certainly it, and not the more prosperous river trade in agricultural products, represents the mercantile phase of the Connecticut Valley's industrialization.

As the nineteenth century's markets opened up, the obsession with

---

[7] H. H. Manchester, *The Story of Silk and Cheney Silks*, (Manchester, Conn.: privately printed, 1916). Manchester does not give sources, but since he is listed as Managing Editor, "The Library of Original Sources," it may be presumed that he had some. I am indebted to Y. V. Fultz for some preliminary research in this area.

[8] The legends of the Connecticut Yankee pots, pans, and peddlers seem to be derived mostly from travelers' reports, (de Chastellux, Dwight, Kendall) and from Lathrop's 1909 Yale dissertation, "The Brass Industry in Connecticut" (see Bibliography I, C).

New England's Early Industrialization

alloying, gilding, and plating continued. The unalloyed metals—iron, copper, zinc, tin—were imported in bars. In the Waterbury area brass, fused by a newly invented English technique, became the leading metal, and rolling was added to molding and forging as a technique to form semifinished or final products. Hardware, buttons, hooks, German silver, and finally, plated silverware became important forms of industry, each in a special town. The climax of this tinkering with the chemistry of molten materials came at last in the vulcanization of rubber by Charles Goodyear, a native of Hamden, after incredible trials and false starts as he moved between Philadelphia and Boston in the years 1835–1845.[9] To the proliferating brass works, silverware makers, and hardware shops of the Naugatuck Valley, parallel to the Connecticut, a significant industry in rubber wares was added by 1850. Both industries depended on imported materials out of which inventiveness, craft skill, and ingenious marketing techniques had created a complex group of Yankee manufactures.

Mechanical advancement in the story begins not with the metal wares, but with wooden clocks.[10] Already by 1790, clocks with wooden works were something of an archaism, dating back to fifteenth century Germany. Clock makers of Paris, London, Boston, and Philadelphia were making works of brass, expensive and finely crafted. Wooden clocks were a country product among German farmers in Pennsylvania, and they exercised the whittling skills and materials of some Yankee farmers. But the works were adapted to the peddler system of distribution; they could ride on a peddler's back or in his saddle bag. Crude and sturdy, they could hold up with little or no adjustment, and could be relied on to tick when unloaded after a long, bumpy journey and fitted into a locally made case. And here the market called out the clock industry's counterpart to Whitney, Eli Terry, a skilled craftsman himself (more so than Whitney) who trained many more clock makers in his shop. This entrepreneur-craftsman appears, no less than

---

[9] On Goodyear, see Ralph F. Wolf, *India Rubber Man: The Story of Charles Goodyear* (see Bibliography I, E). Wolf draws on Goodyear's own account, which he published on rubber pages in a rubber binding, and the account, with references, in *The Dictionary of American Biography*.

[10] Recently, a doctoral student, Scott Molan, has gone over this ground, originally worked over by J. J. Murphy, and carried the investigation further. Among these and the other references in the Selected Bibliography I, B, below, special mention should be made of the early reminiscences of the New Haven entrepreneur Chauncey Jerome, *History of the American Clock Business for the Past Sixty Years, and Life of Chauncey Jerome Written by Himself, including Barnum's Connection with the Clock Business*. This has been reissued by C. H. Bailey, the Managing Director of the American Clock and Watch Museum at Bristol, Conn., 1983. I am indebted to Mr. Bailey for a useful discussion and his interesting exhibition in the museum.

Whitney, to have conceived the idea of batch production of the parts of mechanisms for later assembly.[11]

Terry, in a shop in Plymouth, Connecticut, with a few assistants, announced in 1807 his intention of producing 3,000 clocks in a year. He not only carried out his intention, but even more surprising to his fellow craftsmen, he found no trouble in disposing of his products, at prices much lower than for custom-made clocks, in shops as far south as New Haven and through peddlers into the South and West. Almost nothing is known of Terry's actual methods or machinery, but hand- and water-powered lathes were part of the clock shop's equipment by the 1820s, and it was in the set up and repetition of operations on such machine tools that most cost saving could occur. Terry's quality as a man is shown in his openness and generosity; many clock makers—Seth Thomas, Chauncey Jerome—were trained in his shop. And when his fortune was made by the 1830s, he retired from the competition to spend his last years in his shop making clocks for the sheer pleasure of the craft, by the old craftsman's methods. It is interesting that in the Connecticut Valley the art of woodworking—which in other rural cultures was expressed in statues of saints or ornamentation of doors, thrones, and bedsteads—found its highest expression in the production in scale of moving wheels and balances for the useful purpose of rendering a count of the passing hours. And this art itself, developing from custom hand work to machined batch production and assembly, fell victim at last to the logic of the market. The panic of 1837 is said to have bankrupted dozens of the overstocked and overextended shops, and recovery found the brass clock, by virtue of improvements in the rolling and stamping of brass wheels, taking over the market; well-designed brass shelf clocks could sell for a few dollars, as against the 12–18 dollars required to cover the cost of a wooden grandfather's clock. By 1840 the metal clock and watch industry in Waterbury and nearby towns had begun its long and profitable career.

[11] It is important to emphasize the difference between these methods and those that later produced truly "interchangeable parts." In both early guns and clocks, parts could not be made by hand tools or even by primitive machines and inexact measuring instruments so as to be indiscriminately interchangeable. Once made in batches, they needed shaving or filing to be made to fit. In wooden clocks, the fit could be much looser than in a gun, and the pressures to which the mechanism was subjected were in no way comparable. Moreover, clocks were stationary while guns had to stand up to very rough treatment. Terry's and Whitney's invention, and that of other gun makers before 1830, was to make parts in batches to save set-up costs on tools or such machinery as was used. By and large, the parts still needed some planing, grinding or filing to fit an assembly. Hounshell's recent remarkable book, *From the American System to Mass Production, 1800–1932* (see Bibliography I, E), goes a long way in bringing the story together.

Brass and other hardware, rubber wares, clocks—all were important parts of the Connecticut Valley's development. They came early in the region and stayed late. But firearms hold a central place in the history in several respects. As the industry began to develop, the private demand proved to be vast. A gun was part of nearly every farmhouse's equipment, on the frontier to protect from Indians, and there and within the frontier to hunt game and keep down rodents. But repeatedly, in the 1790s and again at the times of the Mexican and Civil wars it was the large block orders from the federal government and the state militias that pushed the craftsmen into machinery and the machinery into mass production. The two armories created by the Act of 1795 followed European mercantilist precedent, but the ease with which personnel, orders, and techniques passed between the Springfield Armory and the private manufacturers—Whitney, North, and Colt—showed how little the distinction between public and private meant for the final result. More striking is the difference in efficiency and progressiveness between the Yankee armory and that at Harpers Ferry, Virginia. Merrit Roe Smith's splendid study of the latter shows the two cultures in sharp contrast.[12] Springfield, especially under Roswell Lee, was a highly organized, efficient body of skilled workmen, not only building guns but also creating gun-making techniques, instruments, and machinery. The shops of Whitney, in Hamden, and North, in Middletown, stimulated by large federal contracts, worked not competitively but in considerable harmony with the armory. In the Virginia establishment, the only efficient portion, devoted obsessively to the ideal of interchangeability and large-scale production, was the rifle works under the mad Maine Yankee, John Hall. Hall maintained his pursuit of exact calibration, by means of a huge array of fixtures and gauges at every step of rifle production, with the obsessive determination of Captain Ahab chasing Moby Dick. His survival in the antagonistic, highly political local culture of the armory—a culture of influence and family connections that extended from Harpers Ferry to Congress and the White House—has the sound of a morality play of shining Puritan virtue and Virginian vice.

Following Whitney, North, Lee, and Hall, Samuel Colt—a man of many talents—carried the industry into midcentury. Notable in both the clock makers and the gun makers is a curious combination of skill, ingenuity, and showmanship. Whitney, North, Terry, and Hall were craftsmen of high purpose and enlightened devotion to a technological and engineering ideal that formed a basic part of the Yankee culture.

[12] Merritt R. Smith, *Harpers Ferry Armory and the New Technology* (Ithaca; Cornell University Press, 1977).

But Whitney and Jerome in clocks, and in guns, Colt above all, remind us that P. T. Barnum was also a Yankee.

From Machinery to Machine Tools

Production of machine-made textiles requires machinery, and the machine shop was an important feature of a textile mill—the principal feature, one might say, of some of the large ones. This aspect of New England's industrialization—the trail from textiles to textile machinery to the machine and power tools to build textile machinery, thence to generalized machine tools for a variety of basic operations on metal—has been chronicled in a well-known article by Nathan Rosenberg. He has dubbed the phenomenon, "technological convergence": the simultaneous demand for standard lathes, milling machines, shapers, drills and the like, needed to make machinery of the most varied and specific sorts and so to turn out the stupendous array of final goods in an industrial economy. Obviously a large market for machinery itself is required to offer the scale economies in the serial production of standardized machine tools. But textile machines could offer only a small part of a market of requisite scale.

Important, and somewhat unobserved, is the nature of the Connecticut Valley manufactures—clocks and guns that were themselves machines. They had to be made by craftsmen who, if they began to employ batch production and powered tools, became machinists. Unlike the skills in textile manufacture, the skills used to make a gun or a clock were readily transferable to making a machine that would replace hand operations as surely as a gun would replace a stone axe or a bow and arrow.

The New England machine tool industry grew, then, from the two machinery centers: the area west and south of Lowell, from Worcester to Providence, and the Connecticut Valley from Middletown as far up as Windsor, Vermont.[13] Of the two areas, the latter became more central, but in both areas, intensified industrialization shaped cost and market conditions for machine tools and production line in manufacture and in both areas, that intensification corresponded to two phases of especially intense pressure from New England's external environment.

The first was the hothouse effect of the break with England. British restrictions and British competition disappeared from 1776 to 1783,

[13] Carolyn Cooper's recently completed Yale dissertation (see Bibliography I, E) studies the role of one Yankee inventor—machinist-entrepreneur, Thomas Blanchard—in this development. Blanchard invented a lathe for copying gunstocks, shoe lasts, and other complex patterns; a machine for bending wood; and a number of other devices. The dissertation studies not only the technical but also the legal and social aspects of Blanchard's efforts and career.

and competition fell away again in the Napoleonic Wars and the War of 1812 and was finally limited by a protective tariff—moderate by later standards, but significant in that it was established, against the initial resistance of the merchants, as a part of the American tax structure and foreign economic policy. Trade with the British Caribbean had been temporarily interrupted, but the breakup of Spain's mercantilist empire in South America gave new outlets. More important, the Southern plantations, having completed a cycle of growth based on tobacco, were in 1800 just entering their violent and brilliant history based on slaves and cotton, creating both vast supplies of the raw material and a rapidly growing population, black and white, to be clothed and shod. The antebellum period had its boom decades, particularly the 1830s and 1850s, which gave an unevenness to the movement of demand for purchased farm capital items, and which affected the financial conditions of the manufacturing firms. But so long as the agricultural population continued its strong and steady increase, and new lands continued to be both abundant and rich, the continued growth of the textile and the shoe industries was assured.

The second phase—rather different and more complex than the South's effect on New England's industrialization—derived from the influence of the Old Northwest. In this case the first effect was not by way of markets, but by way of migration. Just as the Connecticut lands had pulled families from Massachusetts in the seventeenth century, so central New York pulled settlers from Connecticut and Vermont in the early nineteenth. The opening of the Ohio country produced several waves of migration; beginning even before 1820, the Connecticut Western Reserve in northern Ohio began to fill up. The effect of the westward movement on the various New England farming and industrial areas was not uniform. A movement north was going on at the same time as the first movement west, so the period of Vermont and New Hampshire's agricultural decline was delayed by a few decades. Already the mill towns in eastern and central Massachusetts had begun to reach as far away as Quebec and Ireland to attract farm labor into industrial employment. Connecticut was the hardest hit. From 1800 to 1840 the state's population increased only about 5 percent per decade, compared to 35 percent for the United States as a whole. Migrants in numbers equal to the whole of the natural increase, that is, to at least the size of the state's population in 1800, had gone west by 1840.[14]

[14] Clarence H. Danhof, *Change in Agriculture: the Northern United States, 1830–1870* (Cambridge: Harvard University Press, 1969). Chapter 1 has a good account of the Midwest/Northeast interaction, from the viewpoint of the areas receiving the immigrants.

We have at this point little data, only much theory to guide the story, but the migration must have had a triple-edged effect on Connecticut's young metals and machinery industries, influencing the market, labor supplies, and technology. The migrants, with their companions from other states, must have been mostly farmers moving to where, on new lands, they could farm much the same array of crops and livestock products as they had been, but more productively. Almost from first settlement on, the "East" was to a large degree fed, particularly in meat, from successive "Wests"—from central Massachusetts, from central New York, from central Ohio, and finally from Illinois and, after 1860, Texas and the Plains. The return shipment of the western surpluses, after construction of the canals of the 1820s and the trunk lines of the 1840s, caused the abandonment of farms and basic shifts in the crop mix. The shift from grains reduced labor requirements in northeastern farming, although dairying and hay production to some degree balanced this off. But the sale of those surpluses enlarged the market among farms, small towns, and growing cities in the Old Northwest for just the kinds of consumer durable goods that the manufacturers had begun to turn out. At the same time, with cheaper food and growing markets as a lure, the migrations surely caused Connecticut some loss in its artisan and skilled work force. The occupational composition of the migration stream is not known, but even if it was composed mostly of farmers, the supplies of artisan labor in Connecticut clearly did not increase. The occupational lists of the Midwest towns show a good complement of smiths, mechanics, millers, and the like. Surely many of these had been among the migrants.

Some economic historians, with a fascination for economic theory, have averred that under the market tension in which they existed, and relative to obviously not very ample supplies of something called "capital," American farming and industry nevertheless suffered a chronic shortage of labor, particularly of unskilled labor.[15] The textile industries met this challenge by adopting the machinery that with some

[15] Two recent contributions to this topic are: the papers by Alexander Field, "Land Abundance, Interest/Profit Rates, and Nineteenth-Century American and British Technology," *Journal of Economic History* 43, no. 2 (June 1983) and John James and Jonathan Skinner, "Labor Scarcity in Nineteenth Century Manufacturing," University of Virginia Seminar Papers (Charlottesville 1983). The topic has been complicated by the need to take England as a standard of comparison, although until Field's paper, it was never clear whether the ratio of capital costs to wages was higher or lower there, since America seemed to be both "labor-scarce" *and* "capital-scarce," relative to land. The problem appears to be full of internally created puzzles, and will probably drift out of sight before a "solution" that conforms both to logic and to the facts can be achieved. The papers of James and Skinner, Atack, and David in the present volume are at least tangentially related to the issues involved.

use of mechanics' labor and working capital, greatly increased the productivity of unskilled or semiskilled, formerly domestic labor. Women and children and unskilled farm labor from French Canada and ultimately from Ireland could be used on the machines. In western farming, too, the story is of the labor-augmenting inventions, with modest capital requirements, and farm-fed horsepower, that could keep a farmer fairly busy over wide enough acreage to yield a large commercial surplus. The sewing machine, shoe machinery, the typewriter—all had the effect of substituting an item of manufactured "capital" for unskilled or moderately skilled labor, stretching the labor force to handle a larger volume of resources or semifinished materials. Overall, this is no doubt the way American industrialization worked, and this circumstance may have given a peculiarly strong "labor-saving bias" to the inventive effort and to practice in shop and field. Interchangeable parts, the systematization of work, the ever-finer division of labor, then, may be seen as having been adopted not to utilize idle supplies of unskilled labor, but to economize on unskilled labor by furnishing processes and machines for the tasks to be done—even de-skilling artisan labor in the abandonment of hand processes, with a substitution of system and machinery for skill.

The situation of the Connecticut manufacturers fits this model, but with a difference. Here not unskilled, but skilled labor was the problem and the products by and large were themselves machines for the final market. The same workers who could produce these machines could with a little retraining, devise and work out machines to produce these machines, that is, machine tools. Wages and the array of skills were presumably much the same at both levels of production. No doubt, at some absurdly high rate of interest it would have been too expensive to withdraw workers from current production to form this capital. But it would not have taken much for the value of machine tools to rise enough to swamp the additional interest charges. And once the production structure was thus lengthened, with the maintenance of active competition among producers, not only the cost of consumers' capital, but the cost of producers' machinery and of machine tools themselves would fall. This process cheapened real capital, and so stretched the stock of savings or financial capital, thus making possible an extensive, and often even wasteful use of machinery to augment labor and, ultimately, almost to replace skilled craft labor in most jobs.

From this background arose, in the years 1820–1870, the notable specialized machine-tool producers: Brown and Sharpe, Jones and Lamson, the Whitin works, and others. Alas that records no longer exist to detail just where, when, and at what wages and prices and profit margins the industry was drawn into being![16] Accompanying its development was the application, as in guns, of great mechanical ingenuity. The milling machine is claimed by several of the great names, notably Whitney and North. More certain in attribution are the stocking lathe and Blanchard's other patented and much sued-over inventions. The development of gauges, calipers, jigs, and fixtures is all part of the history of this precision machinery. The manufacture of parts for assembly required precision in the tools, and once such precise and controllable machine tools were made, with flexibility and adaptability to varied purposes, they could be used in turn to make copies of themselves. The generative organs of machine industry were thus created, in Connecticut, and in the Northeast's other machine tool center, north of Philadelphia—as later also in the areas of Cincinnati, Cleveland, and Detroit.

In the second half of the nineteenth century, the industrial tendencies displayed before 1850 were further played out. Textiles, machinery, brass, hardware, guns, machine tools—all continued to grow and flourish while New England's agricultural decline spread to the northern tier of states and out-migration continued. A new element was the financial wealth that was accumulated in New England and moved west, into land, minerals, and railroad and industrial stocks and bonds. But migration alone—of labor, of capital funds, or even of the stream of light machinery the region exported—did not by itself accomplish the industrial miracle of the Middle West. There a late New England Puritan culture mixed with the migration streams from Pennsylvania and the southern uplands to the point where, in the presence of vast new mineral resources, the frame of a complete industrial structure began to arise, encompassing light industry, heavy industry, an agricultural base, and an overlay of commercial and legal institutions and practices. With the midwestern industrial development, the story moves into the twentieth century and New England's own industry moves into the cycle of abandonment, conversion, and partial revival that has characterized its life in recent decades.

III. Some Reflections on the History

"A la fin de toute pensée, il y a un soupir"—attributed to Paul Valéry

Between 1790 and 1840, along its rivers, its streams, and its coastline from Salem to Bridgeport, southern New England experienced a no-

[16] J. W. Roe, in *English and American Tool Builders* (see Bibliography I, D) draws on his earlier Yale dissertation and on such records as were available then. I have found no general treatment published since then to replace it, despite good specialized studies of the Saco-Lowell mills, the Whitin Works, and others.

table Industrial Revolution, lagging behind England's by only a few years. By the standards that economic historians—bourgeois or Marxist—apply, New England's Industrial Revolution was a success. To a Marxist it appears to have broken the mold of a precapitalist form: that of the patriarchal family economy, or some would hold, that of an equally ancient form, the communal village. A new class, merchants and petty capitalists, exhibiting its ingenuity in workshop and marketplace, replaced priests and a king—the latter by a violent revolution that proved to be the forward shadow of its great counterpart in France. In Yankee counting houses and mills, American business civilization was born and began its giantlike growth into a bourgeois society still today awaiting its proletarian revolution. To the bourgeois historian, lacking the Marxist's story line, New England represented a region of that Britain whose culture and Industrial Revolution it shared, or a second "follower country" to which, in its technology, society, economy, religion, and politics, the impulse of the British "modernization process" had been transmitted.

The history of success is, of course, a bore when repeated over and over again. Tolstoy is reputed to have said that all happy families are alike; only the tragic, miserable ones have stories. Without a personal involvement, one's interest in the history of a locality or a firm or even a business leader is short-lived. To paraphrase a notorious comment, when you've seen one textile mill, or one machine shop, you've seen 'em all—unless you are yourself in the trade. Even the scale on which the history is played out does not much matter. Textile mills for cottons or woolens, brass foundries and rolling mills, iron works, machinery and machine-tool shops—the industrial histories have much the same form, with small idiosyncrasies arising from peculiarities in the raw material, the market, or the circumstances of the specific time when the growth began to occur.

In the industrialization of a region, these industries are interlinked. At that level of learning and generalization, the observation of those linkages, implicit in the relation of the technologies to one another and to the human culture out of which they all arise, may arouse interest, curiosity, and even admiration. Town histories come to life when, brushing over them as if rubbing a tombstone, one sees geographical patterns that have an economic or sociological explanation. Histories of firms, arranged by industry and stacked up according to the structure of production, show the "Rosenberg effect." Such a generalization could have been deduced, even "predicted," from a knowledge only of the technologies and the rate of market growth. But it almost certainly would never have been thought of without a large measure of induction—an examination of the actual histories of firms and inventions.

The dynamics implicit in the physical nature of the environment and materials—the "challenge" side of the Toynbee thesis—is one-half of the story that lies hidden like a sea monster, immanent in the history and rewarding long hours of observation. The other half of the story—the "response" by which that potential dynamic is activated—lies in the all-pervasive and complex character of what one is forced to call, for want of a more precise and suggestive term, culture: culture of the artisan, the mechanic, the merchant-entrepreneur, the self-educated inventor; but a culture also of rationality, of close observation. That culture channeled its creativity along practical, material lines, combining with its materials a spiritual sense of both self- and civic righteousness—a self-confidence and self-esteem that came to the descendant of the Puritan, often all too easily, when the sense of sin dissolved into the consciousness of probable election. Foremost in this culture was a sense of pride in what was deemed worthy achievement: the determined effort, by farmer, workman, and businessman alike, to make money in clever trade and in useful production. It is this culture that forms the substratum from which every town history, the history of every firm, of every locality or industry in the region draws its life.

Just as the history of towns or firms in a region, endlessly repetitous, gains meaning only when seen in a pattern, so the histories of one industrializing region after another in the nineteenth century exhibit similar traces of monotony. The textile industries march, endlessly it seems, in wearying parade from one new industrial region to another, like the dove sent out from the ark finding no spot on which to land. In North America, New England's textile phase finds repetition in the southern Piedmont nearly a century later. There the interesting question is a negative or hypothetical one: why did that development fail to occur there earlier as part of the boom times of slavery and cotton? In the Middle West, New England's mechanical industries grew not on harsh soil but on a rich agricultural base, and pre-industrial mentalities, if they ever existed in early colonial times, had long since been replaced by concepts of endless horizons and illimitable wants as a permanent and self-renewing "challenge." At the same time, these light and craftlike mechanical industries fuse after 1870 with the rather different history of the Pennsylvanian heavy industries to form at last, with the huge discoveries of coal, ore, oil, and gas, and in the presence of lake shipping and the railroad, the great belt of American heavy industry. The industrialization of the American economy, broken down by industries and regions, thus forms a history whose outlines are found also in those of the other self-contained national economies of the nineteenth and early twentieth centuries—the German, the French and eventually the Russian, and even, with the exception of the agricultural base, that of Britain. Smaller economies—Switzerland, Sweden, northern Italy—exhibit the entire history in miniature or pieces of it, depending on their positions in world trade.

Still to American and world industrialization, the New England episode is commonly said to have contributed at least one new idea, exhibiting it in so exaggerated a form as to make it the unmistakable and readily imitated hallmark of modern industry. This contribution was the ideal of interchangeable parts and of the thorough organization of machines and labor to create "mass" production. This was the trait of American manufacture that attracted attention from visiting Englishmen and at international fairs at midcentury. Obviously it was not simply in application to guns and clocks that the idea made history. It was the thoroughness, zest, and variety with which the principle was applied when the truly mass market and massive resource base opened up the Midwest that turned Yankee ingenuity into Taylorism and Fordism. The history of the New England episode, taken in the whole context of antebellum American economic history, should illuminate the reason why this development could take such firm hold in so individualistic a society and could prove so fruitful.

Where in colonial history is the origin of this taste for organization to be discerned? One hesitates to trace so complex a history back into the thorny brambles of the Puritan mind. But for the history of American industrialization in the nineteenth century, an anomalous, if not self-contradictory, aspect of the American character, or at least of our view of it, remains to be resolved. How did a nation of individualistic farmers and small-time entrepreneurs move with such ease into the modern world of the large corporate organization? Part of the answer may lie in the traditional sympathy of nineteenth century Americans, noted by de Tocqueville, for community organization. The small town was itself controlled by a neighborly vigilance as close as that of the factory foreman's eye. Evident, too, is the tendency of Americans, spanning a continent, to think big and to admire bigness. Something of the self-assured Texan permeates our picture of the enterprising part of the population everywhere. Yet community life and large settlement projects were organized essentially on a small scale. Nothing about them shows the nature of a centralized corporate management. Waste, not rationalized efficiency, was the hallmark of their success. What trait is it that the individualistic entrepreneur and the production manager and engineer have in common? How do we move so easily from the world of F. J. Turner to the world of A. D. Chandler?

The answer, I suggest, may be found in seventeenth century English Puritanism, as its doctrines and sensitivities were transformed by selective evolution into the several New England varieties and thence into the secular ethic of a nineteenth century Protestant population. For the trait in question is simply an unusual insensitivity to human society, considered as a structure of emotion, vitality, and artistic sensibility. The Puritan drew energy and intolerant self-confidence from the knowledge not merely that he might be saved, but that in Christ he had found the only way to salvation. And he drew direction and guidance from an equally firm belief in the rationality both of the covenant with God and of God's revelation of His order in the physical universe. By the eighteenth century, faith in science was as strong as faith in God; indeed the two faiths had fused together. The Puritan could proceed, then, to inflict order on the world with the same passion with which he had tortured his individual conscience in its search for faith and grace. In the language of two other cultures, the Yogi and the Commissar were two faces of the same soul.

But to convince puritanical economists of the essential Puritanism of their own ways of looking at the world is not the task for a short paper, or for a short life, particularly for one born in that culture and sharing its values, its inspiration, and its blindness.

Selected Bibliography

Even a history that has been relatively neglected has managed over the years to acquire an appreciable bibliography. The list that follows gives only sources on the industrial history itself. A bibliography of sources on (1) social and legal history (including village studies and travelers' accounts, mostly well-known to American historians) and on (2) the relevant intellectual history—the usual sources on Puritanism—and the issues of economic theory raised in section III is available on application to the author.

I. MANUFACTURING INDUSTRY

A. Textiles and Shoes

Burgy, J. Herbert, *The New England Cotton Textile Industry*. Baltimore: Waverly Press, 1932.

Faler, Paul G. *Mechanics and Manufacturers in the Early Industrial Revolution: Lynn, Massachusetts, 1780–1860*. Albany: State University of New York Press, 1981.

Goldin, Claudia, and Sokoloff, Kenneth. "Women, Children and Industrialization in the Early Republic: Evidence from the Manufacturing Censuses." *Journal of Economic History* 42 (Dec. 1982), pp. 771–774.

Hazard, Blanche Evans. *The Organization of the Boot and Shoe Industry in*

Massachusetts Before 1875. Cambridge: Harvard University Press, 1921. Reprint. New York: Johnson Reprint, 1968.

Lincoln, Jonathan Thayer. "Material for a History of American Textile Machinery: The Kilburn-Lincoln Papers." *Journal of Economic and Business History* 4 (Feb. 1932).

Oxley, Jeffrey. *"The Textile Manufacturing Corporation and Boston Capital, The Era of the Lowell System, 1813–1860."* Doctoral dissertation, Stanford University, in process.

Shlakman, Vera. *Economic History of a Factory Town: A Study of Chicopee, Massachusetts.* Smith College Studies in History, vol. 20 (1934–35).

Ware, Caroline F. *The Early New England Cotton Manufacture.* Boston & New York: Houghton Mifflin, 1931.

B. Clocks

Barr, Lockwood Anderson. *Eli Terry's Pillar & Scroll Shelf Clocks* (n.p., 1952).

Camp, Hiram. *Sketch of the Clock Making Business, 1792–1892.* New Haven, 1939 (pamphlet) in the Yale University Library.

Hoopes, Penrose R. "Early Clockmaking in Connecticut." In *Historical Publications of the Tercentenary Commission of the State of Connecticut*, vol. 3. New Haven: Yale University Press, 1933–1936.

Ingraham, Edward. *Connecticut Clockmaking* (pamphlet). Reprinted from 56th Annual Report, the Connecticut Society of Civil Engineers, in Yale University Library.

Jerome, Chauncey. *History of the American Clock Business for the Past Sixty Years, and Life of Chauncey Jerome.* New Haven: F. C. Dayton, Jr., 1860. Reprint. Bristol, Conn.: American Clock and Watch Museum, 1983.

Molan, Scott. *The Origin of the Machinery Industry in the United States: A Regional History of Mechanization in the Nineteenth Century.* Doctoral disseration, the New School of Social Research, in process.

Murphy, J. J. "Entrepreneurship in the Establishment of the American Clock Industry." *Journal of Economic History* 26 (June 1966). pp. 169–186.

Roberts, Kenneth D. *The Contributions of Joseph Ives to Connecticut Clock Technology, 1810–1862.* Bristol, Conn.: American Clock and Watch Museum, 1970.

C. Brass and Brasswares

Bridgeport Brass Co. *Seven Centuries of Brass Making.* Bridgeport, 1920.

Brecher, Jeremy, Jerry Lombardi, and Jan Stackhouse, eds. *Brass Valley: The Story of Working People's Lives and Struggles in an American Industrial Region.* Philadelphia: Temple University Press, 1982.

Coyne, Franklin, E. *The Development of the Cooperage Industry in the United States, 1620–1940.* Chicago: Lumber Buyers, 1940.

Greeley, Horace, et. al. *The Great Industries of the United States,* Hartford: J. B. Burr & Hyde, 1872.

Howell, Kenneth T., ed. *History of Abel Porter & Company from which Scovill Manufacturing Company is the Direct Descendant.* Waterbury, Conn.: Scovill Manufacturing Co., 1952.

Lathrop, William Gilbert. "The Brass Industry in Connecticut." Dissertation, Yale University, 1909.

———. "The Development of the Brass Industry in Connecticut." In *Historical Publications of the Tercentenary Commission of the State of Connecticut*, vol. 6.

Marburg, Theodore F. *Management Problems and Procedures of a Manufacturing Enterprise, 1802–1852; a Case Study of the Origin of the Scovill Manufacturing Company*. Doctoral dissertation, Clark University, Worcester, Mass., 1942.

Scientific American. Articles on brass manufacture: vol. 41 (Dec. 13, 1879), p. 380; vol. 42 (May 1, 1880), p. 277.

Scovill Manufacturing Co. *Brass Roots*. Waterbury, 1952.

Scovill Manufacturing Co. pamphlets—economic aspects, in Yale University Library: (1) *Scovill Manufacturing Co.*, Waterbury, 1802–1935; (2) *Historical Analysis of Scovill Manufacturing Co.*, "Scovill World-Famed Pioneers in Brass," E. L. Newmarker, 1928; (3) *Scovill Manufacturing Co., The Oldest Brass Company in America*, (reprint from *The Metal Industry of NYC*, Aug. 1923) Publications of the Scovill Manufacturing Company: (1) *Scovill Manufacturing Company* (n.p., n.d.); (2) *The Mill on Mad River*, Waterbury, Conn., 1953.

D. Arms

Cain, Louis P., and Paul J. Uselding, eds. *Business Enterprise and Economic Change*. Kent, Ohio: Kent State University Press, 1973.

Cooper, Carolyn C. "Eli Whitney's Armory: Myth, Machines and Material Evidence." *Journal of the New Haven Colony Historical Society* 31 (Fall 1984), 19–34.

Deyrup, Felicia Johnson. *Arms Makers of the Connecticut Valley: A Regional Study of the Economic Development of the Small Arms Industry, 1798–1870*. Smith College Studies in History, V. B. Holmes and H. Kohn, eds., vol. 23 (1948).

Green, Constance M. *Eli Whitney and the Birth of American Technology*. Boston: Little, Brown, 1956.

Mirsky, Jeannette, and Allan Nevins. *The World of Eli Whitney*. New York: Macmillan, 1952.

North, S. N. D., and Ralph H. North. *Simeon North, First Official Pistol Maker of the United States*. Concord, N.H.: Rumford Press, 1913.

Smith, Merritt Roe. *Harpers Ferry Armory and the New Technology*. Ithaca, N.Y.: Cornell University Press, 1977.

Uselding, Paul. "Elisha K. Root, Forging, and the 'American System.'" *Technology and Culture* 15 (Oct. 1974).

E. Machinery and General Industrial History

Bishop, John Leander. *A History of American Manufactures, from 1608 to 1860*, vol. 1 *1608–1789* vol. 2, *1789–1860*. Philadelphia: Edward Young & Co., 1864. London: S. Low, son & Co., 1864.

Blackall, Frederick S., Jr. "Invention and Industry—Cradled in New England!" Address to Newcomen Society of England, American Branch, New York (pamphlet), 1946.

Bridenbaugh, Carl. *The Colonial Craftsman,* New York: New York University Press, 1950.
Clark, Victor Selden. *History of Manufactures in the United States,* vol. 1 *1776–1860.* Washington, D.C.: Carnegie Institution of Washington, 1916–1928. Published for Carnegie Institution of Washington by McGraw-Hill, 1929.
Cooper, Carolyn C. "The Role of Thomas Blanchard's Woodworking Invention in 19th Century American Industrial Technology." Doctoral dissertation, Yale University, Program in the History of Science, Sept. 1985.
Day, Clive. "The Rise of Manufacturing in Connecticut" in *Historical Publications of the Tercentenary Commission of the State of Connecticut,* vol. 5.
Fuller, Grace Pierpont. *An Introduction to the History of Connecticut as a Manufacturing State. Smith College Studies in History,* J. S. Bassett and S. B. Fay, eds., vol. 1. Northampton, mass., (Oct. 1915).
Gibb, George Sweet. *The Saco-Lowell Shops.* Cambridge: Harvard University Press, 1950.
Grant, Ellsworth Strong. *Yankee Dreamers and Doers.* Chester, Conn.: Pequot Press, n.d.
Hounshell, David A. *From the American System to Mass Production, 1800–1932.* Baltimore: Johns Hopkins University Press, 1984.
Hubbard, Guy. "The Influence of Early Windsor Industries Upon the Mechanic Arts." *Proceedings of the Vermont Historical Society, 1921,* pp. 159–182.
———. "The Machine Tool Industry." In *The Development of American Industries,* John G. Glover, ed., Ch. 26. New York: Prentice-Hall, 1936.
Parton, James. *Famous Americans of Recent Times.* Boston: Fields, Osgood & Co., 1869.
Roe, J. W. "Connecticut Inventors." In *Historical Publications of the Tercentenary Commission of the State of Connecticut,"* vol. 4.
———. *English and American Tool Builders.* New Haven: Yale University Press, 1916. Reprint. New York: McGraw-Hill, 1926.
Rolt, L. T. C., *A Short History of Machine Tools.* Cambridge: M.I.T. Press, 1965.
Rosenberg, N., ed. *The American System of Manufactures.* Edinburgh: Edinburgh University Press, 1969.
Steeds, W. *A History of Machine Tools, 1700–1910.* London: Oxford University Press, 1969.
Taylor, George Rogers. *The Transportation Revolution, 1815–1860.* New York: Harper & Row Torchbooks, 1968.
Tryon, Rolla Milton. *Household Manufactures in the United States, 1640–1860.* Chicago: University of Chicago Press, 1917.
Weaver, Glenn. "Industry in an Agrarian Economy: Early Eighteenth Century Connecticut." *The Connecticut Historical Society Bulletin* 19 (July 1954).
Williamson, Harold. *The Growth of the American Economy,* 2nd ed. New York: Prentice-Hall, 1951, Chs. 9–12.
Wolf, Ralph F. *India Rubber Man: The Story of Charles Goodyear.* Caldwell, Idaho: Caucton Printers, 1939.
Zimiles, Martha, and Murray Zimiles. *Early American Mills.* New York: Bramhall House, 1973.

II. RESOURCE INDUSTRIES AND TRADE

Bailyn, Bernard. *The New England Merchants in the Seventeenth Century.* Cambridge: Harvard Univeristy Press, 1979.

Bidwell, Percy W., and John I. Falconer. *History of Agriculture in the Northern United States, 1620–1860.* Washington, D.C.: Carnegie Institution of Washington, 1925.

Cowles, Alfred A. "Copper and Brass." In *1795–1895. One Hundred Years of American Commerce.* New York: D. O. Haynes & Co., 1895, Ch. 47.

Cronon, William. *Changes in the Land.* New York: Hill and Wang, 1983.

Defebaugh, James Elliott. *History of the Lumber Industry of America*, vol. 2. Chicago: The American Lumberman, 1907.

Innis, Harold A. *The Cod Fisheries: The History of an International Economy.* New Haven: Yale University Press, 1940.

———. *The Fur-Trade of Canada.* In University of Toronto Studies, History and Economics, vol. 5, no. 1. Toronto: Oxford University Press, 1927.

Johnson, Emory R., T. W. Van Metre, G. G. Huebrier, and D. S. Hanchet. *History of Domestic and Foreign Commerce of the United States*, vol. 1. Washington, D.C.: Carnegie Institution of Washington, 1915.

Johnson, Richard R. *Adjustment to Empire: The New England Colonies, 1675–1715.* New Brunswick, N.J.: Rutgers University Press, 1981.

Martin, Margaret E. *Merchants and Trade of the Connecticut River Valley, 1750–1820.* Smith College Studies in History, W. D. Gray, H. Kohn, and R. A. Billington, eds., vol. 24 (Oct. 1938–July 1939).

Moloney, Francis X. *The Fur Trade in New England, 1620–1676.* Cambridge, Mass., 1931. Reprint. Hamden, Conn.: Archon Books, 1967.

Pabst, Margaret Richards. *Agricultural Trends in the Connecticut Valley Region of Massachusetts, 1800–1900.* Smith College Studies in History, vol. 26 (1940).

Parker, William N., and Judith L. V. Klein. "Productivity Growth in Grain Production in the United States, 1840–60 and 1900–10." In *Output, Employment, and Productivity in the United States after 1800.* New York: National Bureau of Economic Research, 1966, 523–582.

Rothenberg, Winifred. "The Market and Massachusetts Farmers." *Journal of Economic History* 51 (June 1981), pp. 283–314.

Weaver, Glenn. "Some Aspects of Early Eighteenth-Century Connecticut Trade." *The Connecticut Historical Society Bulletin* 22 (Jan. 1957).

3
Industrial Labor Market Adjustments in a Region of Recent Settlement: Chicago, 1848–1868

PAUL A. DAVID

> "The provision of an assured quantity of labor service is a difficult matter in a market as imperfect as that of the United States in its first half century."
> —Stanley Lebergott, *Manpower in Economic Growth* (1964)

Ongoing industrialization and the accompanying transformation of a society's economic organization create demands for new kinds of labor services. Often these demands cannot readily be satisfied and are reflected in the appearance of chronically high premiums paid for skilled industrial labor. It is widely observed that in economically backward countries, and in countries still in an early stage of industrial development, margins between the wages of skilled and unskilled urban workers are relatively wide, whereas in countries with long industrial experience they are relatively narrow.[1] Shortages of skilled labor frequently are cited as a serious potential limitation, if not a binding constraint, upon the pace at which industrialization and economic development can proceed.

Conversely, the creation of favorable labor supply conditions has been emphasized as critical to the successful historical transition to manufacturing activity made by certain regions and nations. Recent studies of the early nineteenth century industrial development of the northeastern United States have focused attention upon just this point. Alexander Field has stressed the significance of the formation of "reservoirs" of labor available for factory employment, as a consequence of the imperfect response of rural labor markets to the stagnation of

---

Most of the material for this essay has been drawn from my unpublished manuscript: "Factories at the Prairies' Edge: Industrialization in Chicago, 1848–1898." The indebtedness, intellectual and otherwise, that I accumulated in connection with the latter work (referred to hereafter as FPE) is much too extensive to be surveyed and properly acknowledged here. Stanley Lebergott perhaps can recall the many conversations which he, Melvin Reder, and a very young assistant professor had on the occasion of his first visit to Stanford, and may find in these pages some concrete measure of their formative influence upon the labor market aspects of my research. The present essay is the better for having passed first before the eyes of Gavin Wright, although it would undoubtedly be better still if I had been able to follow all of his advice.

[1] See, e.g., Colin Clark, *The Conditions of Economic Progress*, 3rd ed. (London: Macmillan Co., 1957), p. 525, for the classic statement.

agriculture in antebellum New England.[2] The rather different formulation offered by Claudia Goldin and Kenneth Sokoloff does not put the argument in terms of the mobilization of an underutilized regional labor supply; but their presentation stresses nonetheless what, in their view, constituted crucial labor supply conditions favoring the development of manufacturing industries in antebellum New England, rather than elsewhere in the nation. As a consequence of the nature of the agrarian economy that had been established in the region (specifically its dependence upon grain crops), the relative productivity in the farm sector of women and child workers was low vis-à-vis adult male workers. Hence, according to Goldin and Sokoloff, manufacturing could be expanded profitably by intensively utilizing this relatively cheaper source of labor power, and it correspondingly developed least extensively in a region like the South, where crops such as cotton and tobacco provided tasks in which the productivity of women and children was not notably lower than that of male laborers.[3]

But what of the process of industrialization in the urban centers of regions of ongoing settlement? The rise of midwestern manufacturing activity, even during the antebellum period, has been recognized as a peculiar feature of U.S. historical experience; the close linkage between agriculture and industry in that region is acknowledged as differentiating the course of northern economic development from that of the South. Yet it is plain that the early growth of manufacturing centers such as Cincinnati, St. Louis, and Chicago must have involved a labor market adjustment process quite different from that obtaining in the case of the older, longer-settled regions of the Northeast.

The rise of "factories at the prairies' edge," as it were, obviously had not required the antecedent formation of a reservoir of stranded rural manpower, or woman- and child-power for that matter. The burgeoning cities of the Great Lakes-Plains region and the Midwest arose on the basis of their manufacturing prowess as well as their commercial functions, but in neither did they make such extensive use of women

[2] See Alexander J. Field, "Sectoral Shift in Antebellum Massachusetts: A. Reconsideration," *Explorations in Economic History,* 15 (April 1978), pp. 146–171.

[3] See Claudia Goldin and Kenneth Sokoloff, "The Relative Productivity Hypothesis of Industrialization: The American Case, 1820 to 1850," *Quarterly Journal of Economics* (Aug. 1984), pp. 461–487.

Fred Bateman and Thomas Weiss, *A Deplorable Scarcity: The Failure of Industrialization in the Slave Economy* (Chapel Hill: University of North Carolina Press, 1981), pp. 21–22, points out that the South and Midwest were quite comparable in 1860, in terms of the composition of their manufacturing sectors and the level of per capita industrial production in their urban centers; nevertheless, they argue, the underlying conditions which favored successful industrialization of the Midwest in the postbellum era already differentiated the two regions during the ante-bellum period.

and child workers as did New England prior to the Civil War. This held true despite the fact that the post-1830 agricultural regime of the new region resembled that of the pre-1830 Northeast in its dependence upon grain cultivation. Midwestern urban-industrial centers grew rapidly by becoming magnets for immigrant workers drawn from farther afield and, as I have suggested elsewhere, the pace of their development was proximately governed by their ability to rapidly augment the supply of labor within their own precincts.[4]

My purpose in the present essay is to elaborate upon the epigraph that I have drawn from the writings of Stanley Lebergott. I shall do so by closely examining the nature and consequences of the labor market adjustments that were made along this midwestern route to industrial development—a route that in my view was more characteristic of the nineteenth century path followed by the American economy as an entity than was the course of New England's early emergence as a textile-factory center.

In contrast to the story of industrialization with essentially "unlimited supplies of (local) labor," the initial phases of industrial growth in a region of recent settlement—even in so small an economy as the city of Chicago—were accompanied by generalized short-run resource shortages, including a scarcity of labor. The problem of shortages of *unskilled* labor, however, proved less susceptible to amelioration via demand-side adaptations and adjustments than were the scarcities of skilled labor that appeared during the pre-1871 era. By exploiting flexibilities in industrial product mix, and in hiring standards, manufacturing establishments in the city effectively alleviated pressures on the immediately forthcoming supply of skilled workers, and transferred those pressures to the market for unskilled labor. On the supply side, two conditions operated to check the upward course of skilled wage rates: the relatively high skill content of the migration stream that poured into the Great Lakes-Plains states during the 1850s, and the possibility of resorting to direct recruitment of key industrial workers from older, more developed centers on the eastern seaboard.

But these avenues for supply and demand adjustment were not correspondingly open in the case of unskilled labor. An array of city- and region-building activities were under way, which presented local ventures in manufacturing with intense competition for the services of ordinary laborers and the range of semi-skilled "mechanics." Adjustment of hiring standards, and the consequent upgrading of the less ex-

[4] See Paul A. David, "The Marshallian Dynamics of Industrial Localization: Chicago, 1850–1890" (Paper read at the Chicago Meetings of the Economic History Association, Sept. 1984).

perienced, contributed to shifting the pressures further down the skill ladder during this period. As a result, rapid industrial growth drove the level of money wage rates upwards, but at the same time operated to *compress* skill-associated wage differentials. Skill margins were highly volatile under these conditions, fluctuating inversely with the rate of growth of labor demand in Chicago and its hinterland in the antebellum era, and so remaining remarkably narrow during the very years of the formation of the city's industrial base. Those who conceive of large skill premiums as an outcome, and indicator of accelerated urban-industrial development, and who, accordingly, view such transformations as generating greater inequalities among income-earners, may well find it difficult to fit the experience of this midwestern manufacturing center into their historical preconceptions.

Within the manufacturing sector itself, it appears that the incidence of heightened labor market pressure fell upon industries whose average skill requirements were comparatively low and whose labor inputs constituted a large component in total costs. This pressure was likely to be felt especially keenly in those instances in which the local firms also faced vigorous competition from imported goods. Actually, and none too surprisingly, few important branches of manufacturing that were characterized by the foregoing conjunction of especially disadvantageous circumstances managed to develop in Chicago during the 1850s and 1860s. This points to a two-sided conclusion: the peculiar labor market conditions of this region of recent settlement did not thwart the successful antebellum formation of an industrial base, and thereby confine the city economy to specialization in commercial and financial spheres; however, they are to be numbered among the conditions that caused the pattern of industrial development to proceed along lines diverging from those of the northeastern region of the country.

Colin Clark remarked some time ago that regions of recent settlement may well constitute the exception proving the rule that skill-associated wage margins are large during the initial stages of industrialization and tend to narrow as economic development proceeds.[5] Perception of the contrasts between Chicago's experience and conditions in long-settled but economically backward areas, and the analysis of their implications that is attempted here, will not provide an adequate basis for a full appraisal of Clark's hypothesis. It may, nonetheless, bring us a step closer to a formulation both of the general sequence through which the industrial labor force evolved in nine-

[5] See Clark, *Economic Progress*, pp. 525, 532.

teenth century regions of recent settlement like the United States, and of an appropriate corresponding theory of long-run changes in skill-associated wage differentials.[6]

I. The Labor-Scarce Economy

The labor supply problems of a small economy, when viewed at the aggregate level, appear qualitatively quite different from those immediately relevant to the growth of large, more or less autarkic economies. Where no institutional or legal obstructions impede the migration of labor, the small economy's problem in mobilizing additional workers reduces essentially to the problem of generating the demand for them at the going wage rate; given a highly elastic long-run aggregate supply schedule, formation of an adequate labor force would not appear to be a process that could impose any meaningful independent constraints upon the region's economic growth. Indeed, the conception of a small region's development as taking place under conditions approximating "unlimited supplies of labor" suggests that the important features of the process may be perceived by focusing primarily on the demand side of the story.

Such a conception, however, grossly oversimplifies the mechanism of labor market adjustments during industrialization. For one thing, it abstracts from the range of issues raised by the heterogeneity of labor service. It should hardly be necessary to stress that quality variations exist among the members of the work force assembled in a given occupational category, or that diverse kinds of economic activities require specific and different labor services; by the 1890s, the Chicago meat-packing industry alone distinguished thirty types of labor and paid twenty different wage rates. When the problem is posed in terms of the availability of workers with specific skills and experience, it ceases to be so clear, even in the case of the small open economy, that induced immigration provides quick, automatic solutions.

Moreover, it must be recognized that the environment to which employers adapt is likely to be shaped by short-run as well as long-run conditions; and for any one of a number of reasons, the short-run *aggregate* labor supply schedule may be quite inelastic. In the first place, a small community is not necessarily able to accommodate its employ-

---

[6] See Jeffery G. Williamson and Peter H. Lindert, *American Inequality: A Macroeconomic History* (New York: Academic Press, 1980), Chs. 4, 9, for a recent interpretation of the historical course of skill-associated wage differentials in the U.S., which differs, in both conceptual orientation and empirical basis, from the one offered here.

ers' desired labor force expansion without having local costs of living driven to extravagant heights—thereby necessitating further increments in money wage rates simply to maintain the real income level of the existing labor force. To the extent that employers must find markets for their products outside the community and must, in so doing, compete with producers in regions where living costs are not being similarly raised, they are handicapped by their own region's lack of capacity to meet the needs of an augmented population. If local living costs rise in relation to the prices of exportable goods, an attempt to expand employment will force higher real wage rates, as well as higher money wage rates, upon export producers.

The difficulty just mentioned is in essence transitional, although not on that account trivial. It was undoubtedly extremely serious in Chicago during the inflationary boom of the mid-1830s, when rents and prices of other consumer items—including those traded interregionally—soared in the local market under the impact of the sudden population influx. More generally, problems of this sort would appear to account, at least in part, for the tendency of economic activity to become bipolarized in the classic "boom town": we find, clustered at one extreme, enterprises providing goods and services that are not traded beyond the town's immediate environs and, at the opposite pole, a concentration upon production of an export good in which the town enjoys some strong natural monopoly advantage—such as that conferred upon Chicago by the site's strategic location in the web of inland transportation routes.[7] This phenomenon belongs predominantly to the pre-industrial phase of Chicago's history. Although the underlying problem of short-run capacity shortages in the so-called residentiary activities, along with its manifestation in rapidly rising local prices and wages, would reemerge at later dates, the two will receive little further notice here. There are other complications, however, that command our attention.

While the long-run price elasticity of the aggregate supply of undifferentiated labor may become virtually infinite in the case of a specific city, the response of migration to spatial differences in real income may also be far from perfect. Reaction lags, costs of movement, poor information, all may pose obstacles that result in recurrent labor shortages as short-run constraints upon rapid growth. Furthermore, the attraction that a given town holds for migrating workers is likely to

[7] See Louis P. Cain, "The Theory of Location of Cities—From Mud to Metropolis: Chicago Before the Fire" (Unpublished paper presented to the Thirteenth Conference on Quantitative Methods in Canadian Economic History, Wilfrid Laurier University, March 1984).

depend directly on the size of the community and on the normal number of job openings created by turnover, as well as on the total spatial pattern of real-earnings differences. A small community may consequently find that an excess demand for labor is less readily satisfied by induced migration than is the case in a large city. This is especially likely to be true of excess demands for workers with specialized skills, a point appreciated in 1859 by the author of *Philadelphia and its Manufactures:*

> In fact, in or near large cities only can labor of the first quality be obtained. 'As iron sharpeneth iron, so man sharpeneth the countenance of his friend; and away from the centres of population and competition, the face loseth its sharpness, and the hand its cunning . . . Superior mechanics and dexterous workmen manifest a . . . preference for cities and an abhorrence of isolation; hence, if for no other reason, extensive mechanical or manufacturing operations must be conducted at a great disadvantage in isolated localities.'[8]

While the broad subjects of historical inquiry touched upon in the several foregoing paragraphs certainly have a general bearing upon the role of labor supply conditions in economies undergoing industrialization, one needs to guard against casually casting the narrative of Chicago's experience in the ready-made but ill-suited mold formed by discussions of the labor supply problems of contemporary developing economies. The nineteenth century regions of "recent settlement" like the United States experienced rather unique initial conditions in regard to the comparative availability of unskilled and skilled industrial labor. Furthermore, subsequent alterations in the balance of supplies and demands for different kinds of labor within such regions were dominated by forces that do not figure significantly in the usual stylized accounts of the way industrial transformation and economic progress take place. Both of these points will be apparent in the following brief account of the Chicago labor market during the city's early history.

There can be little question that in its first phases the growth of economic activity in Chicago pressed very closely upon the locally available supply of labor. In this, as in other respects, the city shared the experience that marks the "boom town" everywhere. During the heat of the speculative excitement in the 1830s—still preceding the onset of heavy, enduring migration to the young town and significant local industrial developments—the advertising columns of the local press

[8] Edwin T. Freedley, *Philadelphia and its Manufactures* (Philadelphia, 1859), pp. 30–31.

abounded with symptoms of that condition. The call for labor was not only frequent and widespread, but it solicited services across a broad spectrum of occupations and skills. Notice was given of a need for millwrights, engineers, and common laborers; for tailors' and dressmakers' apprentices; for gun smiths, saddle and harness makers, bakers, boot and shoe makers, brick makers, carpenters and journeymen coopers, cabinet makers, watch makers, chair makers, and wagon makers.[9]

Accompanying the advertisements for labor—skilled, semiskilled, and unskilled alike—came the complaint that the attractions of the land frustrated the accumulation of an adequate pool of artisans and mechanics. Yet the classic observations made by Albert Gallatin and Tench Coxe on the conditions of a previous era in a different part of the country were not simply reiterated in Chicago.[10] In this instance, the distraction created by the abundance of land as a resource to be exploited through farming hardly rivaled that stemming from its potentialities as a speculative asset. Ultimately, it was the large-scale purchases by eastern and southern investors that turned the wheel of the Chicago land casino during the mid-1830s. But in 1834, when the land craze was in its incipient stages, the *Chicago Democrat* disapprovingly noted that appreciation in the value of property acquired through preemption rights and settlers' agreements, or purchased from private parties on credit, caught the imaginations of those who witnessed it and "encouraged mechanics and labourers, on arriving, or soon after, to abandon their appropriate trades and occupations for a bright hope of soon making their fortune under the pre-emption laws."[11]

To Harriet Martineau, who arrived at Chicago in the summer of 1836, "it seemed as if some prevalent mania infected the whole people." The streets were crowded with speculators, hurrying from one sale to another, and as the gentlemen of her party walked the streets, "storekeepers hailed them from their doors, with offers of farms, and all manner of land-lots, advising them to speculate before the price of land rose higher . . . Others, besides lawyers and speculators by trade,

[9] *Chicago Democrat,* June 25, 1834, Sept. 9, 1835, March 30 and April 20, 1836; *Chicago American,* June 27, Aug. 29, Oct. 28, 1835, Feb. 13, March 19, April 2, May 7, June 16, 1836. Cited in Bessie L. Pierce, *History of Chicago,* 3 vols. (New York: 1937, 1940, 1957), pp. 193–194.

[10] See A. Gallatin, *Report on Manufactures,* American State Papers, Finance, II, Report No. 325 (1810); Tench Coxe, *Digest of Manufactures,* American State Papers, Finance, vol. 2, Report No. 407 (1814).

[11] *Chicago Democrat,* June 25, 1834, quoted in Pierce, *History of Chicago,* vol. 1, p. 193. For Chicago land purchases on credit, see ibid. p. 150; J. S. Buckingham, *The Eastern and Western States of America,* 3 vols. (London, 1841?), vol. 3, p. 261.

make a fortune in such extraordinary times."[12] Miss Martineau regarded the situation as patently temporary:

> When the present intoxication of prosperity passes away, some of the inhabitants will go back eastward, there will be an accession of settlers from the mechanic classes . . . the singularity of the place will subside. It will be like all the other new and thriving lake and river ports of America.

The speculative fever did indeed subside. But in the ensuing financial collapse and depression of the late 1830s and early 1840s, the accession of settlers—let alone those drawn from the mechanic classes—proceeded haltingly. The marshals for the U.S. Census of 1840[13] counted no more than 238 commodity production employees in the place, if we generously include the 26 workers in the "lumber yards and trade" within that total. Putting aside the construction trades, an even 100 were engaged in manufacturing proper: 35 in metal working; 26 in agricultural processing activities such as meat butchering and packing, leather tanning, soap and candle making, and flour milling; printing and publishing occupied 19; and another 17 were to be found in the carriage and wagon shops and in furniture making. Add in the 3 workers making "hats, caps and bonnets" and you had it all—a nice, comprehendible economy, but hardly impressive or auspicious.

From 1845 on, however, it was a different story. Baring Brothers contrived a new loan to the state of Illinois, permitting the completion in 1848 of the work on the Illinois and Michigan Canal, which had been suspended in 1842. The land market showed signs of reviving from the moribund condition into which it had sunk during the winter of 1839, and the influx of new settlers catapulted the town's population upward from 8,000 in 1844 to almost 30,000 in 1850.[14] In absolute terms, the additions to the ranks of the industrial labor force, gauged by the number who found jobs in manufacturing and handicraft establishments in the city, was spectacular, especially after 1852. Between 1846 and 1850—when the enumerators of the Seventh Census placed manufacturing employment at 2,081—the increment amounted to something in the neighborhood of 600 workers. According to the local press, another 3,000 were added in the 1850–1854 interval, followed

---

[12] H. Martineau, *Society in America*, 3 vols. (London, 1837), vol. 1, pp. 350, 352–353.

[13] See *U.S. Sixth Census*, "Statistics of the United States of America as Corrected at the Department of State, June 1, 1840" (Washington, D.C.: Blair and Rives, 1841).

[14] See Homer Hoyt, *One Hundred Years of Land Values* (Chicago: University of Chicago Press 1933), pp. 37–52, for Chicago's experience in the Panic of 1837, the following depression, and the recovery in the latter half of the 1840s.

by roughly 5,500 more between 1854 and 1856, bringing manufacturing employment to an estimated 10,573 at the latter date.[15]

This eventual fulfillment of Miss Martineau's prophecy notwithstanding, local comments on the "scarcity of labor" continued to appear. True, newspaper advertising became rather less common a means of recruiting workers for manufacturing ventures during the spurt of industrial growth in the mid-1850s, and the speculative excitement over urban real estate in this period was no longer considered a worrisome snare for wayward members of the labor force. The dimensions, if not the nature of the problem had altered. With the prospects of an eventual solution in the offing, through sustained migration, contemporary Chicagoans were able to derive a measure of satisfaction by lamenting the persistent disparity between the opportunities for profitable employment of labor (and capital) and the currently available supplies: "with all our population and capital we have not half the money nor half the laborers that the commerce, manufactures, and general improvement of the city require."[16] Manufacturing establishments were joyously depicted as working "with all available hands," but then the same thing was said with regard to the utilization of floor space and machinery.[17] Labor "scarcity" was no more than a part of the general resource constraint encountered by the rapidly expanding local economy. As such, it received acknowledgment as one of the facts of life, but little specific discussion.

Virtually no commentary has survived from this period that would suggest the existence either of chronic sectoral bottlenecks in manufacturing owing to shortages of labor or, more significantly, of that *differential* scarcity of skilled workers so frequently considered by current writers as the natural concomitant of the initial phase of industrial development. Three instances in which manufacturing enterprises were reported to have encountered serious obstacles of this kind stand out by virtue of their rarity and comparative insignificance in the total industrial scene. The firm of Doggett, Bassett & Hills, probably the pioneer house engaged in the wholesaling and manufacturing of boots and shoes in the city, started business in June 1846 and had initiated its own

[15] See James W. Norris, *Business Directory and Statistics of the City of Chicago for 1846* (Chicago, 1846), p. 7; J. B. DeBow, ed., *Statistical View of the United States . . . being a Compendium of the Seventh Census* (Washington, 1854), p. 223 (hereafter cited as *Compendium of the Seventh Census*); *Daily Democratic Press,* "Annual Review of 1856," p. 33, for employment in manufacturing.

[16] *Daily Democratic Press,* "Annual Review of 1853," p. 17; cf. also "Annual Review of 1854," p. 23.

[17] See *Daily Democratic Press,* "Annual Review of 1854," p. 29.

manufacturing operation in 1849. According to an account that appeared some twenty-one years later, the firm's early production venture was carried on for a few years under "great discouragements on account of the difficulty of procuring competent workmen"; manufacturing was actually discontinued for a short time, was renewed in 1855, and had ever since "shown a large annual increase in the amount of their goods made." [18] Mr. Hildebrand, proprietor of the Chicago Premium Glove Factory, a small establishment on Lake Street and in 1854 the only firm in the city engaged in making gloves and mittens from oil-tanned skins, found himself unable to meet the demand for his product "owing to the difficulty of obtaining a sufficient number of good glove sewers." Likewise, Gertz and Loder's Brush Factory, the lone representative of that industry during the 1850s, "could have accomplished much more" than its annual production of $8,000 worth of brushes, "but for the want of competent workmen in sufficient number." [19]

This is not to suggest that the rapid growth of manufacturing did not generate excess demands for the services of the skilled. Rather, as will be seen from a closer examination of the situation in the market for skilled labor, a combination of favorable initial supply conditions, entrepreneurial responses to the existence of skill shortages, and flexibilities in the young urban economy (as well as in specific activities) all worked to alleviate the problem. Some of these ameliorating factors were peculiar to Chicago, while others were of a considerably more general character.

II. Quality Versus Quantity

From the point of view of endowing the city with a population well suited to the conduct of skilled industrial pursuits, the source of the original migration to the Chicago region was most propitious. Settlers from the upland South prior to 1830 had clung to the timbered lands of lower and western Illinois and, halting at the southern edges of the prairies, had left the Chicago region virtually untouched. Thus, the task of colonizing the northern area of the Old Northwest Territory in the following decades fell to those who came from the East. After a tour to eastern Illinois in 1835, John Law, of Vincennes, Indiana, wrote to Martin Van Buren, remarking on the then visible portents of the future,

[18] *Chicago Tribune*, "Annual Review of 1870," p. 71.
[19] *Daily Democratic Press*, "Annual Review of 1854," p. 54.

Who are now filling up Ohio Indiana Illinois and Michigan? Who will fill up the New States of the North West? New York and New England. Out of the hundreds emigrating to the West along the Lake Shore whom I met and with most of whom I conversed, I found no one either on foot, on horseback, or in carriages, who was emigrating from the South of New York. They were all from that state or New England. It requires no "prophet or son of a prophet" to predict what their influence is to be.[20]

It was through this movement that Chicago acquired, both directly and indirectly, a significant portion of the Yankee heritage. Of the native-born Americans established in the city at the time of the Seventh Census, 47 percent were born in the Northeast and Middle Atlantic states (particularly New York and Pennsylvania), while another 47 percent were children of the Old Northwest. A substantial proportion of the latter, along with those who had arrived directly from the western New York and northern Pennsylvania counties, could have traced their origins to New England. By contrast, the census takers of 1850 counted a scant 495 sons and daughters of the South among Chicago's residents, mostly Virginians, Kentuckians, and Marylanders.[21]

Some indications of what the New York–New England background of the American born settlers implied for the skill endowment, mechanical abilities, and attitudes of Chicago's early labor force are to be read in stories such as that of the Chicago Mechanics' Institute. Organized as early as 1837, the Institute was propelled by enthusiastic and vigorous local support toward creation of an evening school for apprentices, a reference library of several thousand volumes, and kindred programs "to diffuse knowledge and information throughout the mechanical classes." These efforts won the Mechanics' Institute recognition, by an Act of Congress in 1855, as one of the three leading institutions of its kind in the United States.[22]

---

[20] Letter in the Library of Congress, quoted in F. J. Turner, *The United States 1830–1850* (New York: H. Holt and Co., 1935), p. 265. See A. C. Boggess, *The Settlement of Illinois 1778–1830* (Chicago: Chicago Historical Society, 1908); W. V. Pooley, *The Settlement of Illinois, 1830–1850* (Madison, Wis.: University of Wisconsin, 1908).

[21] *Compendium of the Seventh Census*, p. 399; L. K. Mathews, *The Expansion of New England* (Boston: Houghton Mifflin Co., 1909), pp. 139–69, 198–219. See Pierce, *History of Chicago*, vol. 1, p. 178, n. 20, for references in the local press to the "Yankee" characteristics of the native born. The New England Society, which met on December 22 to celebrate the landing of the Pilgrims, was the first such organization, being founded in 1846. Not until 1848 did the New Yorkers, who outnumbered the New Englanders two to one by 1850, get around to organizing the Excelsior Society. See A. T. Andreas, *History of Chicago*, 3 vols. (Chicago: A. T. Andreas, 1885), vol. 1, pp. 522–523.

[22] See Andreas, *History of Chicago*, vol. 1, pp. 518–521. The Mechanics' Institute

A quantitative assessment of the Yankee-Yorker influence is considerably more difficult to contrive. The 1850 pattern of white illiteracy in the Old Northwest does, however, afford evidence as to the impact of that background upon the initial educational level of the population. Illiteracy rates ranging upward from 10 percent of the white population, or twice the national level, were recorded in the counties of southeastern Ohio along the river, in the northwestern Black Swamp region of that state, in the Ohio River counties of Indiana and in the corridors leading north through the rough driftless area, in the river country of southern Illinois—that is to say, in those part of the Northwest where migration from the upland South had been significant. As one moved northward toward the Great Lakes, white illiteracy declined. It dropped to levels of 2 percent and lower in the broad trail of counties that stretched from New England through central New York and northern Pennsylvania, along the shore of Lake Erie, and into northern Illinois and southern Wisconsin by way of southern Michigan. This high-literacy zone traced the broad path of the eastern migration and the regions where New England stock was predominant at the century's midpoint.[23]

Although the early settlement was the work of native-born easterners, by 1850 more than half of Chicago's population were "foreigners," making the city more cosmopolitan than New York.[24] As the stream of foreign immigrants flowed into the Chicago area during the first surge of industrial development, its composition also had a particularly advantageous effect upon the relative availability of skilled labor.

The Irish constituted the city's largest foreign nationality group in 1850. They had in great measure been attracted first by the employment opportunities created by the resumption of construction work on the canal in the latter half of the 1840s. Yet Hibernian numerical dominance was closely contested by newly arriving Germans. This new

---

thus qualified to receive reports and books of the Smithsonian Institution printed and distributed at governmental expense.

[23] The Seventh Census defined white illiteracy rates as the precentage of the white population aged twenty and older unable to read and write. *Compendium of the Seventh Census*, p. 152; Turner, *United States*, p. 336 and Maps 5 and 9 *ad fin*. The crude illiteracy rate for Cook County in 1850 was 1.6 percent, while the age-specific rate (in the white population aged twenty and over) was roughly 3.0 percent. See *Compendium of the Seventh Census*, pp. 219, 359. The age-specific rate was derived by applying the 1850 age structure of the Chicago white population to the Cook County white population.

[24] *Compendium of the Seventh Census*, p. 399; 45.7 percent of the New York City population was foreign born, compared with 52.3 percent for Chicago.

group congregated north of the Chicago River, between Clark Street and the lake, in what was then the 6th Ward, well apart from the Irish, who tended to cluster along the eastern side of the South Branch in the 2nd Ward. By 1847, the Germans had become sufficiently numerous to warrant the Common Council's passage of a motion to employ a German interpreter to assist collection of the street tax in the North Division. In 1850 they comprised 32.7 percent of the foreign born in Chicago, compared with the 38.4 percent who had come from Ireland. Contrast this with the situation in New York City, where the German-Irish balance at the time was strikingly different: Germans made up less than one-fourth of the foreign-born element, while the Irish formed more than a majority of the Old World's representatives.[25]

The migration of Germans into the Chicago region was particularly heavy during the 1850s and by 1860 they constituted 40 percent of the city's foreign-born, having not only displaced the Irish as the dominant foreign nationality group, but having also increased their importance in the total population despite a slight increase in the overall proportion represented by the native-born.[26] There were but a handful of Austrians and Prussians in Chicago at the middle of the century, so it is likely that the city's early German settlers were drawn predominantly from the Upper Rhine Valley, the major source of German emigration to the United States during the late 1840s and early 1850s.[27] Those arriving in Chicago apparently came equipped with handicraft skills characteristic of Rhinelanders, and well adapted to initial local needs, for they quickly established themselves among the town's butchers, brewers, bakers, shoemakers, and coopers. By the mid-1850s Germans had be-

[25] Ibid. For ward boundaries, see Hugo S. Grosser, Chicago: *A Review of its Governmental History from 1837 to 1906* (Chicago: n.p. 1906); for Germans in the North Division, Pierce, *History of Chicago*, vol. 1, p. 182. According to the Seventh Census (1850), the foreign populations of Chicago and New York were distributed among the leading nationality groups as follows:

	Percentage of foreign born		
	Irish	German	English/Scottish
Chicago	38.4	32.7	15.4
New York City	56.7	24.0	13.2

[26] *U.S. Eighth Census*, 1860, "Population," p. 613. The percentage of foreign born in the total population declined from 52.3 to 50.0, primarily as a result of natural increase in the city. The percentage of Germans rose from 17.0 to 20.0.

[27] See M. L. Hansen, *The Atlantic Migration, 1607–1860*, 2nd ed. (New York: Harper Torchbooks, 1960), pp. 287–295; M. A. Jones, *American Immigration* (Chicago: University of Chicago Press, 1960), pp. 96–97, 110–112, for sources of German emigration prior to 1860. The percentage of Prussians among the German-born in Chicago rose during the 1850s to 26 and reached 42 in 1870, reflecting the rising exodus from the eastern rural areas in those years. (*U.S. Eighth Census*, "Population," p. 613, *U.S. Ninth Census*, "Population," vol. 1, pp. 386–391.)

come conspicuous in Chicago's woodworking industries and especially dominant in wagon-building establishments. Although they diffused themselves throughout the industrial labor force, the Germans were to retain traditional preeminence in these activities as late as 1880.[28]

Despite all the pitfalls surrounding reliance upon occupational statistics as indicators of skill distribution in the labor force,[29] the Census of Occupations, first published for Chicago in 1870, does provide a rough notion of the extent to which the German migration contributed to the development of a skilled industrial labor force. In the first place, Germans appear to have concentrated in commodity-producing occupations—rather than in trade, transport, and service occupations—to a greater degree than did the "Americans" and the Irish. (Included here among commodity-producing occupations are the "manufacturing and mechanical" pursuits embracing the building trades *and common labor*. The latter has thus been transferred from the category of "Professional and Personal Services" where it was placed by the Ninth Census.) Of the 25,778 Germans tabulated, 61 percent listed commodity-producing occupations, compared with 54 percent of the 22,337 Irish and 36 percent of the 39,755 "Americans." This made them the most important nationality, not neglecting the native born, in that occupational group. More significantly, as a group, the German workers tended to concentrate in the more skilled commodity-producing occupations. Whereas 58 percent of the Irish in that major category were unskilled laborers, only 30 percent of the Germans listed their occupation as common laborer.[30] Therefore, although the Germans comprised 20 percent of the city's population and 22.8 percent of the workers listed by the occupational census in 1870, they contributed 30 percent of the workers in manufacturing and mechanical occupations proper.

[28] For Germans in Chicago industry, see R. Fergus, *Directory of the City of Chicago for 1843*, Fergus Historical Series, no. 28 (Chicago, 1896); *Democratic Press*, "Annual Review for 1855," p. 32; *Report of the Department of Health for the City of Chicago, 1880*, "Factory and Workshop Inspectors' Report," p. 43. The latter contains a survey of the labor force showing nationality and industrial affiliation (rather than occupation). Germans accounted for the following percentages of workers in various industries: Bakeries and Confectioneries, 54; Boot- and Shoemaking, 58; Breweries and Distilleries, 72; Wagon- and Carriage-building, 40; Furniture-making, 64; Picture Frame and Molding Factories, 63. Along with the production of Sewing Machines, Tobacco and Cigars, and Tailoring, these were fields of relative specialization by Germans in 1880.

[29] For a judicious discussion of these problems, see National Manpower Council, *A Policy for Skilled Manpower* (New York: Columbia University Press, 1954), pp. 62–71.

[30] *U.S. Ninth Census, 1870*, vol. 1 "Population," p. 782. The following table presents the nationality specialization ratios for gross skill categories in 1870:

Thus, during the first surge of industrialization in Chicago—that of the 1850s—skilled labor was made more available locally both by the characteristics of the area's American-born settlers and by the differential ability of the leading group among foreign-born migrants, the Germans, to assume positions at the higher end of the spectrum of industrial skills.

The favorable skill endowment of the population drawn to Chicago must be regarded as fortuitous to the extent that it resulted from circumstances unrelated to the *specific* labor demands generated by the city's early industrial growth. In these circumstances, it is perhaps not surprising that adjustment of the supply of skilled labor through transfer of workers from other regions was left essentially to the unassisted operation of labor mobility in response to the prospects of higher earnings.

Early in 1854 a group of Chicago businessmen did sponsor a movement to create an office of Commissioner of Immigration, with the proposed commissioner's function to include travel in Germany for the purpose of directing emigration to Illinois. Perhaps not undeservedly, this scheme ran into criticism from the *Illinois Staats-Zeitung* as a piece of "sharp" business (and a political maneuver to boot); it was eventually abandoned.[31] At the municipal level, no governmental action was undertaken to subsidize immigration of any sort, let alone that of skilled workers. No matter how highly the real estate and commercial interests might prize further accretions to the city's population, the effective prevailing attitude is well captured in the *Chicago Times* editorial retort to such a suggestion:

> Chicago does not go to the world. The world comes to Chicago . . . Should Chicago pay people to come hither when it is for their advantage to come? Does Rome pay the expenses of the ever-

	(1) Percentage of workers in all occupations	(2) Percentage of all common laborers	(3) Percentage of all in "Manufacturing and Mechanical" occupations	Commodity production nationality specialization ratios for	
				"Unskilled" (2) / (1)	"Skilled" (3) / (1)
Americans	35.2	19.8	27.0	0.56	0.77
Germans	22.8	24.1	29.9	1.06	1.31
Irish	19.8	37.4	14.4	1.89	0.73
TOTALS	68.8	71.3	81.3		

[31] It did, however, have at least one lasting effect in initiating the formation of the Society for the Production and Aid of German Immigrants. See Andrew J. Townsend, "The Germans of Chicago" (unpublished Ph.D. diss., University of Chicago, 1927), pp. 29–30; Arthur C. Cole, *The Era of the Civil War, 1848–1870*, in *The Centennial History of Illinois*, C. W. Alvord, ed., vol. 3 (Springfield: Illinois Centennial Commission, 1919), pp. 23, 24; Andreas, pp. 669–670.

lasting crowds of pilgrims that are journeying to its shrines? If Rome does not pay, why should Chicago pay?[32]

When transatlantic migration was interrupted and pressure upon the domestic labor supply increased during the Civil War, the *Chicago Tribune* came out in support of Lincoln's message of December 8, 1863, favoring federal assistance for immigration to the United States. With annual population increases in 1861–1863 greater than anything hitherto seen in Chicago—albeit rates of growth were lower than in 1850–1856—it is difficult to judge how heavily local concerns, as opposed to national problems and Republican loyalties, weighed in determining the newspaper's stand. In any case, the *Tribune* went on to disregard externalities and, implicitly viewing emigration purely as an act of private investment, suggested it be financed by bonds, interest on which would be paid from the earnings of immigrants after their arrival in the United States.[33]

The belief that migration to Chicago partook of the nature of a holy pilgrimage, however satisfying, was not essential doctrine. Straightforward appeal to the beneficial workings of perfect factor mobility and conviction—buttressed by Chicago's amazing growth—that existing imperfections in factor markets could not be very serious provided alternative justification for the laissez faire attitude maintained by the city in the pre-Fire era. "The true inducements which are held out to both labor and capital should be widely disseminated and correctly understood. This accomplished the two will arrange themselves in their true proportions without any further trouble."[34]

Quite apart from the question of governmental initiative, direct private recruitment of the skilled from other regions appears to have played only a limited, highly specialized role in satisfying local industrial requirements. Skilled workers were imported typically in small number from the eastern seaboard, or from abroad, in connection with the establishment of a new line of manufacturing activity, especially the introduction of new processes or higher-quality products, and the installation and supervision of unfamiliar equipment.

---

[32] *Chicago Times*, March 25, 1868, quoted in Pierce, *History of Chicago*, vol. 2, p. 34.

[33] See *Chicago Tribune*, Dec. 12, 1863. On Lincoln's message and the movement for federal encouragement of immigration, Charlotte Erickson, *American Industry and the European Immigrant 1860–1883* (Cambridge, Mass.: Harvard University Press, 1957), pp. 6–13. Concern regarding emigration to Chicago antedated Lincoln's message; in the summer of 1863, the Rev. E. G. Tuttle of St. Ansgarius Church was dispatched to Baltimore and other eastern cities to encourage emigration. See Pierce, *History of Chicago*, vol. 2, p. 157, n. 36.

[34] *Democratic Press*, "Annual Review for 1854," p. 29.

Thus, Richard T. Crane related that a few months after setting up a thriving brass foundry in the city, during 1855, he determined to go into the line of brass finishing as well as casting and "sent for two first-class workmen [he] had known in Brooklyn." A decade later, Crane Company imported a man from Newark, New Jersey, then the seat of the malleable-iron–casting trade, to direct the production of malleable-iron pipe fittings in its newly erected foundry, allegedly the first malleable-iron foundry built west of Pittsburgh. The building of a small butt-weld (iron) pipe mill in 1864 occasioned another search for talent by the Crane Company. ". . . The manufacture of pipe being looked on then as a rather high art, and as that business had been kept almost entirely a secret," Crane encountered considerable difficulties in locating someone to supervise the erection and operation of the new Chicago mill. Until 1845, most wrought-iron pipe and fittings had been imported into the United States from England, and there were comparatively few domestic producers, but a man to do the job was finally found in Philadelphia and brought to Chicago.[35]

In 1870, to offer another illustration, the Chicago Vise and Tool Company—upon installing "new and expensive machinery" for making solid box vises, blacksmiths' bellows, picks, mattocks, grub hoes, and related implements—secured the services of "a number of thorough experienced workmen from the celebrated establishment of Peter Wright, England, for the express purpose of manufacturing this line of goods in a superior manner."[36] From such evidence as is available, then, Chicago's experience appears not to have diverged materially from the broad pre-Civil War pattern in the older seaboard centers, which saw selective private recruitment of skilled "foreign" workers for "domestic" industrial establishments.[37]

One may grant, then, that population migration and the possibilities of direct recruitment operated to raise even the short-run elasticity of the skilled-labor supply schedule facing this young city's industries. Nevertheless, it would be misleading to depict supply responses as

[35] R. T. Crane, *The Autobiography of Richard Teller Crane* (Chicago, 1927), pp. 33, 38, 43, 45–47, 50. Crane had apprenticed in a bell foundry in Brooklyn during the 1840s. Newark, New Jersey, appears to have become the center of pipe manufacturing, primarily as a result of the originating work done there by Seth Boyden. Cf. *The Valve World* (Chicago), 1906.

[36] *Chicago Tribune*, "Annual Review for 1870," p. 97. See also the account of an unsuccessful attempt to establish an iron-ship-building yard in the vicinity of Chicago that involved importing skilled workmen (and patterns) from Britain in 1867–68, *U.S. House, Congress, Exec. Doc. No. 27*, 41st Cong. 2d Sess., 1869, pp. lxxxii–iii.

[37] See Erickson, *European Immigrants*, pp. 3 6; Rowland T. Berthoff, *British Immigrants in Industrial America* (Cambridge, Mass.: Harvard University Press, 1953), pp. 48, 62–63; Victor S. Clark, *History of Manufactures in the United States* (Washington: McGraw-Hill for the Carnegie Institute, 1929), vol. 1, pp. 398–401.

having borne the entire burden of the adjustment to the conditions of labor scarcity that materialized. To do so would be to embrace empirical error, and perpetuate the narrow, unrewarding conception of specific labor supply conditions as constituting a rigid "prerequisite" for industrialization. Lest the flexibilities and the scope for substitutions inherent in the process of economic transformation go unperceived, attention must now be turned to the range of adaptations called forth from the demand side of the market for skilled labor.[38]

To designate some members of the labor force as more skilled than others itself implies that jobs are filled through a selection process. In turn, selection requires the existence of a set of minimum standards referred to in hiring and firing workers. Whether such requirements are explicitly stated or not, whether they are formulated solely in terms of objective technical performance criteria or are merely of the "No Irish Need Apply" variety, it is difficult to imagine jobs in an industrial community for which minimum standards do not exist. Yet effective hiring standards are not immutable: in tight labor markets employers tend to lower the minimum upon which they insist as qualification for filling a particular job; when applicants become more abundant the procedure is reversed. Upgrading currently available members of the labor force relative to the skill requirements of the jobs they are to fill is, therefore, an alternative to increasing wages in response to scarcities of workers who qualify under the old standards.[39]

It is virtually impossible to determine just how far hiring standards for "skilled" positions were lowered during Chicago's initial industrial development, or the extent to which the rapid expansion of the industrial labor force was achieved at the cost of accepting less-qualified workers. But there can be little doubt that such adjustments were made. Does one face a difficulty in reconciling the existence of the practice of filling skilled positions by easing hiring standards, on the one hand, and, on the other, the local newspapers' recurrent encomiastic spasms on the quality of workers found in Chicago's factories and shops?[40] It would hardly seem so, even if one refused to apply the test

---

[38] My indebtedness to the writings of Alexander Gerschenkron on the vexing subject of "prerequisites" for economic development is obvious in this formulation. See A. Gerschenkron, *Economic Backwardness in Historical Perspective* (Cambridge, Mass.: Harvard University Press, 1962), esp. Chs. 1, 2, "Postscript," pp. 353–364.

[39] See M. W. Reder, "The Theory of Occupational Wage Differentials," *American Economic Review* (Dec. 1955), 833–835, for a discussion of hiring standards, on which this paragraph is based. Reder notes that jobs for which selection is nonexistent or is made randomly, tend to be those in which there is no guarantee of minimum earnings and no risk of injury to or destruction of capital. The "No Irish Need Apply" tag actually appeared in an advertisement in the *Chicago Tribune,* Feb. 26, 1866.

[40] See, e.g., *Daily Democratic Press,* "Annual Review for 1855," p. 23; also note 41 below.

tacent satis laudant. Successive reductions of standards are not at all incompatible with the presence in the labor force of highly qualified workers; indeed, the latter would become more noticeable as a consequence of the inferior quality of the newest recruits and the supervisory and training roles that the more skilled workers would be likely to assume.

We are not, however, confined to transmuting superficially negative indications on this point. From positive evidence of the areas into which it extended, one might surmise that by modern reckoning the relaxation of hiring standards was carried quite far during the pre-Fire era. Where greater dangers of large property loss or similar calamity attend the incompetence of workers, employers should be especially concerned with maintaining hiring standards in the face of labor shortages.[41] Although not an industrial activity, railroading operations in the 1850s, and locomotive running in particular, would appear to provide as marked an example of such conditions as might be found within the contemporary Chicago manufacturing sector itself. Yet, as one who had served on the Galena and Chicago Union road and its corporate successors during the 1850s later recounted,

> Master mechanics everywhere found it necessary to promote firemen to be engineers as fast as they could possibly be trusted to serve as runners; they did not call them engineers; about all they needed to know was to tell the difference between the sound of steam and of water as they opened the gauge cocks to find out how much water there was in the boiler. (Glass water gauges were then unknown.)[42]

Upgrading employees in this fashion is patently a form of wage discrimination, inasmuch as it circumvents raising the wages of existing fully qualified workers in the process of trying to attract others of the same quality. Where training costs can be kept at a minimum, relaxation of hiring standards is likely to be an advantageous course for employers, even in the face of substantial risks to property. In this instance, training was apparently kept to the bare essentials.[43]

Some attempts by employers to ease hiring standards did not pass

[41] Cf. Reder, "Wage Differentials," p. 835.

[42] Dewitt C. Prescott, *Early Day Railroading from Chicago* (Chicago: David B. Clarkson Co., 1910), p. 20. Note that these remarks were prefaced by Prescott's declaration, "The New England states, New York and Pennsylvania, simply emptied themselves of their choicest men and mechanics; the cream of the mechanical world took Greeley's advice and went west."

[43] Ibid., p. 21. Prescott relates the story of an Irish fireman who, having been promoted to runner of a locomotive, reported that the valves of his engine were in need of refacing. The master mechanic inquired how he knew they were. " 'Well, by Jasus, a farmer came into Garden Prairie and told me so.' "

unprotested. During 1853, the Chicago compositors, recently chartered as Local No. 16 of the National Typographical Union, pressed for further wage increases, only to be met by employers' efforts to induce Miss Harriet Case to form a corps of women compositors to replace the men. This she refused to do unless the women would receive the higher wages demanded by the unionists, who reciprocated by presenting her with a "beautiful ring." Although the compositors appear to have successfully thwarted the owners of the major printing establishments in this bid, some headway was made in the industry; by 1854 the *Free West,* in its appeal to women subscribers, proclaimed that its type was set by "females."[44]

Not until the tightening of labor markets throughout the North during the Civil War was the reduction of hiring standards reflected in any significant entry of females into industrial occupations in Chicago. Despite agitation and an attempted boycott of the *Chicago Times* by the typographers, the newspapers proceeded during 1864 to employ women as typesetters, paying them rates well below the union scale.[45] By 1865, women and girls were engaged in a variety of manufacturing occupations outside the staple needlework trades. These included not only cigar making, baking, confectionery work, and bookbinding, all activities where women had been found in small numbers during the 1850s, but also newer lines such as boot stitching and upholstering.[46] Lowering of job-qualification standards did not, however, lead very far in this direction. Although in the aftermath of the war experience the manufacturing sector's reliance upon female labor had increased, it still remained essentially marginal and was concentrated most heavily in the needlework trades. According to the U.S. Census of 1870, women and girls made up only 12 percent of the gainfully occupied manufacturing work force; 3,174, or 84 percent of them, toiled as milliners, dress and mantua makers, tailoresses, and seamstresses.[47]

---

[44] See Andreas, *History of Chicago,* vol. 1, p. 415; Pierce, *History of Chicago,* vol. 1, pp. 160–161, n. 48.

[45] See Pierce, *History of Chicago,* vol. 2, pp. 162–164. The resolution presented by the Chicago Typographical Union to the General Trades Assembly (Sept. 10, 1864) charged that the use of women was "only a temporary expedient [pending] further importation and employment of disreputable printers, who either because of their incompetence or unworthiness, can obtain employment in offices where they are allowed to work for less than the regular established and fair remuneration." (Reproduced from *The Workingman's Advocate,* Sept. 17, 1864, in Pierce, vol. 2, Appendix 7, pp. 501–03).

[46] *Chicago Daily Democrat,* Oct. 21, 1852; *Daily Democratic Press,* Oct. 20, 1852; *Chicago Tribune,* "Annual Review for 1865," pp. 10–11.

[47] *U.S. Ninth Census* (1870), "Population," vol. 1, p. 782. By 1870 the proportion of those employed in needlework who were women was higher than it had been a de-

Where quality adjustments, rather than price adjustments, can be initiated in a particular labor market at the employer's option, their existence is likely to pass with little comment, save that from organized skilled workers thereby threatened. This is especially so if employers' prospects for relief through labor migration remain open. It is a rather different matter when, as was the case in 1861–1864, labor markets everywhere become tight and skilled hands are withdrawn from the work force. Thus, during the period of the Civil War, the forced, downward revision of hiring standards in industry and the upward pressure on wages of less qualified workers evoked grumbles from employers about costly delays and mistakes caused by "green and obstreperous" hands.[48] More lasting complaints against the alteration that had taken place in the attitudes and technical competence of industrial workers were to be heard not only in Chicago, but throughout manufacturing centers in the North several years after the guns had fallen silent. Most pervasive among the charges brought by employers was the assertion that labor productivity had declined because many who held jobs as mechanics were "not really" mechanics, never having served as apprentices and, therefore, "lacking the skill which normally would have secured them employment."

In 1868–1869, David A. Wells, Special Commissioner of the Revenue, was prompted to investigate the matter.[49] He sent out a questionnaire asking employers whether they thought members of their labor force "perform as much work a day now as formerly." An unidentified Chicago manufacturer's estimate that there had been a decline in labor productivity amounting to 12.5 percent was the most restrained among the preponderance of opinion to the same effect. The response to Wells's query from New York was "They perform from one-fourth to one-third less work [since than before] the war" [sic]; from Buffalo, "Twenty-five percent discount"; from Troy, "No, not by fifteen percent in the average"; from Worcester, Massachusetts, "We don't get as much work as formerly by fifteen percent."; from St. Louis, "No, by fifteen to twenty percent." A Bridgeport, Connecticut, manufacturer held out against the consensus, maintaining that mechanics working with improved machinery produced more than they had before the war, while others produced less.

---

cade earlier. The Census of Occupations for the same year listed a total of 18,300 working women of all ages, 16 pecent of all gainful workers enumerated in the city.

[48] See, e.g., the McCormick Company Letter Press Copy Books for 1863 and 1864, cited in W. T. Hutchinson, *Cyrus Hall McCormick,* vol. 2 pp. 88–89.

[49] See U.S., Congress, House, [David A. Wells] "Report of the Special Commissioner of the Revenue," *Exec. Doc. No. 27,* 41st Cong., 2d Sess., 1869, pp. xxxii–iii.

To be sure, the way Wells's question was phrased catered to the employer's inherent conviction that workers are not what they used to be. But the facts remain that between 1859–60 and 1869–70, average labor productivity *did* decline in eleven out of the seventeen major industries in Chicago, and that for Chicago manufacturing as a whole, labor productivity dropped by roughly 30 percent during the decade, thereby reverting to approximately the level prevailing in 1849–50. According to Robert Gallman's estimates, the latter was also true of average real value added per worker in the nation's manufacturing and mining sector, although the reversion to the midcentury level was accomplished by an intercensal decline of only 13 percent.[50] Pending further research into these intriguing findings, it seems only reasonable to accept the opinions of contemporaries: the widespread cessation of labor-productivity growth most probably reflected reductions of hiring standards, and a consequent deterioration in the average quality of the industrial labor force, brought about under wartime pressures.

The hiring of "poorer quality" workers in lieu of those with greater training and a broader range of technical competence does not, of course, exhaust the set of conceivable substitution possibilities open on the demand side of the labor market. Nor was it in fact the only alternative to wage adjustments that was exploited during the earlier phases of industrialization in Chicago. The establishment of some manufacturing activities itself tended to relieve pressure upon the available pool of skilled workers by altering the overall structure of production in a manner that permitted more productive utilization of the semiskilled. From 1847 through the mid-1850s in Chicago, the rapidly expanding planing mills and door and sash factories were, therefore, quite appropriately viewed by contemporaries as "convenient and economical substitutes for carpenters, a supply of whom equal to the needs of the country might indeed be regarded as an improbability."[51] Some early innovations constituted especially striking adaptations in this connection. For instance, during the 1850s local factories turned out prefabricated sections of houses, including roofing, fitted windows, and already-hung doors ready for shipment and assembly.[52]

Beyond the possibilities of reorganizing the way specified final outputs were produced, flexibilities in the selection of product mix offered other avenues for substitution among skill groups. Perhaps most

[50] See David, "Marshallian Dynamics"; David, *FPE*, Ch. 4 (ms) contains more on the 1860–1870 industrial labor productivity decline in Chicago, and in the U.S.

[51] *Democratic Press*, "Annual Review for 1854," pp. 40–41.

[52] See *Weekly Chicago Democrat*, Oct. 26, 1847, cited in Pierce, *History of Chicago*, p. 195, n. 105.

straightforward were the reduced requirements for highly skilled craft workers effected by concentrating manufacturing efforts on lower-quality, standardized product lines. Fortunately, by catering to the preferences of the rural western market for serviceable, but less finely finished or elegantly styled consumer goods, Chicago's manufacturers could both enjoy some transport-cost advantages and avoid trying to duplicate the skill mix of the labor force employed by their counterparts in the eastern states. As was the case with "common and useful" pine furniture and the heavier grades of boots and shoes made from western leather, so the production of lumber wagons, farm wagons, and trucker's drays called for a different quality of workmanship than did light-carriage building.[53]

It should not be inferred from this portrayal that a western market for higher-quality products did not exist. Quite the contrary. The initial phases of western industrialization may in part be viewed as a process of *selective import substitution* involving, among other things, the gross substitution of less- for more-skilled labor, through continued dependence upon extraregional supplies of certain grades of manufactured goods. Thus, Chicago's manufacturing establishments did not attempt to satisfy all portions of even the local market for some classes of goods, and importation tended to occur at the higher-quality end of those product lines. Fine hardwood furniture and light carriages, for example, were items of import during the 1850s. The same was true of women's shoes, where eastern fashions and lighter leathers combined with more demanding standards of workmanship to create a product that was not attempted by the city's boot and shoe factories until late in the 1870s. The more complicated items of machinery, such as double-cylinder printing presses, typically were imported from the East. Even in the steam engine line, where early local progress had been substantial, Chicago-built engines were to be found, during the 1840s, serving in an auxiliary capacity to the larger, more reliable pieces of equipment from shops such as the Morgan Iron Works in New York City.[54]

[53] See, on quality of furniture, *Democratic Press*, "Annual Review for 1853," pp. 54–55; "Annual Review for 1854," p. 43; "Annual Review for 1855," pp. 34–35. On boots and shoes the above comments do not apply to small custom-work shops, but rather to the larger producers, especially from 1861 on. See *Democratic Press*, "Annual Review for 1854," p. 46; "Annual Review for 1855," pp. 39–40; E. M. Hoover, *Location Theory and the Shoe and Leather Industries* (Cambridge, Mass.: Harvard University Press, 1937), pp. 269–71. On wagons and carriages, see *Democratic Press*, "Annual Review for 1852," p. 13; "Annual Review for 1853," p. 54; "Annual Review for 1855," p. 30.

[54] See *Democratic Press*, "Annual Review for 1855," p. 30.; also, on printing presses, ibid. pp. 79–80; on boots and shoes, Clark, *History of Manufactures*, vol. 2, p. 474–77, *Chicago Tribune*, "Annual Review for 1879,"; on steam engines, *Democratic Press*, "Annual Review for 1853," pp. 45–46.

At this juncture it is particularly appropriate to explicitly consider some features of the market for unskilled labor in the Chicago area during the years before the Fire of 1871. For, if continuing reliance upon imports is recognized, *inter alia,* as the other side of the adjustment of the industrial output mix to conditions of skilled labor shortage, in at least one respect a rather different situation must have obtained with respect to the requirements for common labor.

III. The Market for Unskilled Labor

In nineteenth century American experience, the opening up of a new region typically generated a demand for manual labor that was irrevocably skewed toward the lower end of the spectrum of industrial skills. Construction was relatively more important and, by comparison with most manufacturing industries, the building trades tended to be characterized by a high ratio of wage costs to total costs.[55] Moreover, the requirements for common labor in such activities were relatively large and, in contrast to those for craft workers, remained for the most part unaffected by such changes in technique as the use of prefabricated structural elements. Perhaps most significantly, regional and local needs for housing, transport facilities, storage, and associated capacity in the distributive sector, as well as other items of urban infrastructure like streets and sewers, could not be satisfied by importation. Construction is the demand-localized activity *par excellence.* Thus the gross substitution possibilities opened, in the case of skilled industrial labor by commodity importation, were not available as a means of reducing the pressure upon the supply of unskilled labor in this developing "frontier" region.

The crux of this problem, however, is not that a regional balance-of-payments constraint forces expansion of commodity-exports, produced with less-skilled labor, to pay for the importation of high-skill–content goods. That would certainly make matters more serious. But even with current account deficits financed by generous lending from other areas, as was the case during Chicago's participation in the boom of the 1850s, the high rates of physical capital formation in structures and other nonimportables drove the developing regional economy up against a short-run common-labor constraint.

To these specific historical observations, another and more general consideration must be added. When workers are sought to fill places at the bottom of the skill ladder, it is not so feasible as it is at the upper

---

[55] On the relative size of the construction sector, see David, "Marshallian Dynamics," pp. 15ff.

end to achieve an increase in the supply of labor for the tasks in question through the device of upgrading workers relative to jobs. Putting it another way, the downward flexibility of hiring standards begins to diminish as their level falls. A newly developing region is therefore confronted by fundamental asymmetries in the range of possible demand adaptations to conditions of skilled and unskilled labor scarcity. For the reasons given, some of the main avenues of nonprice response to shortages of unskilled labor are closed off, forcing greater dependence upon attracting common labor from other regions through increased wage rates and direct recruitment on a large scale.

While the impact of internal improvement projects upon western labor markets is scarcely an unfamiliar story, it is not often explicitly related in these terms.[56] Nor do the more general implications of construction labor requirements, with regard to the availability of unskilled workers in industry and other activities, usually receive the recognition they would seem to deserve. Some closer attention to these matters is therefore in order.

The revival of work on the Illinois and Michigan Canal during the latter half of the 1840s quickly pushed up both the money and the real wages of unskilled labor in the region around Chicago. By 1846–47, the standard rate for common laborers on the canal reached $1.00 a day, approximately twice the general labor rate paid in 1843.[57] Local supplies still remained inadequate; during the summer of 1846, the canal project's Chief Engineer, William Gooding, strongly urged the use of agents to recruit laborers among the immigrants arriving at New York, Boston, and Montreal. Gooding's problem was made more complicated the following summer when contractors faced with similar difficulties on the Wabash and Erie Canal in Indiana and on the Michigan Central Railroad scoured labor markets farther west—thereby pro-

[56] However, Harvey H. Segal estimated that in 1840 employment on canal construction alone accounted for roughly 5.5 percent of the total nonagricultural labor force in the ten canal-building states. Segal noted that in such sparsely settled states as Illinois and Indiana, the ratio of canal employment to total employment was undoubtedly much above the ten-state average. See Carter Goodrich, ed., *Canals and American Economic Development* (New York: Columbia University Press, 1961), p. 206. See below for a parallel estimate relating to railroad construction employment in the midwestern states during the boom of the 1850s.

[57] For money wages of common labor see Table 1. Cf. James W. Putnam, *The Illinois and Michigan Canal* (Chicago: University of Chicago Press, 1918), p. 61; Pierce, *History of Chicago,* vol. 1, pp. 194–196, on changes in wages and the cost of living in Chicago during the 1840s. The Chicago daily wage for common labor in 1847 was the same as that prevailing in 1836, when the inflationary boom of the 1830s was approaching its peak. Yet local prices of food, clothing, and rents recovered much less rapidly than wages from their 1841–1843 depression levels.

ducing high turnover rates in the common-labor force that had been secured for the Illinois canal project.[58] It is indicative of the nature of labor market conditions in a rapidly developing region that the Chicago area's first strike was occasioned by the demands of common laborers, not those of skilled industrial workers: Irish laborers on the Summit Division of the canal quit work for two weeks during the summer of 1847 to enforce their demands for $1.25 a day—and more agreeable foremen.[59]

During the decade that followed, engineers in charge of railroad construction in the Midwest faced precisely the same sort of labor problems, only on a larger scale. In the spring of 1853, a year that saw 769 miles of rail line completed in Indiana and Illinois alone, the Chief Engineer for the Illinois Central, Roswell B. Mason, despairingly reported that as a result of railroad building already in progress elsewhere, men were available at neither Chicago nor St. Louis for work on his road. Average wages for common labor rose to, and in some cases beyond, $1.25 per day. Even so, the supply elicited was insufficient. To acquire and maintain a work complement sometimes exceeding 10,000 men, the Illinois Central and its contractors found it necessary to send agents to New York and New Orleans, to place broadsides and newspaper advertisements in eastern cities, and to make arrangements with immigration agents to forward men from New York. On one occasion, agents were dispatched directly to Ireland to secure a large group of laborers.[60]

The years between 1850 and 1860 saw 6,790 miles of railroad line opened in Illinois, Indiana, Michigan, Wisconsin, and Iowa. An average of 991 miles was added annually to the midwestern rail network during the four-year period of most intense activity, 1853–1856. We may crudely estimate that in this interval, average annual railroad construction employment for the five-state area was on the order of 23,000–34,000. That would represent approximately 10–13 percent of the entire nonagricultural male labor force available to the region in

[58] See William Gooding to William H. Swift, July 6 and Aug. 8, 1846, July 22 and Sept. 7, 1847, *William W. Swift Papers, 1843–1865* (mss., Chicago Historical Society). For W. Gooding, W. H. Swift, and their respective roles in the completion of the canal, see the account given by Andreas, *History of Chicago,* vol. 1, pp. 170–173.

[59] See *Swift Papers,* W. Gooding to W. H. Swift, July 22, 1847; Pierce, *History of Chicago,* vol. 1, p. 199; Andreas, *History of Chicago,* vol. 1, p. 171.

[60] See F. L. Paxson, "The Railroads of the 'Old Northwest' Before the Civil War," *Transactions of the Wisconsin Academy,* 17, part I, no. 4 (1911), p. 270, for mileage added in 1853. For the Illinois Central's labor recruitment problems and techniques, see Paul W. Gates, *The Illinois Central Railroad and Its Colonization Work* Cambridge, Mass.: Harvard University Press, 1943), pp. 94–98.

1850, or something between 25 and 36 percent of the male nonfarm common laborers reported by the Census of Occupations at that date.[61] It is hardly surprising that the Illinois Central's contractors alone were forced to bring 5,000–10,000 laborers into Illinois from other regions in the years between 1852 and 1856.[62]

Ultimately, such workers were released from internal improvement projects and became available for other activities, including unskilled industrial tasks. But from Chicago's experience, it is evident that the creation of regional physical overhead facilities and rapid industrial advances tended to occur together in time instead of proceeding in orderly succession. The industrial spurt of 1852–1856 and the subsequent surge of 1869–1873 both took place against a background of accelerated movement of people into the city's hinterland and rapid extension of the region's transport network; both were accompanied by rising urban land values and high levels of nonindustrial construction in the city.

The special attention given here to the competition between construction and manufacturing in the market for unskilled labor appears warranted, if only as a corrective measure. In the writing of American economic history, so much emphasis has been devoted to how the economic opportunities open in agriculture affected wage levels and recruitment of workers for manufacturing that the construction sector's shifting demands upon the supply of unskilled labor have tended to pass almost without notice.[63] The perennial debate on the validity of

---

[61] See Paxson, "Railroads," pp. 267–274; [Henry V. Poor] *Poor's Manual of Railroads, 1867–68*, for rail mileage built. (The Iowa mileage was taken from *Poor's*, the rest from Paxson.)

The five-state employment estimate is of the crudest sort: alternative labor coefficients of 23 and 34 workers per mile of line built were derived from Paxson's mileage data and from the employment statistics given by Gates for the Illinois Central Railroad in 1853 and 1854. (See Gates, p. 96, n. 54, 97–98, n. 61.) These were simply applied to the annual average mileage added in the five-state area in 1853–1856. The lower coefficient is based on employment and mileage in 1854 alone; the higher figure on the 1853–1854 averages.

In 1850 the census reported 255.5 thousand free male nonfarm workers aged fifteen and over in the five-state region, and 93.9 thousand nonfarm "laborers." See *Compendium of the Seventh Census*, p. 128.

[62] See Gates, *Illinois Central Railroad*, p. 97.

[63] See Paul A. David, *Technical Choice, Innovation and Economic Growth* (New York: Cambridge University Press, 1975), ch. 1, for a review of the literature on the consequences of agricultural land abundance and the industrial labor scarcity, following H. J. Habakkuk, *American and British Technology in the Nineteenth Century* (Cambridge, England, 1962), pp. 11–14, on this tendency. Cf., however, [*ibid.*], p. 23 for brief recognition that agriculture was not the only competitor for manufacturing workers. George R. Taylor, *The Transportation Revolution, 1815–1860* (New York: Rinehart and Winston, 1951), pp. 289–290, also notes that ". . . construction

Labor Market Adjustments: Chicago 75

the "safety-valve" hypothesis, in its original and "postmortem" versions, has played a part in keeping interest focused upon the question of competition between frontier agriculture and industry in eastern labor markets.[64]

At the risk of becoming even briefly entangled in the toils of that controversy, it may be remarked that insofar as the trek to western lands was an alternative to townward movement for members of the eastern agricultural population, the westward migration of *potential* workers reduced the eastern industrial labor supply and tended to maintain wages at higher levels than would have otherwise prevailed. One can allow this resurrection of the safety-valve concept, and even extend it to include the general westward movement, while accepting the consensus of historical judgment against the original view advanced by followers of Frederick J. Turner. There is still no reason to think industrial workers were actually drawn out of the eastern urban labor force in significant numbers by the attractions of western farming. Nor does it follow that average earnings in either eastern or western agricultural employment set the immediate floor under the wages of unskilled industrial labor in the East throughout the pre-Civil War period, as many writers continue to insist.[65] Indeed, there are some bits

---

of the turnpikes, canals, and railroads created an unprecedented demand for unskilled workers." But Taylor then proceeds, largely on the basis of Matthew Carey's account of canal laborers in 1831 (a year in which, incidentally, canal investment declined markedly), to stress that they constituted the lower stratum of the free labor force, and to detail the miseries and violence of life in the construction gangs. See Matthew Carey, *Address to the Wealthy of the Land* (Philadelphia, 1831); Goodrich, *Canals,* pp. 173, 208–210.

[64] See, for the safety-valve thesis as advanced by a follower of F. J. Turner, Joseph Schafer, "Was the West a Safety Valve for Labor?" *Mississippi Valley Historical Review* 24 (1937), pp. 299–314.

Carter Goodrich and Sol Davidson, "The Wage-Earner in the Westward Movement," Parts I and II, *Political Science Quarterly* 50 (June 1935), pp. 161–185, 51 (March 1936), pp. 61–116; Murray Kane "Some Considerations on the Safety-Valve Doctrine," *Mississippi Valley Historical Review* 23 (Sept. 1936), pp. 169–188; and Clarence Danhof, "Farm-Making Costs and the 'Safety-Valve': 1850–1860," *Journal of Political Economy* 49 (June 1949), pp. 317–359), are all critical of the original hypothesis, but less so than is Frederick A. Shannon, "A Post Mortem on the Labor-Safety-Valve Theory," *Agricultural History* 19 (Jan. 1945), pp. 31–37; and *The Farmer's Last Frontier* (New York: Farrar and Rinehart, 1945), pp. 306, 356–359).

Later arguments for liberalized versions of the safety-valve theory were advanced by Norman J. Simler, "The Safety-Valve Doctrine Re-Evaluated," *Agricultural History* 32 (Oct. 1958), pp. 250–257, and George G. S. Murphy and Arnold Zellner, "Sequential Growth, the Labor-Safety-Valve Doctrine and the Development of American Unionism," *Journal of Economic History* 19 (Sept. 1959), pp. 402–421).

[65] See, e.g., Habakkuk, *American and British Technology* pp. 12, n. 2, 39–40. This was the view of late eighteenth and early nineteenth century American manufacturers. Although its applicability to the period after 1830 has been widely asserted, a solid

of evidence that would appear directly to controvert that argument and the interpretations of antebellum industrialization based upon it.[66]

To come closer to home, it remains far from clear that the attractions of high earnings in midwestern agriculture were the proximate cause of the unskilled labor scarcities encountered by *western* industry. On the one hand, it can be pointed out that during the 1850s, western agriculture meant farming the prairies rather than the timbered lands of the Ohio Valley, and the typical costs of farm-making had thus been raised sufficiently to exclude those lacking substantial capital from immediate entry into owner-operated agriculture.[67] On the other hand, there existed credit facilities and the possibilities of acquiring—through wage labor or tenancy, or both—the requisite capital and experience with prairie husbandry; these features facilitated entry into agriculture by men of initially slender means.[68] Whether or not tenants and farm laborers typically were able to attain the status of proprietors is patently of secondary significance to this discussion, if a very large number of unskilled workers were occupied trying to climb the agricultural ladder and, hence, unavailable for industrial employment.

Unfortunately, it is too easy to form an exaggerated impression, and virtually impossible to arrive at a precise assessment of the quantitative significance of the agricultural employment opportunities for unskilled labor in Chicago's hinterland during the initial phases of the city's industrial development. On the basis of the statistics published in

---

evidentiary basis for it remains to be established. Cf., the interesting and somehow overlooked discussion on this point in Clark, *History of Manufactures,* p. 392.

[66] For example, in 1832, New England farmers were complaining that factories attracted labor and raised wages so that agriculture became unprofitable. Cf., Isaad Hill [New Hampshire], *"Speech in the U.S. Senate on Mr. Clay's Resolution in Relation to the Tariff,"* cited in Clark, p. 392 and Paul W. Gates, *The Farmer's Age: Agriculture 1815–1860* (New York: Holt, Rinehart and Winston, 1960), p. 271, on the difficulties of obtaining farm labor in New England in the 1840s; also below, n. 72, on farm wage rates, where it is noted that New England wages exceeded rates for farm labor in the Midwest in 1850. These observations may not be unreconcilable with Field's ("Sectoral Shift in Antebellum Massachusetts") contention that the contraction of agriculture in New England had created a source of cheap labor for the region's industries, but absorbing them would certainly complicate that argument. They also cast some doubts upon Goldin and Sokoloff's ("Relative Productivity Hypothesis") interpretation of the low relative wage rates of women and children vis-à-vis men in the Northeast as reflecting the relative strength of demands for these kinds of workers in the region's agricultural sector.

[67] Cf. Danhof, "Farm-Making Costs," *passim.* Danhof estimates initial capital requirements averaged over $1,000 in the 1850s, exclusive of land costs. Cf. also, Allan G. Bogue, "Farming in the Prairie Peninsula, 1830–1890," *Journal of Economic History* 23 (March 1963), pp. 3–29, esp. p. 24.

[68] Cf. Bogue, "Prairie Peninsula," pp. 24–26; Shannon, *Farmer's Last Frontier,* pp. 359–360; Gates, *Farmer's Age,* pp. 96–98, 275, 403–404.

the Census of Occupations, Paul W. Gates concuded that for every two farms in the nonslave states in 1860, there was one "farm laborer" (or tenant). In Iowa and Wisconsin, the ratio of farms to farm laborers, as reported by the census, approximated the average in the free states as a whole, while there were relatively fewer farm laborers per farm in Illinois and Indiana.[69] Yet the farm laborer reported in the occupational census was not simply an independent, mobile, hired hand. This designation also covered the older children of "farmer's" families. From such fragmentary evidence as is available for the period, it appears that the large broods of farm children provided the bulk of the nonowner agricultural work force in the country behind Chicago. Samples drawn from the 1850 U.S. Census rolls of Illinois and Iowa townships show that family members other than owner-operators made up 26–35 percent, and hired hands only 11 percent, of the total male farm-labor force aged fifteen and older; a hired man was to be found on but one in five or six farms.[70] Equivalent data for Indiana farm households in 1850 and 1860 suggest that the use of hired hands, outside the height of the harvest season, was considerably less significant than that.[71] If boldly applied to the total number of males aged fifteen and older who claimed agricultural occupations in Michigan, Iowa, Indiana, Illinois, and Wisconsin in 1850, even the 11-percent figure for hired hands would indicate that the region's 239,175 farms—which during the next ten years almost doubled in number—at that date employed only somewhere in the neighborhood of 48,000 hired male laborers.[72] That equals but a shade more than half the number of male nonfarm "laborers" in the region at the beginning of the 1850s.[73]

This picture of agriculture's comparatively heavy reliance upon younger members of the family, tied to the farm by custom and law so long as they remained minors, jibes well with contemporary rural comments regarding the difficulties of securing reliable wage labor. It was not the prices so much as the quality, or lack thereof, which evoked dissatisfaction. Complaints about irresponsibility, disinclination to put

[69] See Gates, *Farmer's Age*, pp. 272–273.

[70] See Bogue, "Prairie Peninsula," Table 3, p. 22, and also pp. 25–26, where other samples indicate tenancy rates (in the total population of male farm operators) of 7–11 percent in Illinois, and somewhere between 11 and 20 percent in Iowa in 1850.

[71] I am grateful to Mr. Charles Lave for making available transcriptions of the manuscript censuses of Johnson Township, Clinton Co., Indiana. In 1860, hired hands in the sample male labor force (aged fifteen and over) of 263 farm "workers" were proportionately more important than in 1850, but still represented a scant 3.4 percent of the total, or one hired man per twenty farm households.

[72] See *Compendium of the Seventh Census*, pp. 128, 169; *U.S. Eighth Census* (1860), "Agriculture," p. 22.

[73] Cf. above, n. 61.

in a full day's work, and carelessness with equipment were universal among farmers' expressions of dissatisfaction with the character of their temporary hired help.[74] But reduced hiring standards that accepted such traits did not provide a solution to the farm-labor problem. For example during the expansion prior to 1857, midwestern farmers found it increasingly difficult, even in the harvest season, to compete with railroad and other construction contractors who offered well above a dollar a day for the very commonest of common labor. As workers were pulled westward by higher wage rates, and some by the lure of the gold country in Colorado and California, complaints of harvesting delays caused by the scarcity of farm labor arose even in the more densely settled agricultural regions of New York and Ohio.[75] This was, after all, the setting in which the adoption of harvesting machinery made such rapid headway. Of Illinois in 1857 it was said,

> . . . all grain is here cut by machine. Cradles are out of the question. . . . If grain is too badly lodged to be so gathered it is quietly left alone. . . . This work done by machinery is not very much cheaper than it could be done by hand, but the great question is—where are the hands to come from?[76]

If agricultural competition for laborers was not the proximate force pushing wage rates upward during the boom, the agrarian demand for unskilled workers may have helped alleviate the downward pressure on common-labor rates when regional and urban construction was halted and unemployed workers began piling up in the cities. With unemployment on the rise during the winter of 1857 and the spring of the following year, Chicago's newspapers advised those thrown out of work to make the best of their situation by seeking jobs in rural areas to the west of the city.[77] The check thus provided to declining urban wages

---

[74] See Gates, *Farmer's Age*, pp. 272, 275.

[75] See ibid., p. 277, and Stanley Lebergott, "Wage Trends, 1800–1900," in W. N. Parker, ed., *Trends in the American Economy in the Nineteenth Century*, Studies in Income and Wealth, 24 (Princeton, 1960), pp. 452–453, on regional patterns in farm labor wages. According to the results of a special inquiry published in *Compendium of the Seventh Census* (p. 164), monthly farm-hand wages (with board) in 1850 were higher in Wisconsin, Illinois, Michigan, Iowa, and Missouri (in that order) than in New York, Ohio, and Indiana. However, rates for Massachusetts ($13.55), Rhode Island ($13.52), Maine ($13.12), Vermont ($13.00), and Connecticut ($12.72) all exceeded the $12.55 reported for Illinois or the $12.69 reported for Wisconsin.

[76] *The Iowa Farmer* (1857), quoted by Danhof, "Farm-Making Costs," pp. 348–349. On the adoption of mechanical reaping and the growth of the market for farm equipment, see Paul A. David, *Technical Choice, Innovation and Economic Growth* (Cambridge: Cambridge University Press, 1975), Ch. 4.

[77] See *The Weekly Democratic Press*, Feb. 21, 1857; *Chicago Daily Democrat*, June 12, 14, 1858. These urgings continued as late as the winter of 1861. See *Chicago Tribune*, Jan. 9, 1861, cited by Pierce, *History of Chicago*, vol. 2, pp. 156–157, n. 29, 32.

was not, however, very effective. By the spring of 1858, Chicago laborers who had asked between $1.25 and $1.50 a day only two years before were accepting employment at 75 cents, when they could find it. The city government, having cut the daily rate for labor on its streets to 50 cents in an effort to create more jobs, had to limit employment to two days a week per man to spread the work among the applicants.[78] Farmers just did not suddenly abandon substantial fixed-cost techniques adapted to conditions of dear labor in order to take advantage of its temporary abundance.

Furthermore, it is questionable whether farmers had full access to the depression-formed labor reserve. Although seasonal and cyclical movements of workers into and out of agriculture undoubtedly took place, and were part of the normal employment pattern in the lives of some members of the unskilled labor force,[79] lack of information about dispersed job openings on farms and resistance to such occupational and locational changes rendered the overall adjustment process highly imperfect. Despite the exhortations of the local press, the degree of mobility among the mass of unskilled urban laborers appears to have been especially low during periods of depression. They remained idle or employed at low wages in centers like Chicago, while farmers in the surrounding country continued to lament shortages of hands and offered pay rates much in excess of prevailing urban levels. In 1860, for example, with the going daily wage for common laborers in the city at 75 cents to $1.00—still showing the effects of the depression—harvest hands were in such demand that Wisconsin farmers reportedly offered them anywhere from $1.00 to $3.00 a day.[80] Chicago was not alone in this experience: the New York Association for Improving the Condition of the Poor, in its report for 1858, argued that emigration to rural areas was not a suitable depression remedy for its clients, who were "not interested in agriculture" and represented "for the most part the less enterprising and industrious classes of the population."[81]

The burden of the foregoing observations is that, contrary to the impression frequently conveyed, during periods of rapid expansion in a region of recent settlement it was the demand for labor not on farms, but on construction projects in and around growing cities like Chicago

[78] See Pierce, *History of Chicago,* vol. 2, pp. 155–156.
[79] See Gates, *Farmer's Age,* p. 273–274.
[80] See *Chicago Daily Democrat,* July 19, 1860, cited in Pierce, *History of Chicago,* vol. 2, p. 156, n. 31. Farm labor was reported at a premium in Indiana during the winter of 1857, according to the *Philadelphia Public Ledger,* Dec. 14, 1857, cited by Leah H. Feder, *Unemployment Relief in Periods of Depression* (New York: Russell Sage Foundation, 1936), p. 35, n. 1.
[81] Feder, *Unemployment Relief,* pp. 35–36.

that tended to set an immediate floor under wage rates for unskilled workers in manufacturing. In addition, the agricultural demand remained rather ineffective in preventing the collapse of common laborers' wage rates when the expansionary movement was halted. This argument, it should be stressed, concerns the nature of conditions on the existing margin; it scarcely implies that under wholly different circumstances in which, say, hired and family workers were released from agriculture *en masse,* unskilled laborers' wages in manufacturing (and construction) would not have been generally very much lower.

Moreover, as the level of local and regional construction activities changed, the presence of sectoral interdependencies tended to amplify the short-run impact upon the availability and wages of unskilled labor in Chicago's industries. Among these interdependencies, the indirect labor requirements that regional physical capital formation generated in the lumber industry, and in lumber distribution, deserve our notice. As *Hunt's Merchants' Magazine* was quick to perceive,

> Iron, as being the basis of all machinery, and the chief element in the construction of railroads, has been said to furnish, by the extent of its consumption, a true measure of the state of civilization. With equal propriety it may be said that the consumption of lumber, in a State in progress of being settled, is at once both a measure of its prosperity and the degree of its development.[82]

The correctness of this view is affirmed, for the period prior to the 1880s, by the significant correlation that existed between swings in the volume of gross construction in Chicago and fluctuations in the city's net lumber receipts.[83]

Taking into account the timber-cutting, milling, lake-transport, storage, and wholesaling aspects of the business, the involvement of the Chicago members of the "lumber interest" was quite extensive: a contemporary estimated that in 1856–57 approximately 10,000 hands were employed in the business "one way and another."[84] This figure might be juxtaposed with the contemporaneous estimate, cited above, 10,573 as the average number of workers employed in the city's man-

---

[82] *Hunt's Merchants' Magazine* 26 (April 1852), p. 435.

[83] See David, *FPE,* Appendix C, for underlying data. Correlation analysis of year-to-year changes in the volume of gross construction in Chicago and recorded net receipts of lumber in the city yields a correlation coefficient (adjusted for degrees of freedom) of $R^2 = .78$ for the period 1854–1877.

[84] *Daily Democratic Press,* "Annual Review for 1857," p. 22. The writer, William Bross, also stated that apart from the property owned by Chicago lumber merchants in the form of mills and woodlands, their wealth tied up in vessels, stocks, docks, and commercial real estate in the city in 1856–1857 "cannot be less than ten or a dozen million dollars."

ufacturing sector. It would, most likely, have represented a seasonal maximum rather than an annual average. On a rough reckoning, somewhere between 1,800 and 1,900 workers appear to have found employment as crewmen on the lake vessels needed to deliver the (average) 735 million feet of lumber, plus shingles and other subsidiary wood products, received in Chicago in those two years; only 688 sailors had served in the fleet that brought 148 million feet of lumber to the city's wharves in 1852.[85] While employment in this activity therefore expanded rapidly during the boom of the 1850s, the lumber fleet's demands for seamen did not keep pace with the long-run upward trend in lumber received at Chicago, 94 percent of which still came by way of the lake in 1877. The average crew size remained roughly constant at seven to eight sailors per vessel, but, as the vessels became faster and increased in cargo capacity, they were able to transport more lumber before the arrival of winter closed navigation on the lakes. Employment in the lumber fleet was put at 2,500 in 1874, which would suggest that average labor productivity (measured in feet of lumber transported to Chicago per man) had risen at an average compound rate of approximately 3 percent per annum during the preceding twenty years. The estimate of productivity increase probably overstates the gains during the 1850s and understates those after the Civil War, since significant changes in techniques of lumber carriage, such as the use of barges, were a postwar development.[86]

In contrast, there is no evidence to indicate a significant increase in the average productivity of the lumber shovers. On the assumption that none occurred, it may be estimated that in 1856–57 another 3,000 men were needed to transfer, pile, and sort the lumber upon its arrival in Chicago. This suggests that at that date, perhaps slightly more than half of the 10,000 jobs generated by Chicago's total lumber business

[85] See *Hunt's Merchants' Magazine* 26 (April, 1852), 435, for employment in the "Chicago" lumber fleet in 1852. The estimate for 1856–1857 was derived by computing an average input coefficient for sailors per million feet of lumber received in Chicago in 1852, and another such coefficient for 1874 (see following footnote), and interpolating between them geometrically, assuming a steady rate of increase in average labor productivity between 1852 and 1874. The adjusted input coefficient for 1856–1857 was then applied to average lumber receipts in Chicago over the two-year period, taken from the *Chicago Board of Trade Report for 1896*, p. 102. The exact figure obtained was 1,872 sailors.

[86] See *Chicago Board of Trade Report for 1877*, pp. 144–151, for sources of lumber received at Chicago; *Chicago Times,* Jan. 1, 1874, for employment in the fleet carrying lumber to Chicago. Total lumber receipts were adjusted for the fraction not received via the lake to obtain an average labor-productivity figure of 403,000 ft. per man in 1874, whereas the productivity figure derived from 1852 data (see preceding footnote) was 214,000 ft. per man. For changes in the lumber fleet, see Pierce, *History of Chicago,* vol. 3, p. 100.

took the form of employment in the lumber country. Only a third of the estimated 1856–57 city work complement would have sufficed to handle the movement of lumber through Chicago in 1852, but with the collapse of the regional construction boom after 1857, and the decline in the volume of receipts and shipments, as many as 3,000 hands would not be required again until 1864. By the close of 1873, however, some 7,000 men reportedly were needed in the same activity, a level of employment that probably was closely approached during the five preceding years.[87]

In addition to the fluctuating demand for labor created at the docks and yards in the city itself, shifts in the demand for works in "outside" lumbering operations were directly communicated to the Chicago labor market. For, during the 1850s and 1860s, the city served as an important recruiting point for the camps and mills in the northern pineries. This was a quite natural arrangement: not only was Chicago the region's major urban center and lumber mart, but many of the lumber settlements across Lake Michigan were actually colonial enterprises of the lumber dealers headquartered in the city. From Muskegon, Manistee, Big Sauble, and other points on the eastern shore of the lake above Grand Haven, Michigan, vessels carried lumber to the yards along the Chicago River's branches, from whence they made the return voyage in the spring and summer months, ballasted with men and provisions secured by the head offices in the metropolis. Although no direct evidence has come to light, it is highly probable that some of the lumber operations on the opposite side of Lake Michigan, in the Green Bay district, followed a similar recruitment pattern. Outposts of Chicago enterprise, such as William B. Ogden's thriving settlement at Peshtigo, Wisconsin, had been established there too during the pre-Fire era.[88]

---

[87] See *Chicago Times,* Jan. 1, 1874. The estimates for labor requirements in lumber handling in the city between 1852 and 1874 were derived by computing a coefficient for the later date (4.26 men per million feet of lumber *received and shipped* through Chicago) and applying it—assuming no change in labor productivity—to data giving the recorded totals of receipts and shipments at earlier dates, taken from *Chicago Board of Trade Report for 1896,* pp. 102–103.

[88] See, on the lumber dealers and their early interests in Michigan and Wisconsin, Andreas, *History of Chicago,* vols. 2, pp. 689–693, and 3, pp. 365–386; *Hunt's Merchants' Magazine* 54 (Feb. 1866), p. 104; John Moses and Joseph Kirkland, *History of Chicago,* 2 vols., (Chicago, 1895), vol. 1, p. 378–388; Pierce, *History of Chicago,* vol. 2, pp. 103–104. For William B. Ogden and the Peshtigo (Wisconsin) Company, which was formed in 1856 and employed 500 men in its mills and shops in 1871, see Pierce, vol. 3, p. 100; James W. Shehan and George P. Upton, *The Great Conflagration* (Chicago, 1871), p. 376. The far-flung activities of the lumber interest during the post-Fire period are more adequately documented in Pierce, vol. 3, pp. 99–107.

The activities of Charles Mears (1814–1895) were in many respects representative of the early entrepreneurs who flourished in this field. After planting an initial series of lumber settlements in Michigan, in 1848 he set himself up as a lumber merchant on West Water Street in Chicago, and, through a succession of complex partnership arrangements, went on to acquire further holdings of timberland, to start other mill settlements in the Michigan lumber country, to build and acquire a fleet of lake vessels, and to open additional yards in the city before reluctantly retiring from the business in 1883. In 1870 the firm of Mears, Bates & Co. was one of the largest lumber dealers in the city, receiving supplies from their own mills in the Wisconsin and Michigan lumber country and shipping "frequently to New Orleans, Memphis, all over the states of Tennessee, Missouri, Kansas, Nebraska, and to thousands of points in Illinois."[89] As of 1875, Charles Mears had located 40,000 acres of timberland, built five harbors, and set up fifteen mills in the Michigan counties of Muskegon, Oceana, and Mason.[90] His business papers and correspondence shed light upon labor recruitment practices in the northwestern lumber industry during the 1850s and 1860s, about which comparatively little is known, and, in so doing, they also testify to changes in the state of the Chicago labor market.

It was Mears' practice to obtain signed "agreements" from the men sent out from Chicago to the pineries, usually obligating them to work for a year at stipulated total wages, plus board and laundry service. (The predominance of single men at the settlements is clear, if only from the last proviso.) During 1855 and 1856, and even up to the fall of 1857, the typical annual money wages offered a single male for general work at the Michigan establishments was $160. On a monthly basis this amounted to perhaps only slightly more than the wages (with board) paid for farm labor around Chicago, but, unlike most farm situations, work was provided at that rate throughout the year. Higher wages were paid in some instances: Thomas Ocerby, who said he could

[89] *Chicago Tribune,* "Annual Review of 1870," p. 47.
[90] See manuscript handbook, *Charles Mears Collection* (Chicago Historical Society); Andreas, *History of Chicago,* vol. 2, pp. 691–692. Charles Mears was born at North Billerica, Mass. He taught school there and participated briefly in the lumber and provision business at Lowell before moving with his brothers, Edwin and Nathan, to Paw Paw (Van Buren Co.), Mich., in 1836 to set up a general store. In 1837 he went to the White Lake, Mich., area to obtain title to a tract of timberland in advance of a party of eastern speculators, where (being unable to resell it due to the ensuing panic of that year?) he built a dam and a mill at Pentwater, with mechanics and machinery brought from the East, and shipped his first cargo of lumber to Chicago in 1838. During the 1840s and early 1850s, mill settlements were started at Duck Lake, White River, Little Sauble, and Big Sauble, Mich.

also run an engine, signed up in June 1855 for twelve months at $216 plus board and washing.[91] By the summer of 1856, general labor could not be readily secured at those rates. August found Mears writing from Chicago, "I shall send a number of German hands and hope Mr. Lamieux [the manager of the mill at Duck Lake, Michigan] will do the best he can with them as no others come around."[92]

Occasionally, agreements were made with married men whose wives, in return for an additional $40 per annum, were expected to see to it that C. Mears & Co. made good on the clauses promising clean shirts. The wording of the following document is typical of most agreements concluded during the 1850s—except for the provision regarding the wife:[93]

> I hereby agree to work for Charles Mears at his Lumbering Establishments in Michigan with my wife Wilhemine one year for the Sum of Two Hundred dollars with board and washing, and to work as follows; From Sun rise till Sun set when the days are more than Twelve hours long, Twelve hours when there are twelve hours of daylight, and not le[f]s than Eleven hours at any time in the year, also to furnish myself with a good axe, assist in the loading of Vessels if able whenever requested to do so, and do all I can to forward the work, and promote the interest of my employer.
> Witne[f]s   [signed] Theodor Susen

Such compacts were regarded as binding upon the employee but could be broken at Mears's pleasure. Thus, at the end of October 1857, when conditions in the Chicago labor and lumber markets had changed, Mears wrote to the manager of one of the Michigan mills:

> We are now having the hardest times with the most *Gloomy* prospects for Business in Future that this Country has ever seen. . . . I can now hire a plenty of good hands at from 8 to $10 per month by the year and would be glad to pay off all who are willing to give up their contracts [and] leave [and] to have others at the going wages.[94]

He proceeded to give explicit instructions that all those with written agreements, except the engineer and head sawyers and "a very few other of the oldest [and] most reliable hands" would have to make new

---

[91] See *Mears Collection,* "agreements" in Correspondence Folder/I (1838–1857). There are references to annual wages for general workers running as high as $180 a year during this period (see Mears to P. D. Fraser, July 4, 1858), but the bulk of surviving signed agreements were made at $160.

[92] C. Mears to "Friend Moore," Aug. 20, 1856, *Mears Collection,* Correspondence Folder/I.

[93] Agreement dated May 27, 1856, *Mears Collection,* Correspondence Folder/I.

[94] C. Mears to William Sprigg, Oct. 30, 1857, *Mears Collection,* Correspondence Folder/I.

agreements at $120 a year, with board and so on, or leave. The $8-per-month figure was probably rather extreme, for a standard wage rate of $10 per month for "outside" work was established and appears to have been maintained pretty well throughout 1858–1861.[95] Nonetheless, wages for general labor during this period of depression in the lumber trade would appear to have dropped to, if not below, the level prevailing in midwestern agriculture. In 1858, Chicago newspapers carried reports that farmers were offering eight-month contracts to hired hands at wages ranging between $10 and $12 per month in addition to board.[96]

The foregoing survey has shown the competing demands for unskilled workers—in agriculture, construction, and lumbering—that impinged upon the Chicago labor market during periods of rapid regional development and industrial growth in the city. This review has, of necessity, been both cursory and incomplete as a description of the total scene on the unskilled labor market. Yet it is sufficient to call into question approaches to the labor market adjustment process that assume industrialization to take place in a setting of relative abundance of unskilled (vis-à-vis skilled) labor. Clearly, such a conception is considerably less apposite to the case at hand than it would be had development actually proceeded in a tidy sequential fashion, with industrial growth occurring in a long-established agricultural economy already provided with its basic transport facilities and urban infrastructure.[97]

IV. Skill Margins and Labor Market Dynamics

It is now useful to recapitulate quickly some of the more salient features of the industrial labor market adjustment process under conditions of rapid growth. The requirements for skilled workers in the expanding industrial sector had in some degree already been reduced

[95] See *Mears Collection,* Correspondence Folders/II (1858–1859) and III (1861–1866).

[96] See *Chicago Weekly Democrat,* May 1, 1858, cited in Pierce, *History of Chicago,* vol. 2, p. 156, n. 31.

[97] The work of W. W. Rostow, specifically his application to U.S. history of the concept of a "preconditioning period" for industrial "take-off," would appear to have fostered a misleading sequential representation of the typical American developmental experience. Of the U.S. and other regions of settlement in the nineteenth century, Rostow writes, in *The Stages of Economic Growth* (Cambridge, England: Cambridge University Press, 1950): ". . . the process of their transition to modern growth was largely a matter of building social overhead capital . . . and of finding an economic setting in which a shift from agriculture and trade to manufacturing was profitable" (pp. 17–18); ". . . the preparation of a viable base for modern industrial structure requires that quite revolutionary changes be brought in two nonindustrial sectors: agriculture and social overhead capital, most notably in transport" (pp. 25–26). The issue is not whether, but when such changes are brought about.

by longer-run adaptations in the structure of production and in the mix of products with respect to quality; these requirements could also be partially satisfied through limited direct importation of key workers, and, most immediately, through the downward revision of hiring standards. The latter process entailed a search through successively lower skill layers of the existing labor force. It thereby augmented the pressure upon the supply of workers available for less skill-demanding tasks—pressure that resulted from strong concurrent labor requirements in regional and local nonmanufacturing activities, especially those of a highly market-oriented character, like construction. The two sources of pressure tended to quickly drain the initially shallow regional reserve labor pool, creating a situation where the wages of workers in less skilled occupations would be pushed upward more rapidly than those of workers filling positions farther up the skill ladder.[98] In addition, the external factors shaping the composition of emigration to the United States, and to the Great Lakes region in particular, had the effect of maintaining a high long-run elasticity in the supply of skilled workers relative to that in the supply of unskilled labor; therefore, the equilibrating rise in the relative wages of the unskilled would have proceeded further than would have been required by the existence of the opposite conditions on the supply side.

When thus arrayed, the features of the labor market indicated as pertinent to Chicago in the pre-Fire era suggest that early surges of industrial growth would create strong forces compressing skill margins, and containing them within a narrow range. No comparable forces were operating to generate the high premiums for skilled labor one has come to anticipate finding during early industrialization, on the basis of the experience of developing countries endowed with large reserves of unskilled workers.[99] Understandably enough, the absence of high premiums for skill did not strike contemporary observers of Chicago's "boom" as extraordinary. At the beginning of 1854, the *Daily Democratic Press* rather casually noted that the city's rapid economic development had effected an alteration of the position of the ordinary mechanic within the labor force hierarchy:

[98] Under the specified conditions it cannot be *deduced* that the relative wages paid for common, unskilled labor will be raised (cf. Reder, "Wage Differentials," p. 838). But if, as will be seen, this did in fact happen, then the relatively greater drain on the pool of workers available for those tasks would constitute a major explanatory factor. The use of "would be" in the sentence of text is therefore a prime illustration of the subtleties of rationalized narrative—the economic historian's forte—rather than a strict theoretical prediction.

[99] See ibid., esp. pp. 845–884, for discussion of other factors—particularly those bearing on the ease of substitution of unskilled for skilled labor—that affect interregional differences in skill margins in the long run.

In dull times, and in cities which have passed the culminating point of their prosperity, master mechanics can select their journeymen, and do somewhat as they wish. For the last year or two, those that worked by the day or week were the real masters, for good mechanics could command almost any price they chose to ask.[100]

On the other side of the ledger, to carry the argument through more explicitly than the *Democratic Press* ventured, in times of dull trade, and especially when regional and local construction activities were brought to a standstill, the less skilled grades of labor, sucked into the region by ample employment opportunities and relatively high wages during the preceding boom, would be transformed into a temporary reserve army of job-seekers in the larger towns and cities. Urban employers could take advantage of the altered situation to revise hiring standards upward, thereby thrusting workers who had formerly held higher-skill positions into immediate competition with those further down the skill echelon. And there was an added incentive for them to do so: the less broadly trained worker could be employed quite productively in a "skilled" job so long as a strong market permitted production on a larger scale, with finer subdivision and standardization of tasks; however, that worker would prove rather less satisfactory when production runs were shortened and firms had to move workers from task to task to keep them busy.[101]

Under such conditions, not only would money wages generally decline, but skill margins would tend to widen and quickly approach dimensions approximating those traditionally associated with the early phases of industrialization in the less developed economies. Over time, therefore, sharp changes in the pace of economic growth in the city and region would be likely to produce marked instability in skill-associated wage differentials, without any pronounced secular movement, so long as the fundamental supply conditions and demand characteristics described remained unaltered.

The available data on skill margins in Chicago during the years prior to the Fire, although they consist of only intermittent rather than continuous observations, are adequate enough to corroborate the existence of precisely the pattern of changes in skill margins that the foregoing considerations would lead us to expect. Table 3.1 presents the daily money wages of four types of skilled workers—blacksmiths and other ironworkers, machinists, carpenters, and house painters—each as per-

[100] *Daily Democratic Press*, "Annual Review for 1853," p. 45.
[101] It may be similarly noted that product-quality considerations favoring the raising of job requirements would tend to assume greater importance in a buyers' market than they had in a sellers' market.

TABLE 3.1
Percentage Average Skill Margins in Chicago, 1846–47 to 1898

Year	Average daily wage rate for common labor (dollars) (1)	Blacksmiths & ironworkers (2)	Machinists (3)	Carpenters (4)	House painters (5)
1846–47	1.00	—	—	125	—
1852	0.87	143	171	—	171
1853–54	1.25	130	130	140	120
1856	1.37	—	—	127	—
1860	0.87	171	271	186	172
1863	1.87	120	125	113	107
1865–66	1.60	180	203	218	167
1870	1.90	165	172	135	106
1871	1.90	171	170	152	108
1872	1.90	172	168	148	138
1873	1.90	173	169	135	105
1874	1.75	179	172	124	105
1875	1.74	170	170	127	101
1876	1.60	176	177	135	107
1877	1.51	183	183	149	118
1878	1.50	180	181	143	128
1879	1.49	180	180	150	130
1880	1.58	171	173	139	132
1881	1.59	179	172	149	143
1882	1.60	177	174	145	158
1883	1.50	187	184	155	173
1884	1.50	185	183	160	172
1885	1.50	187	184	157	178
1886	1.50	187	183	163	161
1887	1.50	188	191	165	160
1888	1.50	187	190	164	168
1889	1.50	186	190	154	162
1890	1.51	188	189	152	153
1891	1.50	189	190	172	161
1892	1.51	187	188	172	172
1893	1.51	192	193	196	186
1894	1.50	184	183	197	175
1895	1.50	186	186	180	176
1896	1.50	188	186	169	174
1897	1.50	188	187	178	187
1898	1.50	187	187	162	187

SOURCES:
1846–47: Pierce, *History of Chicago*, vol. 1, p. 195.
1852: *Chicago Daily Democrat*, July 10, 1852.
1853–54: *Daily Democratic Press*, "Annual Review of 1853," p. 45; *Daily Democratic Press*, March 16, 1854.
1856: Pierce, *History of Chicago*, vol. 2, p. 155.
1860: *Chicago Tribune*, "Annual Review of 1865," pp. 10–11.
1863: *Chicago Tribune*, Oct. 18, 1863, Jan. 14, 1864.
1865–66: *Chicago Tribune*, "Annual Review of 1865," pp. 10–11; *Chicago Tribune*, Jan. 1, 1867.
1870–1898: U.S. Dept. of Labor, Bull. No. 18, *Wages in the United States and Europe, 1870 to 1898* (Washington, 1898).

NOTE: For "Blacksmiths & Ironworkers," from 1870 to 1898, separate average quotations for blacksmiths, boilermakers, and molders were arithmetically averaged.

centages of the going wages paid unskilled day laborers by manufacturing and other establishments in the city. In offering the statistics for the dates prior to 1870, one can do no less for caution's sake than repeat the warning in one of the contemporary sources from which they have been drawn:

> It must be borne in mind . . . that in all branches the rates paid vary greatly in proportion to the skill of employees, and often during the year have fluctuated in consequence of a pressure upon employers. These figures must then be considered as a close average.[102]

Perhaps the initially most striking aspect of these series is the wide range of variation in the observed margins between the wage rate in a given skilled pursuit and the common-labor wage rate over the period from 1846–47 to 1870. For machinists, the range extended from a narrow differential of 20 percent above the common-labor rate to one of 171 percent; for blacksmiths and other ironworkers, from 20 to 80 percent; for house painters, from 7 to 72 percent; for carpenters, from 25 to 118 percent. A second immediately notable feature is the high degree of correspondence between series in the direction of the margin changes from one date to the next. Because the rapidly shifting relative demands for various types of labor militated against the formation of a highly rigid structure of interindustry (and interskill) differentials, the amplitudes of those changes were hardly so similar.

A more or less persistent wage pattern prevailed in the Chicago market during this period, despite the absence of rigidity in the structure of occupational differentials. This may be seen in greater detail from the sample of money wage rates in fourteen occupations assembled in Table 3.2 for three selected dates in the 1850s and 1860s. Ship caulkers and ship carpenters and joiners, who must in the main have been needed for seasonal repair work rather than ship building, remained among the most highly paid members of the industrial work force. Butchers, too, occupied a comparatively secure position among the labor elite, as distinct from the choppers and packers, and still less skilled workers, engaged in the meat-packing industry at substantially lower wages.[103] At the lower end of the wage scale, as a rule, one could

---

[102] *Chicago Tribune,* "Annual Review for 1865," p. 10. The data for the years 1870–1898 warrant similar caution but have the virtue of being derived from a single source. For each occupation, the underlying figures represent average quotations prepared by the Department of Labor from the payroll records of at least two establishments in Chicago. See Notes and Sources of Table 3.

[103] In Jan. 1853, while packing was in progress, packers' and choppers' wages reportedly averaged $1.62 per day, compared with butchers' wages of $2.00 and the average daily wage estimated for the total meat-packing labor force of $1.50. See *Daily Democratic Press,* "Annual Review for 1852," pp. 8, 45.

TABLE 3.2
Average Daily Money Wage Rates in Chicago for Selected Years
(In dollars)

Occupation	1853[a]	1860	1865
Ship caulkers	2.37	2.25	4.00
Butchers	2.00	3.00	4.50
Ship carpenters and joiners	1.87	2.50	4.00
Stone cutters	1.87	1.50	4.00
Machinists	1.75	2.37	3.25
(Masons and) plasterers	1.75	1.50	3.25
Blacksmiths and ironworkers	1.62	2.00	4.00
Wagon and carriage makers	1.62	2.00	3.50
Carpenters	1.50	1.87	3.50
Cabinetmakers	1.50	1.50	3.00
House painters	1.50	1.50	2.50
Day laborers	1.25	.87	1.60
Tanners	1.12	1.00	2.25
(Blacksmiths') blowers and strikers	0.94	1.12	2.25

SOURCES: 1853—*Daily Democratic Press*, "Annual Review of 1853," p. 45; 1860 and 1865—*Chicago Tribune*, "Annual Review of 1865," pp. 10–11.

[a] Average wage rates during January 1853. The other data are given in the sources as averages for the years cited.

expect to find house painters, blacksmiths' blowers and strikers, and day laborers.

These geneal impressions of ordinal stability in the wage structure are confirmed by closer examination of the data in Table 3.2. Considerable alterations in the relative standings of some occupations notwithstanding, rank-correlation analysis reveals the existence of significant correlations between the overall wage patterns for 1853 (January), 1860, and 1865–66.[104] In light of earlier remarks on the volatility of the demand for workers in construction-associated activities, it is interesting to note that the relative movements of the wages of stonecutters, masons, plasterers were particularly important contributors to the element of disarray among the successive rankings.

Although the series presented in Table 3.1 are broken by many gaps before 1870, it is with the timing of the skill margins' fluctuations during that era that particular interest resides. The margins data exhibit the expected tendency toward low levels during periods of rapid growth

[104] Due to the presence of numerous identical wage quotations for different occupations in the same year, the analysis was performed by employing Kendall's method of treating "tied ranks." See M. G. Kendall, *Rank Correlation Methods* (London, 1955), Chs. 3, 4. For the 1853 and 1860 rankings the analysis yielded (rank correlation coefficient) $t = .639$, and a highly significant score, $S = 2.95\,\sigma$. For the 1860 and 1865–66 rankings, $t = .695$, $S = 3.22\sigma$; for the 1853 and 1865 rankings, $t = .702$, $S = 3.26\sigma$.

or marked labor scarcity, on the one hand, while on the other, they rapidly ascend to high levels during years when pressure in the Chicago labor market had eased. The quite narrow 1846–47 margin shown between carpenters' and common laborers' wages also appears to indicate the relationship, during the expansion of local activity that accompanied the final stages of work on the Illinois and Michigan Canal, between the wages of most artisans and mechanics engaged in manufacturing in Chicago, and the prevailing rates for unskilled general labor.[105] Almost equally low margins—if not actually lower, as in the case of house painters—seem to have prevailed in the expansion of the mid-1850s, as far as can be discerned from the observations for 1853–54 and the 1856 carpenters' differential.

Following the onset of the Civil War, skill margins were once again driven downward in Chicago. In this case the shift in the supply curve was such that even a comparatively weak expansion would encounter a labor supply constraint. But a rebound in local activity—in the recovery from the after-effects of the banking crisis, and the derangement of trade brought on by the collapse in the price of southern state bonds in 1860–61—did play some part, along with the withdrawal of skilled and unskilled workers for military service, in producing the recorded decline in skill margins between 1860 and 1863.[106] In some trades, appreciable inroads were made upon the body of skilled workers by the Union's need of military manpower. As one example, roughly 20 percent of the members of the Chicago branch of the International Typographical Union enlisted in the armed forces, contributing to the situation in which, as has been seen, employers made further headway in introducing female compositors into the printing industry.[107] But on the whole, the wartime drain on the civilian labor force would seem to have been focused especially strongly on the pool of unskilled workers. The latter, were they not carried off by the sudden lure of Canada when a draft seemed impending, were drawn into the forces by the appeals to patriotism and, failing that, by the free whiskey dispensed at recruiting offices in the city.[108] Both inducements to enlist were reinforced by the bounty money offered recruits at various times by the Cook County Board of Supervisors and the Chicago Common Council. Payments were also made by private individuals and groups, like the Chicago Board of Trade and the Mercantile Association, to men

---

[105] See Pierce, *History of Chicago,* vol. 1, p. 195.
[106] See F. Cyril James, *The Growth of Chicago Banks,* 2 vols. (New York: Harper and Bros., 1938), vol. 1, pp. 276–285.
[107] See Pierce, *History of Chicago,* vol. 2, p. 161.
[108] See Hutchinson, *McCormick,* vol. 2, pp. 88–89.

who would stand as substitutes for names entered on the muster rolls or actually drawn in a draft.[109]

The notion of labor shortages literally caused by the drafting of workers patently is inapplicable to the experience of Chicago during the Civil War. No draft was imposed in the city in either 1862 or 1863, and substitute recruits were procured by the draft committees before the time of departure for all save the fifty men actually sent as draftees from Cook County during the entire course of the war; seven among those fifty, as it turned out, were relieved by substitutes procured at Springfield.[110]

Substituting for draftees, like the method of subsidizing volunteers to avoid imposing a draft prior to 1864, had the effect of transfering the bearing of arms to the shoulders of those who were economically less well-off, which by and large meant the less skilled.[111] More generally, the burden of military service fell more heavily upon the less skilled workers in the agricultural sector than upon the urban-industrial labor force. This contention has been advanced with regard to the overall situation in the North; it appears equally justifiable as an assertion about conditions within a predominantly agricultural area such as the state of Illinois.[112]

Turning now to the other side of the pattern of changes in skill margins, Table 3.1 discloses a slight expansion of margins between 1846–47 and 1852, a sharper rise from the low level of margins in 1853–54 and 1856 to those recorded for 1860, and a still more dramatic increase between 1863 and 1865–66. These movements are quite consistent with our hypothesis that skilled labor commanded greater premiums when the growth of aggregate demand faltered and alleviated pressure upon the region's labor reserve.

Although it is rather less firmly established than the other move-

[109] In 1862 the Board of Supervisors voted to levy a tax of $200,000, to be dispersed in bounties of $60 each to volunteers. Prices and fiscal measures changed with the course of the war, even at this level. On September 5, 1864, the Supervisors passed an ordinance authorizing the issue of county script, to the amount of $300,000, and provided that each recruit thereafter sworn in and credited to Cook Co. should receive a bounty of $300. The bounty was later raised to $400 and, by the summer of 1864, three-year volunteers were asking $550–$650. See Andreas, *History of Chicago*, vol. 2, pp. 167–168; Pierce, *History of Chicago*, vol. 2, pp. 272–273.

[110] See Andreas, ibid., p. 168. However, Pierce, ibid., p. 274, n. 92, cities 59 as the number of drafted Chicagoans held to service.

[111] See Pierce, ibid., p. 274, on the makeup of a sample of men replaced by substitutes in Cook Co.

[112] See Emerson D. Fite, *Social and Industrial Conditions in the North During the Civil War* (New York: University of Wisconsin Press, 1910), pp. 197–199, on the North, generally. For the sources of recruits in Illinois, see the statistics compiled in *Illinois Adjutant General's Report: September 1, 1864* (Springfield, Illinois, 1864).

ments in skill margins, the rise from 1846–47 to 1852 was presumably connected with the comparative lull in Chicago's commercial and industrial expansion during 1851 and 1852. Indeed, from one point of view, the widening of skill margins in this interval might be regarded as a fragment to be added to the other bits of evidence indicating that a brief period of retardation intervened between the "canal-completion boom" and the establishment of direct rail contact with the East during the spring and summer of 1852.

As the contrast between the state of the Chicago labor market during the surge of the mid-1850s and conditions in the 1858–1860 slack period has already been sketched, the accompanying relative decline in the wages of common laborers requires no further comment here. It may, however, be noted that the upward movement from 1853–54 (and 1856) to 1860, common to all the skill-margin series presented in Table 3.1, hardly accords with the assertion by Robert Ozanne that skill margins in Chicago were compressed rather than widened by the depression of the late 1850s.[113] Ozanne's 100-year series, based solely on skill margin (expressed as a percentage of the common wage rate) of 200 in that year, his data show a slight decline between 1858 and 1859, followed by a more precipitous drop the next year. This was due to the more rapid recovery of money wages for common labor at McCormick's, as Ozanne himself noted. Obviously, these data cannot show the contrast between margins in the 1857–58 depression at McCormick's and those during the preceding boom. Nor does the experience of a single establishment provide a very firm basis for generalizations about the relationship between skill margins and the overall level of activity.[114]

The closing months of 1865 saw arrivals of demobilized troops in Chicago. Although the Board of Trade expressed its pleasure in welcoming home the soldiers it had outfitted and sent off to defend the

[113] See Robert Ozanne, "A Century of Occupational Differentials in Manufacturing." *Review of Economics and Statistics*, 44 (August, 1962): pp. 292–299; reprinted as Appendix D, in Robert Ozanne, *Wages in Practice and Theory: McCormick and International Harvester, 1860–1960*, Madison, Wis.: University of Wisconsin Press, 1968.

[114] Further, as far as the actual behavior of Ozanne's series following 1858 is concerned, it should be recognized that for the period 1858–1897, his skill margins were computed by using the average of the upper sextile of the production-worker wage-rate distribution at McCormick's to represent the wage of skilled workers. It is this average, rather than wage rates paid workers filling various specific jobs, that is compared with common laborers' wages in computing the skill-differential series. Now, unless the rank ordering of specific occupations by wage level was *perfectly* invariant through time—and on this score of evidence in Table 3.2, above, relating to the wage structure in the pre-Fire era is hardly reassuring—the usefulness of Ozanne's series as an indicator of the behavior of skill margins is seriously jeopardized.

94   Paul A. David

Union, the dislocations and readjustments that accompanied demobilization could not have been pleasant, either for the general business community or for those seeking work in a crowded labor market during the hard winter of 1865–66. The difficult employment situation faced by the returning troops was especially aggravated in Chicago by the fact that meat-packing numbered among those local industries experiencing the national postwar recession. The packers, who normally provided jobs in the autumn and winter months, severely curtailed operations in beef and pork when provision markets became irregular and weakened with the continuing decline in the gold premium. Further, a short hog crop kept the price of raw materials from declining in the industry's dominant branch.[115]

Newspaper comments of the preceding years on the scarcity of labor were now supplanted by warnings, addressed to country boys in search of work, to avoid the city.[116] December of 1865 in Chicago found Charles Mears writing with imperfectly contained glee to the manager of his mill at Pentwater, Michigan,

> Men are very plenty here, and I could send you any number at $20.00 per month, and I think I could do so now by land [navigation on the Lake being closed] if you advise me in time. You should contract with all your men for six months or more to the best advantage, they take any advantage they can of us, and we should protect ourselves when we can.[117]

In this setting, skill margins bounded upward from the depths plumbed during the war, surpassing in many instances the heights reached during 1860.

As Table 3.1 shows, wages for carpenters, and for blacksmiths and other ironworkers were higher relative to the common labor rate in 1865–66 than in 1860. House painters' and common laborers' wages stood in roughly the same relation at the two dates, while the differential for machinists had declined. Supplementary wage quotations available for this pair of dates, reveal that the differential over common laborers' wages had been significantly narrowed, not only for machinists but also for master millers, master brewers, flint-glass blowers, ship carpenters, and other rather less highly skilled occupations in a

---

[115] See *Chicago Board of Trade Report for 1865–66* (1866), p. 9; Pierce, *History of Chicago*, vol. 2, p. 158, on demobilization in Chicago; *CBT Report for 1865–66*, pp. 46, 53, *Chicago Tribune*, "Annual Review of 1865," pp. 30–31, on the recession in the packing industry; Rendings Fels, *American Business Cycles, 1865–97* (Chapel Hill: University of North Carolina Press, 1959), pp. 92–96, on the postwar recession.

[116] Cf. Pierce, *History of Chicago*, vol. 2, p. 158.

[117] C. Mears to H. C. Magg, Dec. 11, 1865, *Mears Collection*, Correspondence Folder/III (1861–1866).

number of trades, like brass working and bookmaking. In contrast, money wage rates for sawyers, cabinetmakers, varnishers, upholsterers, and other wood and furniture workers, for leather tanners, and for candy makers joined those for carpenters and ironworkers in rising at a rate exceeding that of common laborers' wages.[118]

In a number of respects this pattern in skill-margin changes would appear to reflect the short-run impact of alterations taking place within Chicago's industrial structure during the decade of the 1860s. In industrial lines that between 1859–60 and 1869–70 increased their net output more rapidly than the rate at which aggregate manufacturing net output grew—as did leather tanning, confectionery, and wood products, especially furniture—the skill differentials in 1865–66 were higher than those in 1860. Observed declines in skill differentials occurred in flour milling, brewing, and printing and publishing—industries that were growing relatively slowly during the decade.[119] A crude indicator of comparative rates of sectoral growth in the 1860–1865 interval, namely the relative frequency of *successful* firm formations in different major sectors, provides some further information which, if fragmentary, is at least consistent with the pattern already described.[120] Comparing 1860–61 with 1862–1865, the relative frequency of successful firm formations in the nonferrous metal industries dropped sharply, while the reverse was the case in the leather industry and in the ironworking industries. As noted above, skill differentials for leather tranners and for blacksmiths and other ironworkers (except machinists) rose, while the differential for brass molders and finishers declined.[121] Although this does not account for the entire pattern of alterations in skill margins, it seems reasonable to suppose that a single rapidly expanding industry would have been more likely to encounter a relative scarcity of skilled workers than a shortage of unspecialized, unskilled workers; therefore, it ought not be surprising to observe that short-run changes in skill margins tend to be *positively* associated with interindustry differences in growth rates.[122]

---

[118] Based upon wage statistics published in the *Chicago Tribune*, "Annual Review of 1865," pp. 10–11.

[119] See David, *FPE*, Appendix A-II, Tables A-II:12 and A-II:16 for underlying real-output estimates.

[120] See *FPE*, Ch. 2, Table 2, for successful firm-formation rates.

[121] Over the decade of the 1860s, however, net output of foundries and machine shops grew at 10.3 percent per annum, while that of other ironworking establishments, excluding rolling mills and furnaces, grew at 13.0 percent. The growth rate of all manufacturing net output was 12.1 percent. See David, "Marshallian Dynamics," Table 1.

[122] See Melvin W. Reder, "Wage Differentials: Theory and Measurement," in *Aspects of Labor Economics* (Universities-National Bureau Committee for Economic Research, Conference, Princeton, 1962), pp. 257–317, esp. pp. 276–277, on the

Skill margins in Chicago probably remained at high levels during 1866, and may have climbed still further during 1867 and 1868, for employment conditions did not improve dramatically and may have worsened. The volume of construction activity, having rocketed upward between 1862 and 1865, moved up and down erratically in the years immediatley following, while new construction per capita failed to rise at all prior to 1868–69. Although flour-milling output and production of agricultural equipment were on the increase, elsewhere in the industrial sector—in meat packing, distilling, and leather tanning, for example—retrenchments and cuts in production became the order of the day.[123] Continued immigration at the high postwar level brought a flood of newly arrived Scandinavians into the city and combined with the unsettled state of trade and industry to create a serious unemployment problem for both skilled and unskilled members of the labor force, especially during the winter months of 1866–1868.[124]

Yet by 1870, under the pressure of both the rising demand for con-

---

distinction between short-run and long-run theory relating to interindustry wage differentials and skill margins. A fallacy of composition is involved in trying to extend the foregoing (short-run) theoretical argument relating skill differentials and the rate of growth of employment or output in a single industry (whose demand for labor does not bulk large in the aggregate demand for labor) to the general behavior of skill differentials under conditions of rapid (slow) aggregate growth and consequent excess demand for (supply of) labor. There is thus no logical conflict between the argument advanced earlier in the text regarding the effects of aggregate demand conditions on skill differentials, and the short-run *industry* analysis advanced here. The crucial element in the aggregate analysis is that the rate of growth is sufficiently rapid—given supply conditions—to exhaust the available reserve of unskilled. If that does not happen, skill margins need not be compressed but may actually widen during a period of accelerated growth, particularly in the sectors that are leading the expansion. Moreover, the comparison between 1860 and 1865 involves two dates at which there was considerable slack in the Chicago labor market. But, in those circumstances, it is likely that when the rate of growth of those sectors slows down relative to other sectors, skill margins therein will no longer continue rising and may, instead, decrease.

Unfortunately, arguments appropriate to one level of analysis (say, the single industry) are sometimes mixed with those appropriate to the other (say, aggregate demands for labor), and assumptions as to the exhaustion, or nonexhaustion of the labor reserve are left unspecified, creating, at best, considerable confusion in the literature on skill margins. Cf., e.g., David E. Novack and Richard Perlman, "The Structure of Wages in the American Iron and Steel Industry, 1860–1890," *Journal of Economic History*, 17 (Sept. 1962), pp. 334–347, esp. pp. 336–339; Ozanne, "Occupational Differentials," pp. 296–297.

[123] David *FPE*, Appendix C, Table C-3 contains the underlying data. The rise in agricultural machinery production during this period of slack in the Chicago labor market was accompanied by the widening of skill margins at the McCormick Reaper Works between 1866 and 1868. Cf. Ozanne, Chart I, p. 293.

[124] See Pierce, *History of Chicago*, vol. 2, pp. 158–159, for newspaper references to the distress in the latter half of the 1860s, all of which appeared during December and January of those years.

struction labor on local building projects and midwestern railroads, and the marked acceleration of the growth rate of aggregate industrial production in the environs of the city during the great spurt of 1869-1873, skill margins for the occupations shown in Table 3.1 had been forced substantially below their 1865–66 levels.[125] On this occasion, however, with the exception of the differential between house painters' and common laborers' wages, skill margins did not descend to the low levels recorded previously, in 1846–47, in the expansion of the mid-1850s, or in the course of the Civil War. The break with the past in this regard was especially notable in the metalworking trades, as reflected by the margins for machinists, blacksmiths, and other ironworkers. It signified the dawning of a new era in the regions industrial history, characterized by a transformation of its labor market dynamics.

V. Epilogue

The failure of skill margins to revert to their former narrow dimension during the major industrial surge of the late 1860s and early 1870s constituted the first phase of a significant set of developments that differentiated the experience of the Chicago labor market in the next thirty years from that during the preceding two decades. For, between 1870 and 1879, as Table 3.1 reveals, the amplitude of short-period movements in skill margins was severely damped out and, still more notably, the margins displayed a significant tendency to drift upward secularly.

Once the conditions highlighted in the foregoing account had been transformed by the passing of the infrastructure-development boom in the city's hinterland, and by Chicago's growing dominance in the midwestern regional labor market, episodes of rapid expansion of manufacturing production were less and less likely to encounter short-run unskilled labor constraints. The stage was thus set for the emergence of a new pattern of cyclical response in the premiums paid to occupants of "skilled" positions, as well as for the sustained secular widening of those differentials which is evident from Table 3.1. But this second part of the story can be left for telling on another occasion—perhaps at the next celebration of Stanley Lebergott's continuing contributions to the writing of American economic history.

---

[125] Ozanne's data for the McCormick works ("Occupational Differentials", Chart I, p. 293) show margins ceasing to rise in 1869 and narrowing in 1870, but not contracting below their 1865–1866 width until 1871.

4
Postbellum Southern Labor Markets

GAVIN WRIGHT

From an examination of farm-labor wage rates going as far back as 1818, Stanley Lebergott noted "the rapidity with which an integrated national labor market tended to develop," concluding that labor as well as capital was "remarkably mobile . . . sufficiently mobile to bring [an] extensive reduction of differences in rates among states."[1] This is a picture of the American experience consistent with many other interpretations and empirical studies.

But the postbellum South is an anomaly. Southern farm wages after the Civil War were barely half of northern levels, and there seems to have been no tendency toward convergence: indeed, if we compute the real absolute differences, the "cost of staying in the South" was higher after 1900 than at any time in the nineteenth century. The dominant tradition among historians of the South holds that this anomaly merely confirms that southern labor institutions were utterly different from the free-labor markets of the north. Lien laws, debt peonage, vagrancy prosecutions, convict lease—all of them undeniably real—make this interpretation plausible. Jonathan Wiener writes,

> The coercive mode of labor control gave southern agriculture its distinctive character. Sharecropping was a form of "bound" labor, with restrictions on the free market in labor that did not prevail in fully developed capitalist societies such as that of the North.[2]

In a similar spirit, Jay Mandle analyzes the postbellum South as a region dominated by "the plantation mode of production," the salient features of which were "repression and enforced immobility."[3] A study

---

For comments on the first draft of this essay, the author extends thanks to Brian Arthur, Paul David, Stephen DeCanio, Bill Parker, Wendy Rayack, Gary Saxonhouse, David Weiman, Tom Weiss, Warren Whatley, and the participants in the Wesleyan Symposium in honor of Stanley Lebergott. I am grateful to Basic Books for permission to draw on material in *Old South, New South* (New York, 1986).

[1] Stanley Lebergott, *Manpower in Economic Growth* (New York: McGraw-Hill, 1964), p. 134.

[2] Jonathan Wiener, *Social Origins of the New South* (Baton Rouge: LSU Press, 1978), p. 70.

[3] Jay Mandle, *The Roots of Black Poverty* (Durham: Duke University Press, 1978), p. vi.

of an impoverished tobacco county in Virginia also reports that "no free market in labor actually functioned."[4] These descriptions are not confined to the nineteenth century but persist well into the twentieth, as in this 1930 statement by Louise Venable Kennedy:

> Many of them [blacks], too, have been subjected to practical peonage, since in some of the Southern states a tenant farmer is required to stay with his land owner as long as he is in debt to him. By seeing to it that the colored tenant has not been able to secure enough profit from his year's work to pay off his obligations, the owners have frequently succeeded in keeping Negroes on the land even against their will.[5]

It is all clear and consistent: the North was capitalist and had mobile labor; the South was coercive and had bound labor.[6]

Yet readers of Stanley Lebergott's recent book *The Americans* know that this simple formulation encounters certain empirical difficulties. Lebergott documents the active recruitment of southeastern blacks by planters from Arkansas, Alabama, and Mississippi, concluding, "The planter drive for more labor in the new states . . . [kept] the income of sharecroppers in line with the major alternatives open to croppers in the market."[7] Most economists who have looked into the matter have also concluded that a market existed and did function. Stephen De Canio found that the typical sharecropper's earnings were not significantly below the value of his marginal product. Ralph Shlomowitz, in a thorough review of literary evidence for the period 1865–1880, reports that terms of employment did respond to market conditions and to the preferences of the freedmen, and that attempts at collusion by planters quickly failed. "Competitive market forces," he writes, "were ascendant." Historian Michael Wayne looked closely at the postwar pressures on planters of the Natchez District and similarly declared: "The road to the New South plantation ran through the marketplace."[8]

---

[4] Crandall A. Shifflett, *Patronage and Poverty in the Tobacco South: Lousa County, Virginia 1860–1900* (Knoxville: University of Tennessee Press, 1982), p. xii.

[5] Louise Venable Kennedy, *The Negro Peasant Turns Cityward* (New York: Columbia University Press, 1930), pp. 9–10.

[6] Many other examples can be cited. See particularly William Cohen, "Negro Involuntary Servitude in the South, 1865–1940," *Journal of Southern History* 42 (Feb. 1976) pp. 31–60; Daniel A. Novak, *The Wheel of Servitude* (Lexington: University Press of Kentucky, 1978); Pete Daniel, *Shadow of Slavery* (Urbana: University of Illinois Press, 1972).

[7] Stanley Lebergott, *The Americans* (New York: Norton, 1984), p. 259.

[8] Stephen DeCanio, *Agriculture in the Postbellum South* (Cambridge, Mass.: MIT Press, 1974); Ralph Shlomowitz, "'Bound' or 'Free'? Black Labor in Cotton and Sugar Cane Farming, 1865–1880," *Journal of Southern History* 50 (Nov. 1984), p. 572; Michael Wayne, *The Reshaping of Plantation Society*, (Baton Rouge: LSU Press, 1983), p. 149.

Now when two groups of scholars see things so differently, it is always possible that one is flatly mistaken while the other is clearly correct. More often, though, they are thinking about the matter differently, responding to different clues in the sources and to different presumptions about what is normal and what is surprising. The believers in "enforced immobility" have attempted to shore up their position by various arguments, none of which has been compelling. Jay Mandle, for example, after reviewing the evidence that sharecroppers had ample opportunities to move from one landlord to another, concluded that "job denial in the North" must have been the more important factor in confining blacks to the plantation sector; but this would seem to have nothing to do with the "plantation mode of production" nor, for that matter, with "enforced immobility."[9] Another writer under similar pressure suggests that it is not the *fact* of control that matters but the "attempt to maintain control," because planters "would not relax their vigilance over the racial order."[10] If unsuccessful *attempts* at control are considered just as repressive as actual control, then anything goes. These defenses are feeble, and yet the majority of southern historians have a gut feeling that the economists have not quite gotten the story right. Their problem, it seems to me, is that they have accepted the intellectual hegemony of supply and demand: mobility is good, immobility is bad, and institutions are repressive if they interfere with market forces. A well-functioning wage-labor market, economists contend, is an essential feature of economic progress under capitalism, for individuals and for economies. Having accepted these assumptions implicitly, the historical critics of southern society lack a set of economic concepts and vocabulary with which to articulate their indictment. Their problem is illustrated by Gilbert Fite's recent statement: "There was no *meaningful* labor mobility for the great majority of sharecroppers and tenants. . . . Sharecroppers and tenants frequently moved from farm to farm, but they found conditions *about the same everywhere.*"[11] But equivalence of conditions everywhere describes a labor market in equilibrium, exactly the evidence Lebergott viewed as a sign of "remarkable mobility."

A closely related topic deserves to be treated simultaneously, the effects of race and "tradition" on the operation of southern *industrial* labor markets. Writers from W. J. Cash to W. H. Nicholls to Dwight

---

[9] Mandle, *Roots,* p. 35.

[10] Stanley Greenberg, *Race and State in Capitalist Development* (New Haven: Yale University Press, 1980), pp. 108, 116.

[11] Gilbert Fite, *Cotton Fields No More* (Lexington: University Press of Kentucky, 1984), p. 28 (italics added).

Billings have characterized the mill village as a transfer of the plantation to the factory, for whites only. Nicholls portrayed the development problem of the South as a choice between "tradition" and "regional progress." Robert Higgs analyzes the treatment of blacks in the South as a struggle between "competition" and "coercion," the forces of the market and the forces of prejudice and repression. Jennifer Roback has recently argued that the Jim Crow laws passed between 1890 and 1910 served to reduce black mobility both within agriculture and from farms to cities, thereby lowering relative black wages.[12] Was labor mobile between agriculture and industry, and from place to place within industry? Did racial exclusions persist only because competitive market pressures were blocked or thwarted? Most of the issues in the plantation debate have parallels in the industrial labor context.

This paper reexamines the functioning of postbellum southern labor markets, beginning with the presumption that a labor market is a real institution to be examined and analyzed, and not just a name for abstract, invisible forces. Real labor markets did exist in the South, but their effects were not what economists usually take them to be. There was a well-functioning wage-labor market, but it was a regional market rather than a branch of the national labor market. Most southern workers, however, did not want to be in this market and did their best to escape from it. Many of the repressive institutions of the South did not block the operation of the labor market but made it work better. Furthermore, the pressures of the market did not usually undermine racial segregation and inequality but strengthened and reinforced preexisting patterns. If these propositions sound peculiar, it is testimony to the strength of certain of our intellectual organizing tools, which more often than not serve to blur our perceptions of both the thinking and the behavior of nineteenth century Americans.

I. There Was a Regional Labor Market

All accounts agree that in the early postwar period, the freedmen moved from place to place in large numbers, driving hard bargains in an atmosphere of labor scarcity. Their preferences clearly were a factor in the early rise of the decentralized sharecropping system on the plan-

---

[12] William H. Nicholls, *Southern Tradition and Regional Progress* (Chapel Hill: University of North Carolina Press, 1960); Robert Higgs, *Competition and Coercion* (Chicago: University of Chicago Press, 1980); Dwight Billings, *Planters and the Making of a "New South"* (Chapel Hill: University of North Carolina Press, 1979); Jennifer Roback, "Southern Labor Law in the Jim Crow Era: Exploitive or Competitive? *University of Chicago Law Review* 51 (Fall 1984), 1161–1192.

TABLE 4.1
Labor Turnover on Four Plantations in Adam County, Mississippi, 1871–1874

Year	Total tenants	Remaining from 1871 Number	Percent of tenants	Percent of original	Percentage new in each year
1871	42				
1872	38	23	61%	55%	39%
1873	64	20	31	48	56
1874	62	19	31	45	27

SOURCE: Wayne, *Plantation Society*, p. 208.

TABLE 4.2
Turnover on Leatherwood and Camp Branch Plantation, Henry County, Virginia, 1871–1877

Laborers	Years on plantation
21	1
7	2
4	3
7	4
3	5
3	6

SOURCE: Nannie May Tilley, *The Bright-Tobacco Industry, 1860–1929* (Chapel Hill, 1948), p. 99.
NOTE: Median, 1.1 years; mean, 2.4 years.

tations.[13] But did these rates of mobility continue as labor markets softened in the 1870s and as the new institutions settled into reasonably stable grooves? Michael Wayne examined four sets of records from plantations in Adams County, Mississippi, for 1871–1874, and found that less than half of the original group of tenants were still on hand two years later (Table 4.1). Between 27 percent and 56 percent of the total tenant force was new each year. The picture was similar on the Hairston Farm tobacco plantation in Virginia between 1871 and 1877. Of the 45 laborers who worked during this time, 21 stayed less than

[13] Wiener is quite clear on this: "By creating a 'shortage of labor,' the freedmen defeated the planters' effort to preserve the plantation as a single, large-scale unit worked by gangs." See "Class Structure and Economic Development in the American South, 1865–1955," *American Historical Review* 84 (Oct. 1979), 976. The view that the planters "literally were dragged kicking and screaming" into the sharecropping system is pressed by Ronald L. F. Davis, *Good and Faithful Labor* (Westport, CT: Greenwood Press, 1982), p. 190. A more balanced account is Roger L. Ransom and Richard Sutch, *One Kind of Freedom* (New York: Cambridge, 1977), Chs. 4–5.

TABLE 4.3
Turnover on Waverly Plantation, Worth County, Georgia, 1871–1876

Years after arrival	Percent of all tenants still working				
	1871	1872	1873	1874	1875
1 year	53%	58%	44%	38%	24%
2 years	42	37	50[a]	19	
3 years	16	42[a]	19		
4 years	32[a]	16			
5 years	11				

SOURCE: George W. Bryan Papers.
[a] Includes tenants returning as wage laborers.

one year (Table 4.2). On the Waverly plantation of Worth County, Georgia, only 11 of an original cohort of 19 tenants who started in 1871 were still there a year later (and two of these left during the course of the 1872 crop year).[14] By 1876 only two of the original group were left (Table 4.3). Reporting on his 1878 visit to the South, Sir George Campbell observed that "the negroes frequently change about from one estate to another," and many similar remarks may be cited.[15] It does not seem, therefore, that many freedmen were "bound" to their plantations in the 1870s.

The evidence does not look different for later periods. In 1910 the Bureau of the Census carried out a complete tabulation by race and tenure (but unfortunately not age) of responses to the question, "How long have you lived on this farm? ─── ───."[16] Summary figures for share tenants (not differentiated from "croppers" in 1910) are presented in Table 4.4. Although the turnover rate for whites appears to have been slightly higher than that for blacks, the most remarkable fact revealed by the table is that the median share tenant, white or black, had been on his present farm *one year or less*. More than 80 percent of black share tenants had been on their present farms less than five years. Share tenancy was a high-turnover system. If freedom means the right

---

[14] George W. Bryan Papers, Southern Historical Collection, University of North Carolina.

[15] Sir George Campbell, *White and Black: The Outcome of a Visit to the United States* (New York: 1879), p. 151. For further citations, see Shlomowitz, "'Bound' or 'Free'" p. 573.

[16] *U.S. Thirteenth Census*, 1910, "Stability of Farm Operators, or Term of Occupancy of Farms." This report, prepared under the supervision of John Lee Coulter, has remained little known because it was not published with the census and it never even received a bulletin number. After a search of several government documents libraries on the West Coast proved fruitless, a copy was obtained from the National Agricultural Library in [Beltsville,] Maryland. Paul Rhode was instrumental in locating this document.

TABLE 4.4
Term of Occupancy of Share Tenants, 1910

Years on farm	South Atlantic region		East South Central region	
	White	Black	White	Black
Less than 1	37.9%	33.9%	45.6%	39.9%
1 year	17.8	17.4	17.8	15.9
2–4 years	28.1	31.5	24.8	28.1
5–9 years	10.0	10.5	7.5	9.7
10 years and over	6.2	6.6	4.1	6.2

SOURCE: U.S. Census Bulletin, "Stability of Farm Operators" (1910). Figures are percentages of those answering the question. Response rates were South Atlantic white, 94.8%, black, 94.4%; east South Central white, 95.1%, black, 93.1%.

to quit, these farmers were free; if the employers were trying to stop them, they failed.

Perhaps a more economically appropriate way to determine the existence of a labor market is by looking not at quantities but at prices, that is, the wages of farm labor. If "vigilance over the racial order" were the device for restricting labor mobility, we ought to observe that the wages of blacks were below those of whites. Yet the testimony is that black and white farm laborers were paid the same. Planters told the Industrial Commission in 1900, "I think we give them about the same thing. . . . If there is any difference I don't know it."[17] If these statements are regarded as suspect, it should be recalled that these men were not necessarily proud of the equal-wage policy they reported. One of them said,

> We have white men working on the farms. We frequently have applications every day. But when the white men come and are willing to work we have to say: We cannot afford to pay you any more, because I can get a negro for 60¢ a day; if you are willing to work at that price the first vacancy we have you can have it. We occasionally put a white man in that way.[18]

It was a pretty good description of a market process at work. The best survey on this point is the First Annual Report of the North Carolina Bureau of Labor Statistics (1887), which posed the question of racial wage differences to landlords *and* to tenants and laborers in 95 counties. In 94 of the 95 counties, the landlords responded "no," and in 77

[17] *Report of the Industrial Commission on Agriculture and Agricultural Labor,* vol. 10 (1901), pp. 71, 471.
[18] Ibid., p. 446.

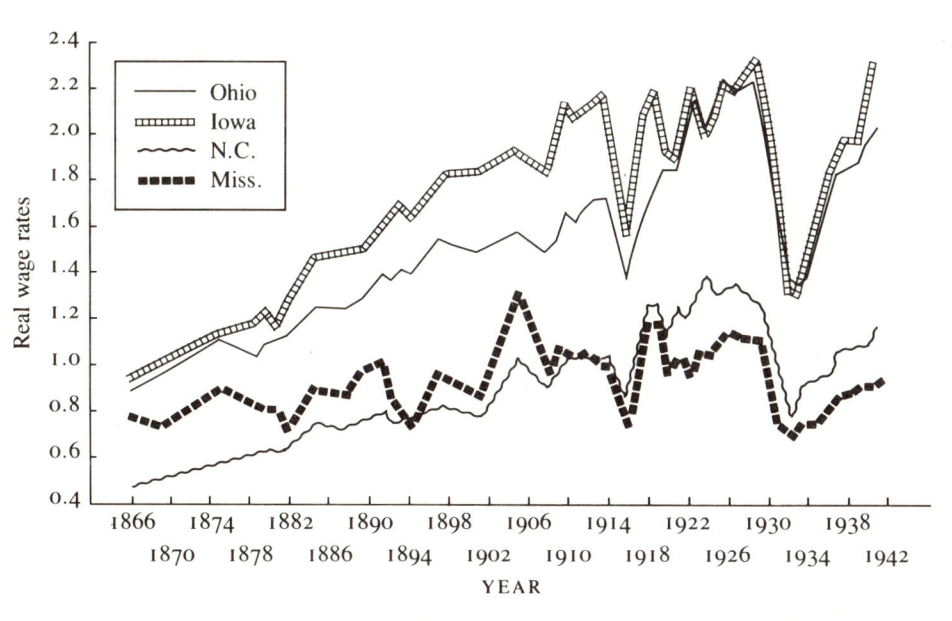

FIGURE 4.1 Farm-Labor Wage Rates Per Day, Without Board, Deflated by WPI, 1866–1942, Selected States

of the 95 counties, the tenants and laborers also indicated that no differences prevailed in wages paid to white and black.[19]

If labor market pressures were strong enough to equilibrate black and white wages, then we ought to be able to observe the effects of economic forces on the course of farm wages over time. The daily farm-wage rates graphed in Figure 4.1 for Mississippi and North Carolina do show some responsiveness to dips and advances in prosperity. Even more clearly, they show exactly the kind of east-west convergence that Lebergott found for the northern states. A similar pattern may be seen between the states of Ohio and Iowa between the 1880s and the 1920s. What is striking, however, is that wages converge within the North and within the South, but the two regions do not converge toward each other. Thus, there is evidence for a labor market in the South, but it is a southern regional market, not a branch of the national labor market.

This interpretation is consistent with the evidence on interstate mi-

[19] Robert Higgs, "Racial War Differentials in North Carolina: Evidence from North Carolina in 1887," *Agricultural History* 52 (April 1978), 308–311. An earlier study by Higgs is "Did Southern Farmers Discrminate?" *Agricultural History* 46 (April 1972) pp. 325–328. See also the exchange between Higgs and Roberts in *Agricultural History* 49 (April 1975), pp. 441–447.

gration. According to the census of 1900, blacks living outside of their state of birth were more than 2½ times as likely to have moved to another southern state than to a state in the North or West. (For whites that ratio was nearly double.) The predominant direction of movement was from the low-wage southeast to the high-wage southwest, but in fact there was substantial migration in both directions (as one would expect in a unified labor market area) within the South.[20] This is the important element of truth in Mandle's contention that the "denial of black jobs" was the key ingredient in keeping black wages low. But unskilled southern *white* wages were nearly as low, and it seems more accurate to say that both were part of an isolated regional labor market.

How can we explain this isolation? It is not a simple matter of cultural distance and transportation cost. Millions of unskilled and often non-English-speaking European immigrants were coming even longer distances to take northern industrial jobs. Clearly the origins of the separation are historical. Labor market information among poorly educated unskilled workers tends to follow informal channels, by word of mouth, passed on through ethnic and kinship connections. The risks and costs of long-distance moves are reduced by the presence of relatives and friends, who can provide fares, temporary housing, and access to employers. Statistical studies confirm that the existence of a first wave of migrants from a country is the single most important factor in generating a second wave.[21] But these are not "nonmarket" elements that "interfere" with market mechanisms. As Michael Piore argues in *Birds of Passage,* long-distance migrants often display a distinctively high responsiveness to pecuniary incentives, because many of them are trying to raise as much money as possible in a short time, and they do not have attachments to particular localities in the host country. Hence they are willing to work long hours, move from place to place, and accept demanding or even demeaning tasks that they would not consider doing at home.[22] Within the South, the same sort of behavior is reported among easterners migrating temporarily to the southwest. Virginians in Texas, the "sons of the best families," worked

[20] Joseph A. Hill, "Interstate Migration," in Bureau of the Census, *Special Report: Supplementary Analysis and Derivative Tables* (1906), p. 305. See also William O. Scroggs, "Interstate Migration of the Negro Population," *Journal of Political Economy* 25 (Dec. 1917), 1040. On the rates of gross migration, see Philip E. Graves, Robert L. Sexton, and Richard K. Vedder, "Slavery Amenities and Factor Price Equalization," *Explorations in Economic History* 20 (1983), pp. 156–162.

[21] James A. Dunleavy, "Regional Preferences and Migrant Settlement," *Research in Economic History* 8 (1983), p. 221.

[22] Michael Piore, *Birds of Passage* (New York: Cambridge University Press, 1979), Chs. 1–4.

at menial jobs that they said they would never accept at home, because (as one put it) it would be "too much of an affliction for my family, and I should lose caste with my lady friends."[23] Thus, historical beginnings and timing can define the scope of the market and its direction of expansion.

These initial conditions are reinforced by the incentives facing labor recruiters in high-wage areas. It was cheaper for individual northern employers to tap into the existing national and international market, or to extend it marginally, than to invest in the large fixed cost required to open a new flow from an isolated area. But southern recruiters found that they could not use this market, because the "footloose" European migrants usually stayed only a brief time before departing to the high-wage North. So southwestern planters had to buy in an area where wages were even lower than they were paying, one with cultural and racial linkages as well. The mechanisms of channeling are illustrated in these reflections by a black congressman from North Carolina:

> When the people have been left to themselves to emigrate, it has been largely to the North and East, and somewhat to the West; but where the agents with oily tongues come about and offer flattering inducements they have gone from one Southern state to another.[24]

Economists and philosophers may debate the precise meaning of being "left to themselves to emigrate," but it is clear from the description that it was much easier for a poor southern farm laborer to follow functioning market channels within the South than to set off alone to an unfamiliar region.

Although this paper concentrates on the pre-World War I period, it may help solidify the argument to contrast this era with what came after. Regressions relating the southern farm wage to the price of cotton and to the national wage for unskilled urban labor show a significant change after World War I.[25] Because of similar fluctuations in nominal levels, all coefficients are significant in both periods, (with the exception of Virginia) so the R^2's are not meaningful. (We do not have an annual series of regional price levels, and this problem would in no way be solved by dividing all the variables by the same deflator.) But we can examine the percentage of the explained variance attributable to the national wage in the earlier and in the later periods (Table 4.5).

---

[23] Virginius Dabney, *Virginia: The New Dominion* (New York, Doubleday, 1971), p. 419.
[24] *Report of Industrial Commission*, vol. 10, p. 428.
[25] Farm wages are taken from USDA, *Crops and Markets* 19, no. 5, and 18, no. 11, using the daily wage without board. Unskilled urban wage taken from Jeffrey Williamson and Peter Lindert, *American Inequality,* (New York: Academic Press) pp. 319–320.

TABLE 4.5
Percentage of Explained Variance in Southern Farm Wage Attributable to U.S. Industrial Wage, 1890–1915 and 1916–1930

Region and State	1890–1915	1916–1930
DEEP SOUTH		
Alabama	9.7%	28.0%
Arkansas	6.6	14.5
Georgia	13.4	18.0
Louisiana	6.4	23.6
Mississippi	5.2	30.1
South Carolina	9.7	28.0
BORDER		
North Carolina	27.8	24.7
Tennessee	17.9	14.2
Virginia	23.6	4.1

In every one of the six states of the deep South, this percentage increased, and in Alabama, Louisiana, Mississippi, and South Carolina, the jump was marked. The border states of North Carolina, Tennessee, and Virginia seem to have already had some labor market links with the North, and do not show such a change. This result is consistent with testimony that a small but persistent flow of migrant labor from these states did predate the war, particularly from Virginia to New Jersey, Pennsylvania, and New York.[26]

The point of this exercise is not to argue that the country had leapt into full labor market integration by the 1920s, but only to use the decisiveness of the change that did occur as a way of underscoring the nearly complete isolation of the deep South before 1916. This was less true for the border states, but they too were well off the beaten track of the rest of the country. The process of integration in the twentieth century could well do with further analysis. In what follows, however, I take regional labor market isolation for granted and concentrate on developments within the South in the period before World War I.

II. Tenancy and Wage Labor in the Plantation Belt

It is ironic that historians should have tried to argue the exploitive nature of southern farming institutions by denying the existence of labor

---

[26] *Report of the Industrial Commission,* vol. 10, p. xxii; Hill, "Interstate Migration," p. 305; La Wanda F. Cox, "The American Agricultural Wage Earner, 1865–1900," *Agricultural History* 22 (April 1948), p. 99.

mobility, because contemporary opinion in the nineteenth century held that mobile laborers were tramps and vagrants, people with no ambition or character, who would never get anywhere because they kept going everywhere. A correspondent to the Industrial Commission wrote,

> The tendency is for laborers to move about more and more. : . . This is very bad for laborers and employers. Our young negro men are becoming tramps, and moving about over the country in gangs to get the most remunerative work.[27]

The commission summary stated that "one of the most frequent complaints made against the negro laborers of the South is that they go about from place to place whenever they have an opportunity of bettering their condition."[28] Economists reading statements like this usually discount them as the self-interested complaints of employers who disliked competition. But even sympathetic observers saw the only hope of progress as coming from "those Negroes who have been able to *attach themselves to the land.*"[29] Speaking in 1897, the black educator W. H. Crogman deplored frequent migration as destructive to home, education, and church life:

> It strikes at the roots of those things by which only any people can hope to rise. . . . Over and over again have I known persons to leave their native state and after wanderings through several others to return finally to the very spot whence they had started, having in that time gained nothing, acquired nothing, except that which is a property common to all bodies once set in motion—a tendency to keep moving.[30]

These statements too may be discounted on one ground or another, but I think they ring true. It was a common saying in the postbellum South that "three moves equal a fire." Gilbert Fite feels that this is excessive, but having experienced both a move and a fire in the past five years, I feel it is a pretty good estimate.[31]

Staying in one place has real productivity effects in a number of ways. A farmer may develop a feel for the responsiveness of the soil and terrain over time. He may learn to recognize local weeds and the idiosyncrasies of the microclimate. But the important benefits are social in nature, as the observers quoted above obviously had in mind.

---

[27] *Report of Industrial Commission,* vol. 11, p. 95.
[28] Ibid., vol. 10, p. xix.
[29] T. J. Woofter, *Negro Migration: Change in Rural Organization and Population of the Cotton Belt* (New York, 1920), pp. 22–23, 89–90 (italics added).
[30] Quoted in Scroggs, "Migration of the Negro," pp. 1042–1043. The statement is identified as having been made "twenty years ago."
[31] Fite, *Cotton Fields No More,* p. 46.

By staying in one place you get to know the people, you find out who is likely to have good advice when problems arise, you develop a working relationship with teachers and ministers and lawyers and merchants, as well as with other farmers. What this sort of familiarity gave a farmer above all was *creditworthiness,* which was an intrinsically localized attribute in the nineteenth century. Southern farm workers aspired to careers not as farm laborers but as farm proprietors. They hoped to accumulate wealth over time so as to buy their own implements and mule, working their way up the ladder to landownership and independence. The way to do this, according to the critics of mobility, was to stay in an area and develop a local credit reputation. A reliable, well-known person could get better credit terms, not just for the annual subsistence needs, but for the purchase of longer-term items like implements, workstock, and land.

The local character of credit relationships was a pervasive fact of life in the nineteenth century. Robert Wiebe notes that farmers made a practice of returning again and again to the same storekeepers for the purpose of ensuring credit; the needs of credit exerted a "powerful bias toward continuing relationships."[32] The practice of branch banking, for example, was barely in its infancy in the early twentieth century, and almost no discussion of the concept had taken place before that time.[33] But even branch banking and other forms of geographic capital flow would not have changed the fact that local farmers had to establish creditworthiness with a local lender or at least a local representative. The high value, liquidity, and portability of slaves provided the major exception to this rule, but it was the exception that proves the rule, because emancipation was followed in short order by a localization of credit relationships in the South.[34] The worst fear of a lender was that the borrower would "pull up" and leave the area, and this was by no means rare. No wonder they gave preference to deferential "clients" with strong local connections and track records, preferences that modern observers interpret as "patronage capitalism" and an exercise in "political hegemony." Crandall Shifflett writes of Lousa County, Virginia, for example, that a tenant had to "deal with and gain the favor of several different patrons, . . . a matrix of creditors, suretors, trustees, and landlords," even for "the manifestly simple matter of buying a

---

[32] Robert H. Wiebe, *The Opening of American Society* (New York: Knopf, 1984), pp. 151, 299, 301, 324.

[33] Eugene Nelson White, *The Regulation and Reform of the American Banking System 1900–1929* (Princeton: Princeton University Press, 1983), pp. 14–15.

[34] The best account is Harold Woodman, *King Cotton and His Retainers* (Lexington: University Press of Kentucky, 1968), Chs. 22–23.

horse."[35] A telling example indeed. A horse was a $100 item, and 80 percent of Lousa County's farmers had less than $100 in personal property. Would *you* lend $100 to an impoverished but highly mobile wage laborer in a declining county? To buy a horse?

The importance of local reputation and its incompatibility with high geographic mobility is reflected in the process by which southern farm labor tended over time to sort itself into two broad categories—locally established and mobile or footloose—and Southern labor market institutions adapted to these categories and institutionalized them. This is the central organizing theme of the plantation model formalized by Warren Whatley for analysis of mechanization in the twentieth century.[36] Most plantations divided their acreage into plots assigned to tenants and sharecroppers and a portion retained by the owner to be cultivated with wage labor. The compelling incentive toward sharecropping, from the landlord's standpoint, was the seasonality of labor requirements and the uncertainty of obtaining and retaining wage labor for the "critical" periods, especially the harvest. It was, in P. K. Bardhan's expression, a "labor-tying" contract, a way of ensuring labor supply for the full year by creating both an obligation and a built-in incentive to stay with the crop to the end.[37] Except in the vicinity of towns, where a regular supply of transient labor could be obtained on short notice, wage labor was uncertain. When John Dent of Georgia complained to one of his hands about "idleness," "his reply was 'I can leave you and I will do so.'—and such all do who are hired for wages, leaving you at a time when their services are most needed."[38] Hiring wage labor on an annual basis was no guarantee either, as this bitter complaint in the Waverly plantation records shows (after four hands had left without notice):

> Received wages for whole year, but left without permission and broke their contracts. They were in the wrong, had become tired

[35] Shifflet, *Patronage and Poverty*, p. 37.

[36] Warren Whatley, "Institutional Change and Mechanization in the Cotton South" (Ph.D. thesis, Stanford University, 1982); "A History of Mechanization in the Cotton South," *Quarterly Journal of Economics* 100 (Nov. 1985), pp. 1191–1215.

[37] Pranab K. Bardhan, "Labor-Tying in a Poor Agrarian Economy: A Theoretical and Empirical Analysis," *Quarterly Journal of Economics* 98 (Aug. 1983), pp. 501–514; *Land, Labor and Rural Poverty* (Delhi: Oxford University Press, 1984) Ch. 5. Bardhan writes, "Historically, agrarian labor-typing brings to mind the blatant cases of obligatory service by the tenant-serf to the lord of the manor (as in the classic instances of European feudalism) or those of debt-peonage to moneylender-cum-landlord. . . . This is to be distinguished from the case where the laborer voluntarily enters long-duration contracts with his employer and reserves the right to leave at the end of the specified period." (pp. 501–502).

[38] Dent Journals, vol. 15, March 3, 1883, quoted in Charles L. Flynn, Jr. *White Land, Black Labor* (Baton Rouge: LSU Press, 1983), p. 78.

of work, and ran away. They have the right to leave without slipping off. They were treated well.[39]

The only remedy was a postharvest payment with incentives, which is the essence of what sharecropping accomplished for the landlord. It is easy to see how this sort of arrangement could resemble "bound labor," but the testimony is now virtually unanimous that sharecropping emerged as a partial response to the *laborer's* desire for work autonomy and a family-based system. For married men, the preference was not merely a psychological desire for day-to-day freedom, but a way of gaining full control over the labor of the members of his own family. As one planter recalled,

> The negro first became a cropper on shares as a step toward what you call in your inquiry, "a more independent" position. Also largely due to the fact that if he hired himself out for wages he alone drew wages and his family such pay for day work as they might secure on the farm. By running the crop himself he could put his wife and children in his own crop.[40]

As this suggests, wage labor was uncertain to the workers as well as the landlords, especially those limited to a local area. When the cotton was "laid by," the wage-labor force might be "laid off" for two months in the summer, and again after the harvest in winter.[41] Of course, a mobile, unattached male worker, willing to pick up and move to wherever work could be found, might be able to put together a reasonably full work year, but such was not much of a way to live for a man with a family.

The division of plantation acreage between tenant plots and owner's land thus corresponds broadly to the division of the labor force into married men who had "settled down" in an area and unmarried young men who were more mobile on a short-term basis. In the Whatley model, the planter strikes a balance between the terms on which local tenants could be attracted and the wages of short-term labor, plus recruitment and transport costs. The local-tenant market clears by adjustment of the plot size, while the wage is set in a wider geographic market. Ideally, the balance would be struck so as to use all the available acreage and equalize the marginal net revenue of the two systems, as in the stylized Figure 4.2. A more complex analysis could incorpo-

---

[39] George W. Bryan Papers, Scotland Plantation Record, 1872.

[40] Quoted in R. P. Brooks, *The Agrarian Revolution in Georgia* (Madison: Bulletin of the University of Wisconsin #639, 1914) p. 46.

[41] See the testimony in the *Report of the Industrial Commission*, vol. 10, pp. xix, 455.

Postbellum Southern Labor Markets 113

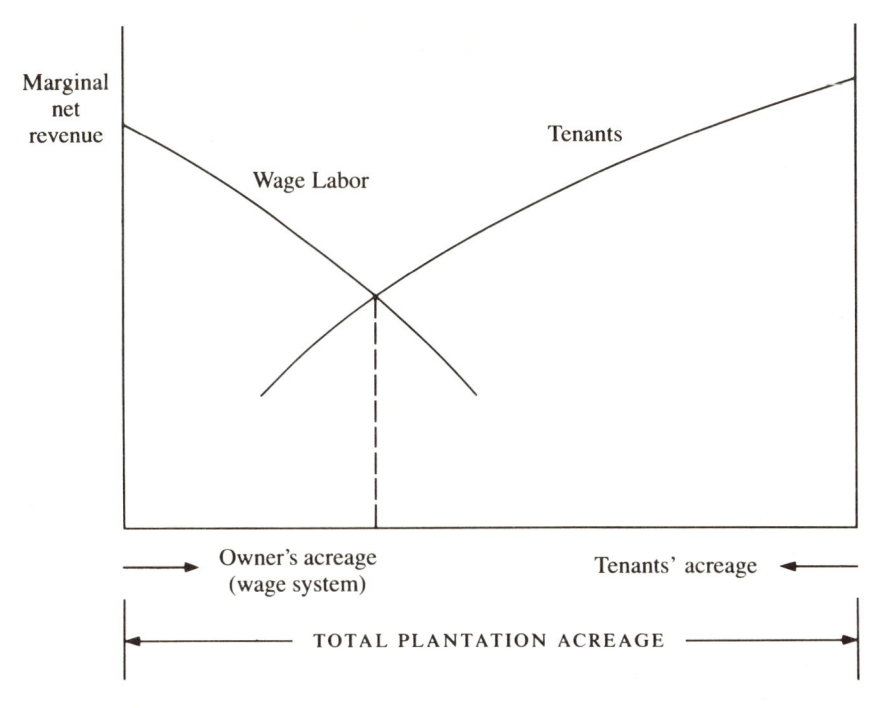

FIGURE 4.2 Allocation of Plantation Acreage between Wage System and Tenant Plots

SOURCE: Adapted from Warren Whatley, "Institutional Change and Mechanization in the Cotton South" (Ph.D. diss., Stanford University, 1982), p. 60.

rate explicitly the uncertainty properties of the two forms, but the essence of the balance described in Figure 4.2 would remain.

The correlation between wage labor and young, unmarried males was by no means perfect, but it was close enough for many southerners to take it for granted. Census data indicate that in 1890, three-fourths of the male farm laborers were unmarried and two-thirds were under the age of 25 (Table 4.6.). A Tennessee Valley planter observed in 1883,

> We employ families generally where we can. Those who are unencumbered with families are disposed to seek the public works. They have gone off to the Mussel Shoals from my neighborhood, and some have come down to Birmingham, and some have gone elsewhere. It is usually those that are encumbered with families that stay at home.[42]

[42] Testimony of P. N. G. Rand, Colbert County, Tennessee, in U.S. Senate Committee on Education and Labor, *Testimony as to the Relations Between Capital and Labor* (1883), vol. 4, pp. 145–146.

TABLE 4.6
Male Farm Laborers, 1890

State	Percent black	Percent married	Percent less than 25 years old
Alabama	63.3%	22.4%	74.2%
Arkansas	44.8	19.7	70.9
Florida	59.8	26.4	65.3
Georgia	63.7	25.7	72.0
Louisiana	74.3	34.4	61.1
Mississippi	75.9	22.9	71.3
North Carolina	49.6	26.0	69.4
South Carolina	73.1	28.1	71.1
Tennessee	37.0	27.0	66.4
Texas	31.9	15.9	70.6
Virginia	52.1	33.2	57.9
11 STATES	57.1%	25.6%	68.7%

SOURCE: 1890 Census, "Occupations," Table 116.

Twenty years later the man who complained about the migratory tendencies of farm laborers went on to say, "The older men, with families, who live at one place, still crop on shares."[43] W. E. B. Du Bois wrote that young black laborers "sink to the class of *metayers*" when they married, which may have been an exaggeration but which reflected the valid point that they did give up something when they settled down in an area.[44] Sharecroppers moved from farm to farm too, but the great majority of their moves were local.[45] The long-distance wage-labor market was always there to take advantage of, but those above a certain age and with a family one wouldn't want to be in it.

Over the course of time, this process of self-selection between the two markets came to be reflected in the expectations of employers, and therefore in the terms offered, including not just wage levels, but types of work, supervision, housing arrangements, and so on. Increasingly in the late nineteenth century, we encounter the firmly stated view that only "shiftless, inferior" types can be found in the wage-labor market. George K. Holmes, wrote in an article entitled "The Peons of the South," published in 1893: "The blacks prefer a tenancy to selling their labor for wages, and in some regions, at least, the white owners who

---

[43] *Report of the Industrial Commission*, vol. 11, p. 95.

[44] W. E. B. DuBois, "Of the Quest of the Golden Fleece," in *The Southern Common People*, Edward Magdol and Jon L. Wakelyn, eds. (Westport, CT: Greenwood Press, 1980), p. 265.

[45] That at least is the impression of Ronald L. F. Davis on the basis of store accounts and plantation records for the Natchez district: "The average black family moved from place to place, from landlord to landlord, and even from supplier to supplier without ever leaving the neighborhood." (*Good and Faithful Labor*, p. 179).

cultivate their farms find that only the inferior laborers can be hired, because the superior ones prefer tenancies."[46] Obviously, he was not equating "peonage" with its literal sense of "bound labor." But if planters felt this way, then the better workers would be well advised to move out of the wage-labor market early on. Some years later, T. J. Woofter wrote, "In fact, most large farm units which remain today are 'mixed,'—the owner hires as many laborers as he can and farms the remaining land with tenants. The result is that only the lower types can be hired for wages."[47]

The line of thought that associated wage labor beyond a certain age with failure and inferiority was widespread in nineteenth century America, but in the South it was closely followed by another irresistible association: wage labor was black work because all laborers were black, but because those that were white had lowered themselves to the level of blacks. When asked why he used black labor exclusively, a Tennessee planter replied, "Well, the class of white men that offer for hire out there as a rule are a very sorry class of men."[48] A Virginia correspondent complained that the best laborers had been drawn off—"what remains is very inefficient, the white being as bad as the colored."[49] Some years later an authority reported it as a "well-known fact" that practically all common laborers working for wages on plantations were negroes.[50]

But "market signaling" effects like this should not be seen as "interference" with the free functioning of market forces. The dominant intellectual tradition in economics views "institutions" as things that stand in opposition to "market forces" and impede their operation. But real, functioning markets are themselves institutions, and one of the ways in which markets deepen, expand, and improve is through the emergence of well-known terms and categories, much the way trading communities tend to converge on one, or at most, a few standard instruments of payment. In this way, the bewildering array of contract forms of the immediate postwar era gave way to a few reasonably standardized packages, crystallized around the distinctions that were usually most important. The recognition of sharecropping as a distinct legal category contributed to this crystallization, but even here one can

[46] George K. Holmes, "The Peons of the South," *Annals of the American Academy of Political and Social Science* 4 (Sept. 1893), p. 71.

[47] Woofter, *Negro Migration*, p. 34.

[48] *Report of Industrial Commission*, vol. 10, p. 477.

[49] J. R. Dodge, "American Farm Labor," Part 3 of *Report of Industrial Commission*, vol. 6, p. 93.

[50] C. O. Brannen, "Relation of Land Tenure to Plantation Organization," USOA Dept. Bulletin no. 1269 (1924), p. 22.

argue (as Brooks did in 1912) that "the law only crystallizes the actual economic facts as they have worked out."[51] The distinction between cropper and tenant, for example, seems to have been reflected in unofficial census enumerators' categories even before the Civil War.[52]

Within this set of packages, wage labor was the class that asked least and expected least. When we say that farm labor was undifferentiated or common labor, we do not mean that there were no differences among workers, but only that the institutional setup *made* no distinction among workers, and the jobs and the pay were designed accordingly. The term "public works" quoted earlier is instructive, since it has such a different meaning today. It did not refer to TVA projects or municipal water plants, but to factories or mines or sawmills that would take virtually anybody. For jobs like that, market forces were powerful and effective, and labor mobility was a real strength. But if you wanted to get anywhere with your life, you had to escape from that market. Southerners, especially blacks, put quite a bit of effort into escaping from the labor market, but they did not always succeed.

III. The Agricultural Ladder

The way to get out of the labor market was to get married, settle down in an area, establish a reputation as a reliable worker and dependable borrower, and accumulate the wealth to move up the "ladder" to higher tenancy levels and ultimately to farm ownership. A surprisingly large number of black as well as white farmers were able to accomplish this, but contrary to the presumptions that have underlain the debate over "bound or free" labor, at each step along the way progress was associated with *reduced* levels of geographic mobility. The same 1910 census report that documents the high turnover of share tenants also shows an equally remarkable *positive relationship between length of tenure on a given farm and status on the agricultural ladder* (Tables 4.7 and 4.8).

It would be a mistake to interpret this pattern as a simple cause and effect, certainly not in the judgmental fashion of the contemporary critics of "tramps" and "vagrants." But the tables are sufficient to show that writers like Jennifer Roback, who take geographic mobility to be a self-evident indicator of economic opportunity, have completely misspecified the issue. The same relationship may be seen in the avail-

---

[51] R. P. Brooks, *The Agrarian Revolution*, p. 459. Important works on the legal history of sharecropping, include Harold Woodman, "Southern Agriculture and the Law," *Agricultural History* 53 (Jan. 1979), pp. 319–337; Gerald D. Jaynes, *Branches Without Roots* (New York: Oxford University Press, 1986).

[52] This is the argument of Frederick Bode and Donald Ginter, *Farm Tenancy and the Census in Antebellum Georgia* (Athens: University of Georgia Press, 1986).

TABLE 4.7
Years on Present Farm by Tenure Class,
South Atlantic Region, 1910

Years on farm	Owners free	Owners mortgaged	Part owners	Cash tenants	Share tenants
WHITE FARMERS					
Less than 1 year	3.8%	9.5%	9.6%	28.9%	37.9%
1 year	4.4	9.2	9.3	17.6	17.8
2–4 years	16.4	25.5	25.0	31.7	28.1
5–9 years	19.0	20.8	20.6	12.9	10.0
10 years and over	56.5	34.8	35.5	8.8	6.2
BLACK FARMERS					
Less than 1 year	2.2%	4.8%	7.5%	17.7%	33.9%
1 year	3.0	6.0	7.4	13.5	17.4
2–4 years	16.1	22.8	23.9	34.9	31.5
5–9 years	20.9	22.8	21.6	18.3	10.5
10 years and over	57.9	43.8	39.6	15.7	6.6

SOURCE: U.S. Census Bulletin, "Stability of Farm Operators" (1910). Figures are percentages of those answering questions.

TABLE 4.8
Years on Present Farm by Tenure Class,
East South Central Region, 1910

Years on farm	Owners free	Owners mortgaged	Part owners	Cash tenants	Share tenants
WHITE FARMERS					
Less than 1 year	4.7%	12.2%	14.3%	36.7%	45.6%
1 year	5.1	10.6	11.1	17.1	17.8
2–4 years	18.9	28.5	26.9	29.8	24.8
5–9 years	20.3	20.7	20.2	10.4	7.5
10 years and over	51.1	28.1	27.4	6.0	4.1
BLACK FARMERS					
Less than 1 year	2.5%	4.6%	9.4%	19.1%	39.3%
1 year	3.4	5.3	7.7	12.4	15.9
2–4 years	16.4	22.8	25.7	34.5	28.1
5–9 years	21.9	25.2	23.0	18.4	9.8
10 years and over	55.7	42.0	34.2	15.6	6.3

SOURCE: U.S. Census Bulletin, "Stability of Farm Operators" (1910). Figures are percentages of those answering questions.

able plantation records: tenants who did well and owned their own implements and workstock were those who stayed on year after year. Those who were "in and out" at different times, or who appeared only briefly, did not get generous terms based on the bargaining power their mobility gave them; on the contrary, they were at the bottom of the heap, earning less, consuming less, producing less. Is this a surprising and paradoxical pattern? Not when we recall that virtually all of the social-mobility studies for the nineteenth century find essentially the same result. Accumulating savings, acquiring assets, and moving into jobs demanding skill or responsibility, all served to reduce the likelihood of leaving town.[53]

The voluminous new literature on sharecropping and other tenancy forms is now so complex and variegated as to daunt any normal reader. But by my reading of it, both theoretical and empirical work now show a certain convergence around the view that credit markets are the key. Tenancies were not just labor contracts: they were "interlinked" land, labor, and credit arrangements.[54] In the plantation belt, croppers and tenants usually obtained the "furnish" from the planter himself. Sometimes the credit came from a local merchant, but in that case it was part of a three-way relationship, the connection between landlord and tenant serving either formally or informally as a basis for the loan. And even where the landlord was not the money lender, there was a "credit" aspect to the contract, in that the laborer had access to the land for a year and over that period could maintain the land or neglect it with lasting effects. Thus, creditworthiness was an essential requirement for progress up the agricultural ladder.

Despite the specification of some theorists, a fixed-rent tenancy was not riskless to the landlord and could not be made so. As T. J. Woofter wrote, "Landlords have to be very particular about renting outright because they risk losing considerable sums on drifters, unsuccessful and dishonest tenants, and they also risk greater depreciation on their land."[55] The basic issue is trustworthiness, but as in most credit markets, this subjective criterion was generally measured by an objective one, namely wealth, or the possession of specific assets. The price of a mule was the main barrier between share and fixed-rent tenancy. In their recent survey of contractual mix in southern agriculture, Alston

---

[53] The best-known studies are Stephen Thernstrom, *Poverty and Progress* (Cambridge, Mass.: Harvard University Press, 1964), and *The Other Bostonians* (Cambridge, Mass: Harvard University Press, 1973).

[54] P. K. Bardhan, *"Interlinkage of Land, Labor and Credit Relations," Oxford Economic Papers,* 32, March 1980, pp. 82–98; Aviskay Braverman and Joseph Stiglitz, "Sharecropping and the Interlinkage of Agrarian Markets," *American Economic Review* 72 (Sept. 1982), pp. 695–715; Gerald Jaynes, "Production and Distribution in Agrarian Economies," *Oxford Economic Papers* 34 (July 1982), pp. 346–367. The agricultural ladder as the key to sharecropping and other tenure forms has been emphasized by Joseph D. Reid, Jr., in a somewhat different formulation. See Reid, "Sharecropping and Tenancy in American History," in *Risk, Uncertainty and Agricultural Development,* J. A. Roumasset, J. Boussard, and I. Singh, eds. (New York: Agricultural Development Council, 1979).

[55] Woofter, *Negro Migration,* p. 86.

and Higgs express skepticism about this proposition on the grounds that landlords could and sometimes did advance the credit for the purchase of mules and implements. But their empirical study of tenure on plantations in Georgia confirms the proposition clearly. As they write, "the ownership of work stock by laborers appears to have been decisive in determining whether fixed rental agreements would be used."[56] Certainly contemporaries took the correlation to be so close as to practically define the rungs of the ladder. Writing in 1913, economist Benjamin Hibbard described a "well-defined caste system among the tenants":

> The lowest class is represented by those who furnish little equipment and receive half, or less, of the crop; above this comes the group whose independence is measured by the possession of a mule and a plow and the means of subsistence till harvest time; the highest class consists of those who can be trusted to deliver a certain quantity of crop or possibly a sum of money, and who are by that fact emancipated in the main from the directing authority of the landlord.[57]

Note the blend of objective and subjective criteria, in which the state of being trusted to deliver a certain sum or quantity is regarded as a "fact." The ownership of a mule or plow was of course no more than a rule of thumb in a particular time and place, and it remained true (as Alston and Higgs emphasize) that any two parties were free to change the arrangement if they both wished to. But such rules of thumb can become influential in their own right, when they come to be known and adopted over a wide area. Most individuals have little basis for differentiating themselves from others and adapt their behavior to the categories of the credit market as they do to the labor market.

The relationship between tenure status and asset ownership is illustrated in the records and contracts of the Lewis plantations of Hale and Marengo counties in Alabama.[58] For one of these plantations, called the Hermitage, an almost continuous record of labor contracts and

---

[56] Lee J. Alston and Robert Higgs, "Contractual Mix in Southern Agriculture Since the Civil War," *Journal of Economic History* 42 (June 1982), p. 351 (regression results on pp. 348–349). The very quotation cited by Alston and Higgs, referring to advances for "the hire of mules and other credits," in fact confirms the notion that credit-worthiness is the key, since it is an admonition against the practice of entering into fixed-payment tenancies "without adequate means."

[57] Benjamin Hibbard, "Tenancy in the Southern States," *Quarterly Journal of Economics* 27 (1913), p. 486.

[58] Lewis Plantation Records, Southern Historical Collection, University of North Carolina. I am indebted to Warren Whatley, both for calling these records to my attention and for providing me with a 26-year tabulation of the Hermitage labor contracts, personnel, and settlements.

settlements from 1875 to 1900 has survived. We have no way of judging the representatives of this one case, but it offers an unusual chance to follow work careers of individuals over a long period, and some of the events are striking. From all indications, the accounts were honestly kept; certainly the Lewises did not contrive to keep the tenants perpetually in debt, and in this respect they were not unusual. Indeed, despite the oft-repeated expression "debt peonage," virtually all specific testimony or survey evidence indicates that most tenants and croppers did not typically end the year in debt in normal times.[59] More importantly, when they did end the year in debt, there is no indication in the Hermitage records that this condition was a barrier to mobility. Tenants with hopelessly large debts were *more* likely to leave, often before the year was over. In the Hermitage records as well as in those of Waverley discussed earlier, there are many cases where an end-of-year debt was given to the tenant in the hopes of retaining him for the following year. When Nelson Bryan (black) finished the season of 1874 at Waverley owing $20.16, for example, the account book read "balance due $20.16, which is given to him, provided he continues to act rightly." Abel Anderson (black) ended 1874 owing $73.28, of which $50.02 was written off, "provided he stays with me in 1875 and does well." One could construe these statements as attempts to use the end-of-year debt to reduce mobility. But it was not the legal control provided by the debt that mattered, because the debts were written off or down as an inducement to stay. And in the six years of the Waverley records, there are no less than nineteen cases of departure by the tenant directly after failing to break even for the previous years.

At the Hermitage, despite the continuing turnover of marginal laborers, a core of more or less permanent tenants emerged over time. Of an original cohort of twenty-five in 1875, only sixteen remained in 1876; but fourteen of these were still on hand five years later in 1881. Through the 1880s, almost half of the tenants in any given year had been at the Hermitage five years or more. Up to 1886, all tenants

---

[59] Even in 1930, hardly a prosperous year, only 13.4% of cotton croppers finished the year in debt, according to one survey (Mandle, *Roots,* p. 49). The black congressman George Henry White, telling of cases where debt had accumulated over time, made a point of saying: "I do not want to impress you that this is the rule . . ." (*Report of the Industrial Commission,* vol. 10, p. 419). W. E. B. Du Bois, describing Dougherty County, Georgia, in 1898, where 175 of 300 tenant families ended the year in debt, also made a point of noting this was a "year of low-priced cotton," and that "in a more prosperous year the situation is far better" ("Golden Fleece," p. 260). Ransom and Sutch do employ the term "Debt Peonage," but they make it clear that they are *not* referring to the case where laborers are legally bound to a particular employer because of debt.

worked on a standard share contract, although there are occasional wage credits for "work on the dick" [dike]. In typical years during the 1880s, most of the experienced hands cleared their accounts and made a profit. In 1883, for example, there were twenty-five tenant accounts, sixteen of which were settled in full at the end of the year. But of the fourteen who had been there five years or more, all but one paid out fully; almost all the "balance due" cases were newcomers who tended not to stay long. Sometimes, it is true, an older tenant would pay out only by putting in a few days wage work after the crop settlement. But even in this case, it is clear that it was not the debt that caused the tenant to sign on for another year.

Indeed, after a series of reasonably good years, we find the following in the records after the settlements for 1885:

> A. C. Lewis, hereby, agrees to rent to JNO DUNN about 60 acres on the Hermitage Place for the year 1886 for 1850 lbs. of lint cotton to class low middling. . . . JNO DUNN agrees to cultivate the land well and thoroughly to keep it up. (January 6, 1886)

This new rental contract is followed by nine others of exactly the same form, with some variations in numbers of acres and rents. All ten of the renters were tenants who had been at the Hermitage for some years. By 1890 there was a tenant force of about twenty, only four of whom were still working on shares. A solid core of fourteen was still there in 1893. But in that disastrous year, every single account ended in deficit, and eight of the fourteen (some of whom had worked on the place for a decade or more) were gone by 1894. The "settlements" are pitiable, tenants forfeiting mules, wagons, cattle, and still not covering their debts. In the following year, every hand worked either for shares or wages; the accounts now refer to "work done on the wages squad" and contain "a list of hands belonging to the road." The Lewises wrote out the terms of a share or a wage contract, and each of the twenty-eight laborers stipulated the option of his choice. No coercion was involved, nothing but free choice, the market at work, and a highly mobile labor force.

This is the clearest evidence I know that the choice between shares and fixed rent was determined by the financial status of the tenant, and that the flourishing mobility of the adult male wage-labor force was a sign of defeat for these people in their efforts to escape this market, rather than an indication of the contribution of market forces to their welfare. Of course, times did improve after the collapse of the mid-1890s, and we know from the research of DeCanio, Higgs, and Margo that blacks made significant progress in wealth accumulation before World War I.[60] Roughly 25 percent of black farm operators were owners by 1900. But the great majority of these owners had small plots of inferior land. The continued growth of this class is a clear sign that they did not see the labor market as their opportunity or their protection. By cultivating a small plot of land, they may well have given up something in standard of living, and they surely gave up their mobility. Thus, those who had the chance chose to escape the labor market, but many others were not that fortunate.

IV. Race and Industrial Labor Markets in the South

Many writers who use the term "labor immobility" may still feel that the foregoing discussion is not fully relevant, because what they really mean is that blacks were confined to the agricultural sector. Cotton textiles, the most dynamic and successful southern industry during the nineteenth century, employed whites almost exclusively. Other industries like furniture and printing excluded blacks to a greater or lesser degree, and even where blacks were employed in industry, segregation usually prevailed, by room, building, job, and status. Such outright racialism would seem to defy both rational profit-seeking behavior and the pressures of marketplace competition. Indeed, Robert Higgs has interpreted the entire period "as an interplay of two systems of behavior: a competitive economic system and a coercive racial system." Although blacks were coerced and suppressed by segregation laws and unequal law enforcement as well as by discrimination in the market, "the region's economic development increasingly undermined the foundations of its traditional race relations."[61]

In fact, the racial practices of employers were almost completely unregulated by law in the southern states. In the wake of C. Vann Woodward, so much has been written about the rise of Jim Crow at the end of the nineteenth century that many people find this hard to believe. One distinguished authority on black labor, Herbert Northrup, has written the following:

---

[60] Stephen J. DeCanio, "Accumulation and Discrimination in the Postbellum South," *Explorations in Economic History* 16 (April 1979), pp. 182–206; Robert Higgs, "Accumulation of Property by Southern Blacks Before World War I," *American Economic Review* 72 (Sept. 1982), pp. 725–737; Robert A. Margo, "Accumulation of Property by Southern Blacks: Comment and Further Evidence," *American Economic Review* 74 (Sept. 1984), pp. 768–776.
[61] Robert Higgs, *Competition and Coercion,* p. 13; "Race and Economy in the South, 1890–1950," in *The Age of Segregation,* Robert Haws, ed. (Jackson, Miss.: The University Press of Mississippi, 1978), p. 30.

Of course, a jim crow system is designed to make the employment of Negroes more costly. It is an historic fact that jim crow laws were a product of the era following the Civil War Reconstruction when there occurred the rise to power in the South of the poor white working and farmer classes. Their objective in "putting the Negro in his place" was, among other things, to curb the competition of Negro labor. By confining Negroes to segregated living areas, they were removed from job information and factory locations; by jim crow transportation, Negroes found it more difficult and unpleasant to get to work; by jim crow schools and training, Negroes were made less qualified to compete for work; and by requiring employers to segregate Negroes as employees and to provide duplicate eating, sanitary, and locker-room facilities, these laws made it more expensive to employ Negroes.[62]

This is completely erroneous. G. T. Stephenson's exhaustive 1910 survey *Race Distribution in American Law* contains no reference to employment or workplace regulation. As late as 1940, the legal authority Charles S. Mangum wrote only that "the problem of industrial segregation in the South is usually left in the hands of private business," noting only three exceptions. The only industrial segregation laws of any importance, a North Carolina statute requiring separate toilets and a South Carolina law requiring segregation in cotton textiles, were not passed until 1913 and 1915, respectively, and were not imitated in other equally segregationist states.[63] Explicitly racial laws enforced segregation in public conveyances, marriage, schools, and places of public accommodation, amusement, and burial, but not in employment. Yet segregation prevailed.[64]

It might be thought that patterns of industrial segregation merely re-

[62] Herbert Northrup and Richard Rowan, *Negro Employment in Southern Industry* (Philadelphia: University of Pennsylvania Press, 1970), pp. 77–78. There are no footnotes attached to this paragraph.

[63] Gilbert Thomas Stephenson, *Race Distinctions in American Law*) (New York: D. Appleton & Co., 1910); Charles S. Mangum, *The Legal Status of the Negro* (Chapel Hill: University of North Carolina Press, 1940) pp. 174–175.; Pauli Murray, ed., *States Laws on Race and Color* (Cincinnati: Methodist Church, 1950), pp. 341–342, 414–415.

[64] These distinctions are unfortunately muddled by Jennifer Roback in her recent study. She coins the term "Jim Crow labor laws," referring to statutes on enticement, vagrancy, contract enforcement, emigrant-agent registration, and convict lease, conveying the impression that these were racial laws associated with the campaigns for disfranchisement and legal segregation after 1890. But the labor laws were not Jim Crow laws and long predated disfranchisement. Indeed, as Roback's own evidence shows, the majority of states had abolished the convict-lease system by 1890. The wave of antivagrancy laws passed between 1903 and 1907 was clearly a response to the tight labor market conditions of those years, conditions that affected workers of both races and helped to integrate regional labor markets more fully than in previous years. Roback's evidence for a differential effect by race is indirect and of doubtful relevance;

flect the racial percentages in industry locations, but this too is not so. Segregation followed industry lines rather than geography. The population of the Piedmont, where cotton textiles concentrated, was about one-third black. In the state of North Carolina, there were all-white cotton mills and furniture factories but also heavily black tobacco factories. Tobacco manufacturing was a major black employer even though it was concentrated in white-majority states like North Carolina, Tennessee, and Kentucky. The industrial patterns hold true even in particular towns and cities. White cotton and black tobacco coexisted in places like Danville, Virginia, and Durham, North Carolina. In Birmingham, where iron and steel workers were two-thirds black, the Avondale cotton mill was 98.1 percent white.[65]

Theoretical work on the economics of discrimination is nearly as confusing as the work on the theory of sharecropping. Yet amidst the bustle, one fairly clear conclusion has emerged: the motives that give rise to racial distinctions are much more likely to generate job segregation than wage discrimination in a competitive market setting. Suppose, for example, that whites require a premium to work with blacks. Then total wage costs are

$$C = w_A(W/L)W + w_B B , \qquad w_A' < 0 \qquad (1)$$

where W is the number of white workers, B is the number of black workers ($W + B = L$), w_B is the black wage, and w_A is the white wage, which is a downward-sloping function of the proportion W/L. It follows at once that a profit-maximizing employer will compare $w_A(1)$ with w_B and hire either all whites or all blacks, whichever is cheaper. If all employers behave this way, the full-employment equilibrium must be

$$w_A(1) = w_B \qquad (2)$$

and there is no discrimination.[66] Similarly, if some employers dislike hiring blacks, others will find it advantageous to hire nothing but

---

she does not examine the available wage data. See Roback, "Southern Labor in the Jim Crow Era," pp. 1165, 1184–1191.

[65] Paul Worthman, "Black Workers and Labor Unions in Birmingham, Alabama, 1897–1904," *Labor History* 10 (Summer 1969), p. 392.

[66] The illustration comes from Kenneth Arrow, "The Theory of Discrimination," in *Discrimination in Labor Markets,* Orley Ashenfelter and Albert Rees, eds., (Princeton: Princeton University Press, 1973). See also Joseph Stiglitz, "Approaches to the Economics of Discrimination," *American Economic Review* 63 (May 1973), pp. 287–295.

blacks. There are so many ways to segregate—by room, job, location, firm—that it would not require any very large number of profit-seeking firms to drive the black wage up to the level of the white wage.

Remarkably enough, this is pretty close to what actually happened in southern unskilled labor markets. A Chattanooga iron company official reported in 1883, "We make it a rule to pay colored men the same as we pay white men for the same kind of labor. . . . We pay men according to the positions they occupy." The head of a brick-making firm in Columbus, Georgia, said, "This is about the cheapest field for labor that I have ever found anywhere . . . white labor of the same class is equally cheap. We pay the same to whites as to blacks, when we employ them."[67] A study of wages in Virginia by Robert Higgs finds that wage differentials by race within firms were rare, and even the modest differences in state occupational average wages for blacks and whites were attributable to variations in wages among localities rather than to discrimination within a local labor market.[68]

More important than the policies or attitudes of individual employers is the effect of the market. So long as both blacks and whites had open access to the farm-labor market, and so long as labor flowed freely between sectors, industrial wages for unskilled labor could not deviate widely from the farm wage. Allowing for differentials based on age and sex, or for particularly strenuous or dangerous work, the evidence indicates that farm and industrial unskilled labor markets were closely linked. This proposition holds true even where industries, like cotton textiles, were lily-white.[69] The equilibrating pressures of the competitive labor market are strikingly illustrated by the convergence of wages for white males in textiles and black males in tobacco manufacturing, despite the radically racial traditions and policies in the two cases. In Virginia in 1907, the median white male wage in cotton manufacturing was $6.78 per week, while the median black male wage in tobacco was $6.00. The *modal* wage of black males in tobacco was actually higher than that of white males in cotton, $6.60 to $6.00. For

---

[67] Senate Report, *Relations Between Capital and Labor,* vol. 4, pp. 171, 489.

[68] Robert Higgs, "Firm-Specific Evidence on Racial Wage Differentials and Workforce Segregation," *American Economic Review* 67 (March 1977), pp. 236–245.

[69] In "Cheap Labor and Southern Textiles, 1880–1930," *Quarterly Journal of Economics* (Nov. 1981), I reported the following (p. 625):

$$\ln w_s = -2.76 + 0.69 \ln w_f + 3.03 \text{ EXP} + 0.143 \ln \text{SP}$$
$$(2.72) \quad (3.26) \quad (3.96)$$
$$R^2 = .931 \quad D-W = 1.25$$

where w_f is the farm wage, EXP is an age-sex index, and SP is the stock of spindles (*t*-ratios in parenthesis).

females, who did not have access to the intermediating agricultural wage jobs, a 25-percent gap in favor of whites persisted.[70]

Thus, the persistence of segregation did not mean that competitive market pressures were absent, and these pressures seem to have acted powerfully in the market for unskilled labor. Even the most notoriously oppressive postbellum labor institution, the convict-lease system, operated primarily to preserve the competitive market rather than to obstruct market forces. The convicts, of course, were not free; by the 1890s, 90 percent were black, and in the hands of unscrupulous mine operators, they could be brutally exploited. But they were never more than a tiny fraction of the labor force, and even in mining, the primary functions of convict labor were to break up strikes and attempted labor organization. That is to say, convict lease was used to maintain the competitive labor market. Widespread corruption led to the abolition of the system in most states by 1900, but where it was continued and where open bidding for convicts prevailed (as in Georgia after 1899), lease rates rose to roughly the prevailing wage levels in the free-labor market.[71]

But just as the open market for undifferentiated labor was not a vehicle for individual black progress in agriculture, the same was true in industrial employment. The unique wage data collected after 1900 by the Commissioner of Labor for Virginia give a more complete picture. Although Virginia was by no means a typical southern state, it did have representation from almost every important southern industry—cotton and tobacco manufactures, sawmills, iron, building trades, and many smaller ones. By combining the occupational wage data with information about racial-segregation lines, we can estimate overall distributions by race for the largest industries. The parameters for 1907 appear in Table 4.9, and fitted log-normal curves for the same year are displayed in Figure 4.3.[72]

---

[70] These statistics are estimated from data in the Eleventh Annual Report of the Bureau of Labor and Industrial Statistics for the State of Virginia (Richmond: David Bottom Superintendent of Public Printing, 1908), pp. 83–96, 189–195. See also note 72.

[71] Matthew J. Mancini, "Race, Economics and the Abandonment of Convict Leasing," *Journal of Negro History* 63 (Fall 1978), esp. p. 348. A good recent discussion of convict leasing may be found in Edward L. Ayers, *Vengeance and Justice: Crime and Punishment in the 19th Century American South* (New York: Oxford University Press, 1984), Ch. 6.

[72] The result presented covers building trades, saw and planing mills, cotton and tobacco manufactures, iron and machinery, and tanneries. The data come from a larger project in collaboration with Warren Whatley, supported by the National Science Foundation. The figures presented correct slightly those in *Old South, New South*, pp. 183–184. Full details will be presented in a forthcoming publication.

Postbellum Southern Labor Markets

TABLE 4.9
Aggregate Statistics, Virginia Male Wage Earners, 1907

Statistic	White	Black
Number of wage earners	22,914	13,765
Mean wage	$11.79	$7.52
Modal wage	$8.00	$8.00
Median wage	$9.96	$7.50
Standard deviation	5.61	1.90
Standard deviation (LOG)	0.438	0.241

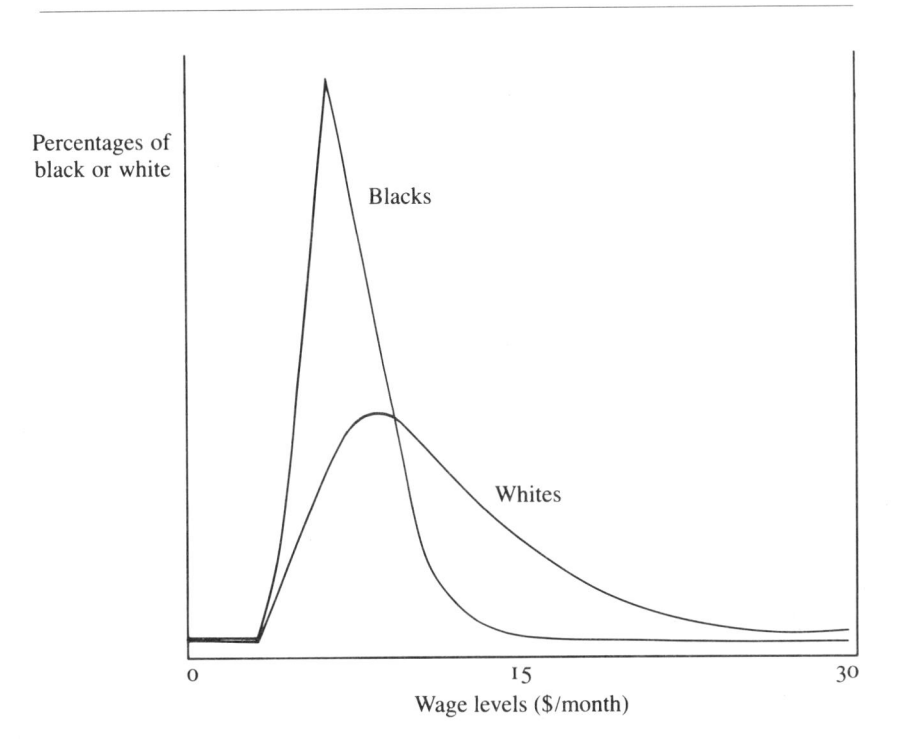

FIGURE 4.3 Aggregate Wage Distributions, Virginia, 1907

This is indeed a horse of a different color. The evidence of equilibration in the unskilled market is clearly there, in that the modes of the two curves are identical at $8.00 per week. But it is *only* in this unskilled labor portion of the curve that equilibration prevails. Means and medians show a 30–50-percent advantage for whites over blacks. Apparently, unequal access to higher-paying jobs was a far more important form of discrimination than wage differentials for the same job. Blacks could get the going wage in the unskilled market, the same

TABLE 4.10
Upward Occupational Mobility by Race, Birmingham, 1880–1914

	1880 group	1890 group	1899 group
5 years			
White	19%	29%	23%
Black	14	9	14
10 years			
White	31	33	40
Black	13	8	17
15 years			
White	34	34	43
Black	14	13	18
20 years			
White	48	34	—
Black	18	16	—
25 years			
White	48	39	—
Black	—	18	—

SOURCE: Paul B. Worthman, "Working Class Mobility in Birmingham, Alabama, 1880–1914," in *Anonymous Americans,* Tamara Hareven, ed. (Englewood Cliffs, N.J.: Prentice-Hall, 1971), Table 9.

wage that was linked to the agricultural market, but they had a hard time going any further.

Studies of occupational mobility by social historians confirm this picture. In Birmingham, for example, with one of the largest concentrations of black industrial labor, more than 80 percent of black workers who stayed for ten years made *no upward progress* during this time (Table 4.10). Separate studies of Atlanta, Georgia, and Washington, D.C., show similar results.[73] The typical white unskilled worker could expect to move up over time, the typical black could not. Blacks were not "confined to agriculture," but in terms of relative position and opportunities for advancement, the prospects for "good blacks" were much more inviting on the farm ladder, precisely because that ladder offered a way of escaping the labor market.

In support of this claim, compare the industrial wage distributions in Figure 4.3 with the distributions of gross farm income by farm, computed from the census of 1900 (Figures 4.4 and 4.5). Although the inequality between the races is visible, it is much less marked than in

[73] Paul B. Worthman, "Working Class Mobility in Birmingham, Alabama, 1880–1914," in *Anonymous Americans,* Tamara Hareven, ed. (Englewood Cliffs, N.J.: Prentice-Hall, 1971); Richard J. Hopkins, "Occupational and Geographical Mobility in Atlanta, 1870–1896," *Journal of Southern History* (May 1968), pp. 200–213; Lois E. Horton and James Oliver Horton, "Race, Occupation and Literacy in Reconstruction Washington, D.C.," in *Towards a New South,* Oliver Vernon Burton and Robert C. McMath, eds. (Westport, CT: Greenwood Press, 1982).

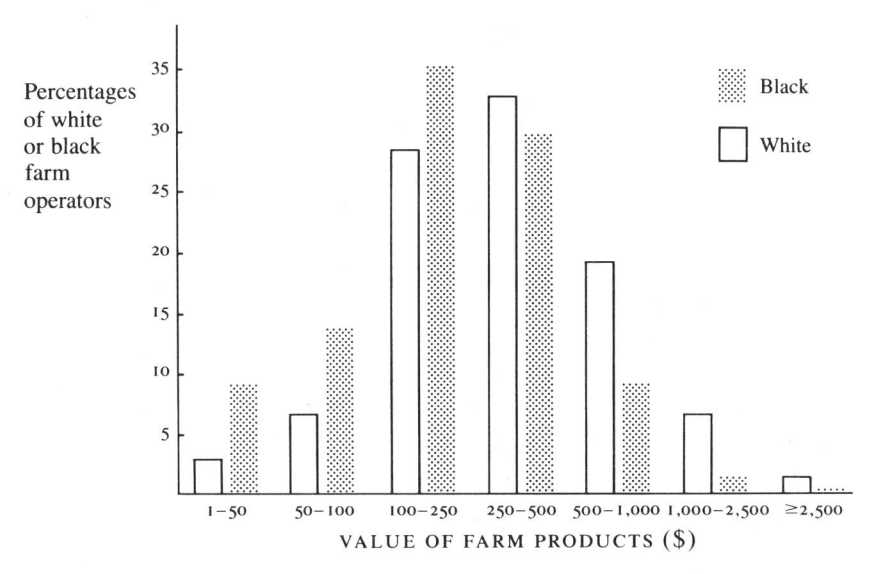

FIGURE 4.4 Gross Farm Income, South Atlantic States, 1900

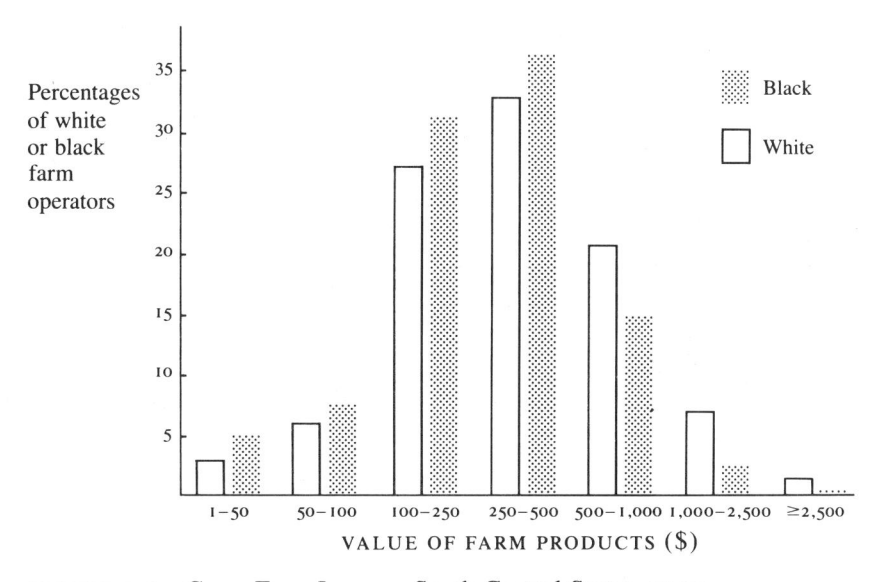

FIGURE 4.5 Gross Farm Income, South Central States, 1900

the industrial curves, and the black distribution has much more of a right-hand tail. It would be a serious error to think that blacks were actually *getting* these incomes in 1900. The figures are gross income, from which one would have to subtract rent and interest payments. But in contrast to the industrial case, these agricultural earnings were in principle there to be claimed by blacks as well as whites, if only they had the persistence and luck to increase their share by accumulation. For many this prospect may have been only a hope, but it was better than what the labor market had to offer for most blacks.

V. Market Pressures and Racial Equity

The vertical segregation represented by the white domination of the right-hand tail of the industrial wage distribution may reflect many possible effects: craft union restriction in the building trades, unequal education and training, employer prejudice, and resistance by white workers to black promotion, to name a few. The key question is, however, whether competitive market pressures were fundamentally hostile to discrimination in the form of vertical segregation. Did segregation represent the power of tradition and prejudice to dominate profit motives, and did industrial progress tend to undermine these patterns? The evidence presented here suggests that, on the contrary, competitive pressures served to strengthen and reconfirm lines of segregation traceable back to antebellum times. The mechanism was on-the-job experience.

Some clues may be found in the testimony by southern employers to the 1883 Senate committee. The ironmen from Alabama and Tennessee, who employed large numbers of blacks, extolled the qualities of black workers, and saw no difference in racial aptitudes. The cotton manufacturers of the southeast, on the other hand, made statements like "I don't think the negro would be fit to work in a cotton factory" and, "He is not adapted to the management of intricate machinery . . . their fingers are too stiff."[74] It is possible that the ironmasters were merely enlightened while the cotton-mill men were merely ignorant and prejudiced. More likely, blacks really were good ironworkers because they had experience. They were or would have been poor cotton-mill workers because in 1883 blacks had virtually no experience in that line of work. And they did not gain any in subsequent years. When the use of blacks in the mills came under discussion in the 1890s, the debate was abstract and speculative, as though no one could really imagine what black textile workers would be like.[75]

The origins of the all-white cotton mills go back to slavery. Slaves

[74] Senate Report, *Relations Between Capital and Labor*, vol. 4, pp. 102, 124, 157, 169, 239, 278, 529, 538, 726, 753.

[75] George Tindall, *South Carolina Negroes, 1879–1900* (Columbia, South Carolina: University of South Carolina Press, 1970), pp. 130, 132; "The Negro as a Mill Hand", *Manufacturer's Record* 24 (Sept. 22, 1983).

had been used successfully in southern mills in the 1840s, but almost all were pulled out during the cotton boom of the 1850s.[76] After the war, the core of experienced workers were all white, giving them an "initial advantage" over blacks. If the economic advantages of segregation are compelling, even a small margin may be enough to swing the entire factory labor force. And these advantages *will* be compelling, because as shown in the previous section, segregation is the way by which a competitive labor market avoids the inefficiencies of racial discrimination. Initially, it may have made no great difference, as blacks could take jobs in agriculture that paid just as well.

Over time, however, the textile labor force gained experience and maturity, and this learning process was an important part of the progress of the industry.[77] Under the mill-village system, new generations of workers actually grew up in the mill, becoming socialized to the routines, the noise, and the machinery at early ages. The significance of early industrial experience is well illustrated by the problem of the "cotton-mill drones" or "tin-bucket toters," indolent fathers who allegedly lived off the earnings of their children. On closer examination, investigators generally found that these were men who had grown up on farms and had brought their families to the mill in desperation, but had then found in a short time that their sons and daughters could make more money than they.[78] But as the new young men and women grew up and had families, this problem receded.

The progress of the textile industry, then, made the prospects for black jobs more and more remote as time passed. It is a good example of increasing returns, where a relatively small initial advantage evolves over time into a competitive necessity. A close analogy is the case of the QWERTY typewriter keyboard, analyzed by Brian Arthur and Paul David, in which an early choice became embedded in the training of the labor force and subsequently dictated the design of technology right on into the age of computers.[79] In both examples, the labor market was the vehicle. Identifying and using workers with established skills in relationship with particular machines offered advantages so great that even new entrants quickly followed the prevailing practices. The best evidence that segregation in textiles did not run counter to

[76] Gavin Wright, "Cheap Labor and Southern Textiles Before 1880–1930," *Journal of Economic History* 39 (Sept. 1979), pp. 655–680.

[77] Ibid. pp. 608–613.

[78] U.S. Senate (for the Secretary of Commerce and Labor), *Report on Condition of Women and Child Wage Earners in the U.S.* (1910), vol. 1, p. 453.

[79] Brian Arthur, "On Competing Technologies and Historical Small Events," (Stanford University, Nov. 1983); Paul David, "Understanding the Economics of QWERTY, or Is History Necessary?" *American Economics Review,* 75 (May 1985), pp. 332–337.

economic pressures is what happened when black mills were tried. There were perhaps as many as a dozen attempts between 1895 and 1905; whatever the precise number may have been, it was precisely equal to the number of failures. Reasons for failure were numerous, as always, but it seems evident that even with new machinery and good management, a completely inexperienced textile labor force just was not competitive by the turn of the century.[80]

Could it be that this interpretation is merely a rationalization for what is better viewed as an exercise of white power? Certainly blacks lacked power, but there are cases where previous experience gave blacks preference in employment. An example is tobacco manufacturing, which employed slave labor before the war and black labor continuously thereafter. Many freed blacks worked for years in the same establishments where they had been slaves. As late as 1930, black women still dominated the same phases of work—sorting, picking, stemming, hanging—that black slaves had done.[81] It might seem that this amounts to nothing more than tradition or resistance to change. Alternatively, Dwight Billings has recently portrayed the tobacco manufacturers as pure, profit-seeking, modern business practitioners willing to use black labor, as contrasted with the more traditional cotton-mill men who used white labor.[82]

More plausible than either of these views is the interpretation that the process of labor market development acted to spread and reinforce an initial association between race and specific skills, making the preexisting pattern the cheapest option for employers of all stripes. Tobacco factories employed blacks, but they were firmly segregated. The tasks that blacks performed were commonly classed as "unskilled," but as in most work, there was wide scope for developing dexterity and diligence. There were any number of subtle bits of know-how and "feel" for the leaves and the routine. Certainly workers had to accustom themselves to the smell and atmosphere of the tobacco rooms. These effects were surely of some importance, if we are to explain the otherwise extraordinary behavior of manufacturers in new centers of production who recruited and paid the transport costs of black workers

---

[80] See the useful study by Leonard Carlson, "Labor Supply, the Acquisition of Skills, and the Location of Southern Textile Mills, 1800–1900," *Journal of Economic History* 41 (March 1981), pp. 65–71.

[81] A. A. Taylor, *The Negro in the Reconstruction of Virginia* (New York 1926), pp. 117–118; 1880 Census, vol. 20, p. 40; Alan Bruce Bromberg, "Slavery in the Virginia Tobacco Factories, 1800–1860" (Ph.D. thesis, University of Virginia, 1968); Frenise A. Logan, *The Negro in North Carolina, 1876–1894*, (University of North Carolina Press, Chapel Hill, 1964).

[82] Billings, *Planters and the Making of the "New South,"* pp. 113–120.

with previous experience. An episode recounted by Nannie Tilley is enlightening:

> With the rise of the industry in the coastal plain and the absence of labor accustomed to the work of redrying plants, operators solved the problem by collecting employees in old tobacco centers and paying for their transportation to the new plants. One such employer, Hoge Irvine, experienced considerable inconvenience in 1900 with a group of stemmers transported from Danville to Kinston. Amusement at Irvine's predicament from fellow tobacconists indicates a situation not unusual. "After paying their way here, he was very much chagrined to find that fourteen of them had skipped. He sent over to Greenville in search of his stemming tourists, and returned last night with eight of them, as happy as distillery hogs and in no wise ashamed of seeking to leave him in the hole. Hoge is wrathy, and says he will get the others."[83]

Is it surprising then that segregation persisted on an industry rather than a geographic basis? This was the way by which the labor market spread. Although the story suggests the difficulties of working on the fringe of an established market, over time the process of mobility through the market merely served to reconfirm the existing segregation, whereas an immobile labor force might have required change. This all happened in what was unmistakably a market context, where workers could come and go as they pleased. Employers paid so little attention to their employees' individual identity that when one new worker dropped dead in the Phinnix prize house in Durham, no one knew his name or place of residence.[84]

The problems of racial equity in the industrial labor market, to be sure, were not merely the result of some accidents of historical timing. The lowly status of most blacks reflected the legacy of slavery, the hostility of white workers, unequal educaton, uncertain legal rights, and the lack of political power, as so many students of southern history have rightly stressed. But the black's plight did not arise from labor immobility or an absence of labor markets; the functioning of vigorous, competitive labor markets and the high mobility of the labor force actually reinforced established patterns and made it economically *difficult* to deviate from them. When money-hungry tobacco manufacturers like Julian Carr expanded into cotton textiles, they used white labor. And when northern manufacturers came south in the twentieth century, with few exceptions they followed local segregation

---

[83] Nannie Tilley, *The Bright-Tobacco Industry* (University of North Carolina Press, Chapel Hill, 1948), pp. 320–321, quoting the *Kinston Free Press* for Sept. 25, 1900.
[84] Ibid., p. 321.

practices, until strong political pressures forced a change. In Alabama, with one of the longest industrial histories in the South, a survey of firms in all major branches of the economy found not a single case before the 1960s where management,

> drawing on cost calculations, business norms, or some abstract concept of justice, chose to desegregate the work place or break down job discrimination. . . . Even in retrospect, off the record, within the confines of their own offices, businessmen did not recall that the racial order created any "impediments" or "difficulties" for their enterprises.[85]

[85] Greenberg, *Race and State in Capitalist Development*, pp. 231–233.

5
The Gender Gap in Historical Perspective

CLAUDIA GOLDIN

> "When I was asked to prepare a paper . . . upon the alleged differences in the wages paid to men and to women for similar work, I felt very reluctant to undertake the task. . . . The problem is apparently one of great complexity, and no simple or universal solution of it can be offered."—Sidney Webb
>
> "Should men and women receive equal pay for equal work? This question is in a peculiar degree perplexed by difficulties that are characteristic of economic science."—F. Y. Edgeworth

This paper explores long-run changes in the relative earnings of females compared with those of males and in the variables that determine a ratio known as the "gender gap." The ratio of female to male full-time earnings has been remarkably stable over the last thirty-five years, hovering just under 0.60 (0.66 adjusted for hours of work), with a mild decline in the early to middle 1950s and a recent rise beginning around 1980.[1] But data on the earnings of males and females are readily available only for the period since 1950, with their publication in the Current Population Reports. Has the gender gap remained as remarkably stable over a longer sweep of American history?

Long-run changes in technology, work organization, educational standards, social norms, and life-cycle labor force participation might be expected to increase the ratio of female to male earnings. The labor market's rewards to strength and dexterity should be minimized by the adoption of machinery and the replacement of human physical labor with inanimate power. Formal education, supplied by the employee, should replace on-the-job training, which might be denied individuals who appear less able or have high labor market turnover. As more

---

I would like to thank Moses Abramovitz for his helpful comments. An earlier version of this paper was presented to the United States Commission on Civil Rights Consultation on Comparable Worth, June 1984.

The Webb quote is from "On the Alleged Differences in the Wages Paid to Men and Women for Similar Work," *Economic Journal* 1 (Dec. 1891).

The Edgeworth quote is from "Equal Pay to Men and Women for Equal Work," *Economic Journal* 32 (Dec. 1922).

[1] On recent trends in the gender gap, see James P. Smith and Michael P. Ward, *Women's Wages and Work in the Twentieth Century* (Santa Monica: Rand Corp., 1984); June O'Neill, "The Trend in the Male-Female Wage Gap in the United States," *Journal of Labor Economics* 3 (Jan. 1984), S91–S116.

women enter and remain in the labor market, their experience in jobs and with firms will approach that of the male labor force. Economic progress, it seems, should narrow and eventually eliminate differences in the earnings of females and males. Short-run data for the past three to four decades do not appear to be consistent with this depiction, but are longer-run historical data?

The increase in female labor force participation over the last few decades has been a signal of social change to many, but its significance has been called into question by the persistence of large differences in the earnings and occupations of males and females. Can the absence of change in the gender gap and occupational structure be reconciled with the increase in female labor force participation rates? Furthermore, why have substantial increases in educational norms over the last several decades apparently not been reflected in a narrowing of the gap?

This paper examines four related topics: (1) the history of the ratio of female to male earnings; (2) an analysis of the ratio at various points in time; (3) an explanation for changes in the ratio over time to determine whether economic progress has narrowed the gender gap; and (4) the reasons for the relative constancy of the gender gap over the past thirty-five years. The magnitude of the gender gap at any one point in time has been the subject of a lengthy and inconclusive literature, and it is only briefly discussed here.[2]

The implicit framework in the analysis that follows is one of an evolving labor market in which skills, education, strength, and job experience are differentially rewarded across occupations. The economy is initially an agricultural one, in which the type of crop is a major determinant of the relative productivity of females, males, and children. Females and the young can be quite effectively employed in working crops such as cotton, rice, and tobacco, while they are at more of a disadvantage in others, such as grains.[3] Home production of manufactured goods and crafts can coexist with agriculture, and the close association of the home and the miniaturized factory encourages all family members to acquire various skills.

[2] Contrast, for example, the findings in Donald J. Treiman and Heidi Hartmann, eds., *Women, Work and Wages: Equal Pay for Jobs of Equal Value* (Washington, D.C.: National Academy Press, 1981) with those in Solomon Polachek, "Women in the Economy: Perspectives on Gender Inequality" (Paper presented to the U.S. Commission on Civil Rights, Consultation on Comparable Worth, June 1984).

[3] These ideas are more clearly set forth in Claudia Goldin and Kenneth Sokoloff, "The Relative Productivity Hypothesis in Industrialization: The American Case, 1820 to 1850," *Quarterly Journal of Economics* 99 (Aug. 1984), pp. 461–487; Claudia Goldin and Kenneth Sokoloff, "Women, Children, and Industrialization: Evidence from the Manufacturing Censuses," *Journal of Economic History* 42 (Dec. 1982), pp. 741–774.

The mechanization of factory production affects the relative earnings of females in a variety of ways. The more intricate division of labor and the replacement of human strength with inanimate power favors females and serves to raise their relative productivity. But the separation of home and market increases the costs of acquiring various skills and puts decision making more in the control of the employer and less in the hands of the family and individual. Thus the initial adoption of the factory system would be expected to increase the relative earnings of females, especially in the agricultural areas in which they were initially at a disadvantage. But certain types of industrial skills might now be acquired only on the job, and in these areas we would expect women, with their low labor force attachment, to gain relatively less.[4]

The widening of the economic marketplace, the further introduction of labor-saving technological advances, and the increase in per capita income all lead to an increased demand for more specialized skills, in particular those of clerical workers and professionals. The rise of the tertiary sector is a common feature in the development of all economies, and within this sector, the relative increase in the clerical trades generally precedes that in the professions. The mechanization of the office enables workers to enter a trade for which learning on the job is of less importance than off-job education, and in which the pay and working conditions are relatively good. In the first three decades of this century, the attributes of these positions were attractive to young women whose labor market attachment was relatively weak.[5] The shift in the occupational structure toward clerical jobs would be expected to raise the relative wage of females.

The increase in professional jobs, however, might have had an opposite effect, at least until women greatly expanded their life-cycle labor force participation. Professional jobs combined the attributes of both clerical and craft positions; they required a high level of education but also rewarded job experience. A substantial stock of knowledge

---

[4] Claudia Goldin, "The Changing Status of Women in the Economy of the Early Republic: Quantitative Evidence," *Journal of Interdisciplinary History* 16 (Winter 1986, pp. 374–404), presents data on the changing occupational distribution of widows from 1790 to 1860 as market production moved from the home to larger establishments. Women held numerous atypical positions, either of a craft or retail nature, in Philadelphia in the 1790s, but were far less frequently found in such positions in the nineteenth century. The increased separation of home and market appears to be the reason for the decline of women in these trades.

[5] Claudia Goldin, "The Historical Evolution of Female Earnings Functions and Occupations," *Explorations in Economic History* 21 (Jan. 1984), pp. 1–27, presents evidence on the role of educational advances in the increase in female clerical workers in the first few decades.

had to be brought to the job, but further amounts were acquired in the marketplace.

When the labor force participation of women was low and their attachment to the work force was weak, increases in the division of labor in both manufacturing and the clerical sector brought an increased demand for their services. The reasons for this increased demand are to be found in the theory of human capital, as well as in models of monitoring and supervisory costs. In the 1920s, women first began to extend their labor force participation into the period of marriage. But because change in the labor force is generally accomplished by the movement of cohorts through time—what John Durand has cogently termed a "succession of generations"—and because the 1930s and 1940s were punctuated by economic abnormalities, these changes resurfaced as late as the 1950s.[6]

There are complex interrelationships among technological change, economic development, and the ratio of female to male earnings and isolating each in a simultaneously determined model would be a difficult task. Increases in education, for example, may have resulted from forces outside the labor market, or may have come in response to heightened demands for skilled workers. Similarly, changes in women's life-cycle labor force participation are probably related to the options available for them in the labor market. It may not be an accident that women first began to increase their labor market involvement with the initial rise of the clerical sector and have entered the labor force in even larger numbers after the shift of male workers into the professions. Despite the absence of an explicit formulation of the underlying structure, the framework just sketched will be of use in understanding both changes in the ratio of female to male earnings and the evolution of contemporary issues regarding gender in the labor market.

Current concern with the ratio of female to male earnings is based on the notion that the gender gap in earnings is a measure of discrimination against women. As the quotations that open this essay suggest, the gender gap was a topic of political and academic interest a century ago when the British Fabian Sidney Webb presented an early rendition of the doctrine of comparable worth, reformulated thirty years later by Edgeworth. The list of economists involved in the initial debate over the "women question" is impressive, with J. S. Mill the leading figure, chronologically as well as intellectually, later joined by Bowley, Cassel, Edgeworth, Rowntree, S. Webb, and the three female economists Fawcett, Rathbone, and Beatrice Webb.

---

[6] John D. Durand, *The Labor Force in the United States: 1890–1960* (New York: Social Science Research Council, 1948), p. 123.

The Gender Gap

British concern with gender differences in wages can be traced to a special combination of factors. Economic thought was influenced by Mill, who had personal and intellectual ties to those committed to equality between the sexes. Edgeworth's concern was more with allocative efficiency in the labor market than with egalitarianism. Equality of treatment (not equality of wages or occupations) between the sexes in the labor market was a part of the laissez faire ideology of Adam Smith. British trade unions and discriminatory social norms and customs were impediments, according to Edgeworth, as costly to an economy as protectionism and monopolies.

Across the Atlantic, at the time, there was far less interest in equal pay between the sexes and equality of treatment in the labor market. Equal pay had become an issue during the First World War and was earlier supported by progressive economists, such as Richart T. Ely. Carroll Wright, the prolific Commissioner of Labor, compiled an impressive array of statistics in 1897 on whether women were paid less than were men for jobs of equal productivity.[7] Edith Abbott, among others at the turn of this century, was concerned with the standard of living of working-class women, but not with the ratio of female to male wages. Despite the interest of Ely and Wright, Americans were, by British standards, apathetic on the "women issue."[8]

The current literature on the gender gap has, with few exceptions, neglected the historical record, continuing America's curious history of apathy on the women issue. Stanley Lebergott's exploration of the ratio of female to male wages in the nascent industrial Northeast provides one notable exception.[9] The present paper furthers the task of completing the historical record.

The doctrine of equal pay for jobs of comparable value has already been adopted by several states and municipalities. The remedy is rooted in the frustrations of individuals with the constancy of the gender gap.

---

[7] See U.S. Commissioner of Labor, *Eleventh Annual Report of the Commissioner of Labor, 1895/96: Work and Wages of Men, Women, and Children* (Washington, D.C., 1897). The precise term was "the same grade of efficiency." These data are used throughout this research especially in the analysis of the gender gap in manufacturing employment in the nineteenth century.

[8] The apathy of American academics and policy makers was explained by the British as a product of our general equality of wages, produced by weak trade unions and less rigid social custom. "Custom is presumably less powerful in regulating wages in the United States than in England, and in the United States the proportion which the average earnings of women in manufacturing industry bear to those of men, is . . . considerably higher than in [Britain]. . . . In the United States, on the other hand, where competition has perhaps freer play, women typewriters receive wages equal to men typewriters," Sidney Webb, "On Alleged Differences in Wages," p. 649.

[9] Stanley Lebergott, *Manpower in Economic Growth: The American Record Since 1800* (New York: McGraw-Hill, 1964).

But has the gap remained stable for more than the last three decades? What explains its level, and what has been responsible for its change or constancy over time?

I. Labor Force Participation Rates

Because earnings are so dependent on the degree of labor market involvement, it is instructive, before examining earnings data, to review the historical record regarding the labor market participation of women in the United States. Participation rates affect earnings in various ways, but each case depends on the precise meaning of participating in the labor force, which determines the relationship between labor force participation and life-cycle labor market experience. Expected life-cycle experience determines whether individuals invest in training, both on and off the job, and the stock of human capital influences monetary rewards in the labor market.[10]

Participation rates for a group can be low, but its members can remain in the labor force for long periods of time. If they do, and if they had such expectations early in their working lives, their investments in job training could be substantial. Participation rates for women have increased rapidly over time, particularly over the last four decades. The marketable skills of this emerging labor force depend on the degree to which these women worked in the past, and this in turn depends on the meaning of labor force participation.

A participation rate of, say, 50 percent can indicate that one-half of all individuals are in the labor force and one-half are not. But a participation rate of 50 percent can also indicate that all individuals are in the labor force half time. Combinations of these two extreme cases could also exist. The meaning of labor market participation is further complicated by changes, with the 1940 census, in the procedures used to compile the national labor force participation rate. Before 1940 the "gainful worker" definition was used, and after that date the "labor force" construct was employed. Under the latter definition, individuals were in the labor force if they responded positively to a question concerning the amount they worked in the previous week. Under the former definition, individuals were in the labor force if they stated that they had an occupation. Because there was no clear notion of what it meant to have an occupation, it is difficult to assess the precise mean-

---

[10] Solomon Polachek, "Differences in Expected Post-School Investment as a Determinant of Market Wage Differentials," *International Economic Review* 16 (June 1975), pp. 451–470, estimates such a model and finds that it explains almost all of the earnings gap, or about twice that of other models.

ing of the "gainful worker" data. Fortunately, other data sets provide the necessary information to distinguish between the two extreme views of labor force participation.[11]

Labor force participation rates for women have varied markedly by age, marital status, nativity, and race. Table 5.1 presents labor force participation rates by race and marital status for 1890–1980.[12] The starting point for these data, 1890, is dictated by the availability of labor force statistics in published format, although other data can be used, particularly for young single women, for the period before.[13]

The labor market involvement of white married women was very low until well into the twentieth century. Rates for single women increased steadily over time, although they were quite high in most industrial and urban areas throughout the nineteenth century. For much of American history, the labor force participation rates of all adult women were low but began to expand during the 1920s. These rates rapidly increased after 1950, first for women over age thirty-five and later for those under thirty-five.

When the data on labor force participation for adult married women are arrayed by birth cohort, the increase in participation rates over time is reflected in average labor market life-cycle experiences. For every cohort of women within their married years, participation rates rose with age, with younger cohorts of women having progressively increased participation rates.[14] Some cohorts, such as those born around 1906–1915 and 1946–1955, had larger increases in participation rates than those preceding them. But all cohorts experienced similar changes across their own life cycles and had participation rates that were higher than those before.

Three aspects of these data, together with the relationship between

[11] Claudia Goldin, "The Female Labor Force and Economic Growth in the United States, 1890 to 1980," in *Long Term Trends in the American Economy,* Stanley L. Engerman and Robert Gallman, eds. (Chicago: University of Chicago Press, forthcoming), contains a discussion of the two labor force concepts; see also Durand, *Labor Force in the United States,* on changes in the definition of the labor force. Claudia Goldin, "Life-Cycle Labor Force Participation of Married Women: Historical Evidence and Implications," National Bureau of Economic Research Working Paper no. 1251 (Cambridge, Mass., Dec. 1983) includes estimates of life-cycle labor force experience for married female cohorts.

[12] More complete data on female labor participation rates can be found in Goldin, "Female Labor Force and Economic Growth."

[13] Goldin and Sokoloff, "Women, Children, and Industrialization," presents a "manufacturing labor force participation rate" for young, single women for 1832–1880 in the American Northeast.

[14] For a more detailed description and analysis of the cohort labor force data see Claudia Goldin, "The Changing Economic Role of Women: A Quantitative Approach," *Journal of Interdisciplinary History* 13 (Spring 1983), pp. 707–733.

TABLE 5.1
Female Labor Force Participation Rates, by Marital Status, Race, and Nativity, 1890–1980

	Sixteen years and over					Fifteen years and over			Sixteen years and over	
	1890	1900[a]	1920	1930	1940	1950	1960	1970	1980	
Total[b]	18.9%	20.6%	23.7%	24.8%	25.8%	29.0%	34.5%	42.6%	51.5%	(49.9%)
Married	4.6	5.6	9.0	11.7	13.8	21.6	30.7	40.8	50.1	(49.2)
Single	40.5	43.5	46.4	50.5	45.5	46.3	42.9	53.0	61.5	
White	16.3	17.9	21.6	23.7	24.5	28.1	33.7	41.9	(49.4)	
Married	2.5	3.2	6.5	9.8	12.5	20.7	29.8	39.7	49.3	(48.1)
Single	38.4	41.5	45.0	48.7	45.9	47.5	43.9	54.5	64.2	
Nonwhite	39.7	43.2	43.1	43.3	37.6	37.1	41.7	48.5	(53.3)	
Married	22.5	26.0	32.5	33.2	27.3	31.8	40.6	52.5	59.0	(60.5)
Single	59.5	60.5	58.8	52.1	41.9	36.1	35.8	43.6	49.4	
Foreign born	19.8			19.1						
Married	3.0			8.5						
Single	70.8			73.8						

SOURCES: Claudia Goldin, "Female Labor Force Participation: The Origin of Black and White Differences, 1870 to 1880," *Journal of Economic History* 37 (March 1977), Table 1. The 1980 data are from U.S. Department of Labor, *Marital and Family Characteristics of the Labor Force*, March 1970, Special Labor Force Report 130 (Washington, D.C., 1982), and are the Current Population Survey figures. Those in parentheses are from U.S. Bureau of the Census, 1980 *Census of Population*, vol. 1, Ch. C, "General, Social and Economic Characteristics," part 1, United States Summary (Washington, D.C., 1983), and are the Population Census figures.

[a] The 1910 labor force figures have been omitted. See text for a discussion of the overcount of the agricultural labor force in that year.

[b] When adjusted for unemployment, by subtracting out the unemployed, the totals are 19.0% for 15-to-64-year-olds in 1890 and 55.4% for 16-to-64-year-olds in 1980.

participation and life-cycle labor force experience, affect the ratio of female to male earnings and changes in the ratio over time. Because participation rates for adult women were low until the relatively recent past, most women and their families would not have found it profitable to invest in job training. Therefore the earnings and occupations of these women could be expected to have differed considerably from those of men, even when these women were young and had high participation rates.

With increased participation rates for adult women, it is necessary to understand whether more women were participating or whether the same women were participating more. It appears that while some combination of these two extreme views is most accurate, a large proportion of women who participated in the labor force when young continued to do so. As participation rates increased over time, women with little labor force experience entered the market, joining those who had accumulated more experience.

Furthermore, because each cohort's participation rates exceeded each previous one's, all women may have had difficulty predicting their own future labor force participation rates. Each cohort, when young, may have extrapolated from the experiences of their elders and thereby underestimated their own future labor force participation rates. The implications of these suggestive remarks are explored further below.

II. Earnings of Females Relative to Those of Males

The Agricultural and Manufacturing Sectors, 1815–1970

The story of relative earnings can begin with data from the agricultural and manufacturing sectors of almost two centuries ago. Earnings ratios for the entire economy, however, can be constructed only for the last century, and only with caution for the pre-1950 period.

The wage of females relative to males was fairly low in the northeastern states prior to industrialization but rose quickly wherever manufacturing activity spread.[15] Around 1815, the ratio of female to male wages in agriculture and domestic activities was 0.288, and it rose to about 0.303–0.371 among manufacturing establishments at the inception of industrialization in the United States in 1820. By 1832 the average ratio in manufacturing was about 0.44, and continuing its rise, it reached just below 0.50 in the northeastern states by 1850. Nation-

---

[15] Goldin and Sokoloff, "Women, Children, and Industrialization," and Goldin and Sokoloff, "Relative Productivity Hypothesis of Industrialization," both present evidence on the increase in relative wages with industrialization; the latter paper models the causes of the increase.

TABLE 5.2
Wage Ratios for Males and Females in Manufacturing Employment, 1815–1970 and Across All Occupations, 1955–1983
(Except where noted these ratios are based on full-time, year-round employees.)

Agriculture, 1815	Manufacturing						
	1820	1832	1850	1885	1890a	1890b	
0.30	0.37–0.30	0.44–0.43	0.46–0.50	0.559	0.539	0.538	

SOURCES:
1815–1850: Claudia Goldin and Kenneth Sokoloff, "Women, Children, and Industrialization in the Early Republic: Evidence from the Manufacturing Censuses," *Journal of Economic History* 42 (Dec. 1982), Table 5. The range is for New England and the Middle Atlantic. The (b) results from Table 5 are given, which use Lebergott's male common laborer wage as the base.
1885: Clarence Long, *Wages and Earnings in the United States: 1860–1890* (Princeton: Princeton University Press, 1960), p. 146, from First Report of the U.S. Commissioner of Labor, daily wages.
1890a: Ibid., p. 148, from Dewey, actual wages used.
1890b: U.S. Census Office, *Report on Manufacturing Industries in the United States at the Eleventh Census: 1890*, part I, "Totals for States and Industries" (Washington, D.C., 1895), actual wages used.

Year	Full-time	Actual	Full-time Weekly	Full-time Hourly
1899	0.535	0.536		
1904	0.536	0.535		
1909	0.536	0.537		
1914	0.535	0.534	0.568	0.592
1921	0.536	0.536	0.617	0.653
1925	0.536	0.536	0.592	0.657
1930			0.578	0.635
1935			0.653	0.700

SOURCES: First two columns, Paul F. Brissenden, *Earnings of Factory Workers, 1899 to 1927: An Analysis of Pay-roll Statistics* (Washington, D.C.: Government Printing Office, 1929), Table 33, p. 85. Second two columns, M. Ada Beney, *Wages, Hours, and Employment in the United States, 1914–1936*, National Industrial Conference Board Study no. 229 (New York, 1936), Table 3, pp. 48–51.

Year	Manufacturing Full-time	Manufacturing Total	All occupations Median, year-round	All occupations Median, weekly Actual	All occupations Median, weekly Hours-adjusted
1939	0.539	0.513			
1950		0.537			
1955	0.580	0.526	0.639		
1960	0.559		0.608		
1965	0.532		0.600		
1970	0.540		0.594		
1975			0.588	0.62	0.68
1980			0.602		
1983				0.66	0.72

SOURCES:
Manufacturing: *Historical Statistics*, G 372–415, pp. 304–305. Female earnings for operatives were multiplied by 1.02 to adjust for craft and supervisory positions where such data were unavailable. Male earnings were weighted equally between craft and operative positions, consistent with the labor force percentages.
All Occupations: June O'Neill, "The Trend in the Male-Female Wage Gap in the United States," *Journal of Labor Economics* 3 (Jan. 1985), Tables 1 and 3. The difference between the year-round and the weekly data is primarily the exclusion of teachers and other less-than-year-round workers from the former. Median earnings of weekly workers are higher for women than for men because of the higher-than-average earnings of female teachers. Both sets of data are from the Current Population Surveys.

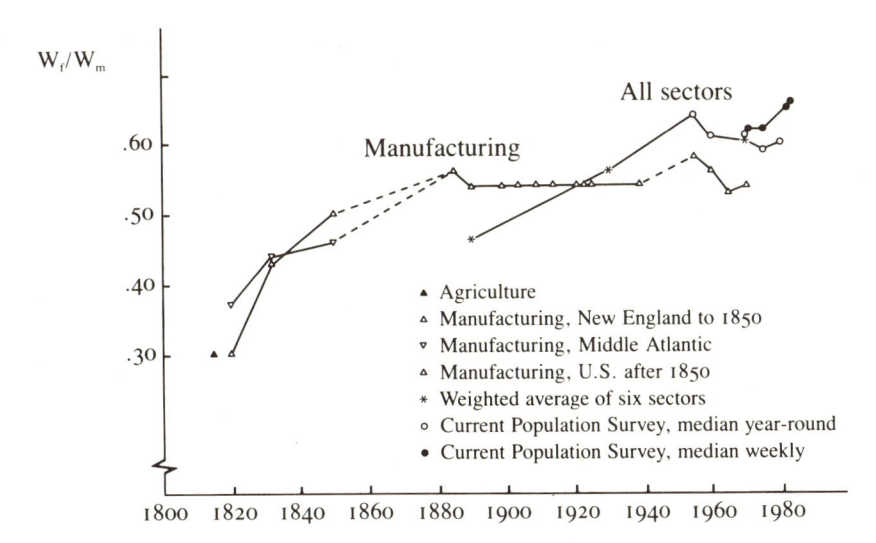

The Gender Gap

W_f/W_m

- ▲ Agriculture
- △ Manufacturing, New England to 1850
- ▽ Manufacturing, Middle Atlantic
- ▵ Manufacturing, U.S. after 1850
- ✱ Weighted average of six sectors
- ○ Current Population Survey, median year-round
- ● Current Population Survey, median weekly

SOURCES: See Tables 5.2 and 5.3.

FIGURE 5.1 The Gender Gap in Historical Perspective: The Manufacturing Sector and the Entire Labor Force, United States, 1815–1983

wide, the ratio rose slowly until about 1885, when it reached its current value of about 0.56 (see Table 5.2 and Figure 5.1). Why the ratio has been virtually stable for the last hundred years is somewhat of a puzzle, but the cause of the earlier rapid and steady increase in relative wages seems clear.

The agricultural sector of the Northeast was primarily a grain-growing area, and its farm work was more strenuous than that in the cotton-growing areas of the South. Adult women and young children rarely worked in the fields of the Northeast, while women and children, both black and white, were often employed in the planting and harvesting of cotton. "In New England," note Bidwell and Falconer from their extensive reading of travelers' reports, "only men as a rule were to be seen in the fields, the women of the family assisting only occasionally in harvest time."[16]

Manufacturing interests in the Northeast took full advantage of the large and relatively elastic supply of female and child labor in their use of machinery and intricate division of labor. The work could be learned quickly and performed by individuals of limited strength. As

[16] Percy W. Bidwell and John F. Falconer, *History of Agriculture in the Northern United States, 1620–1860* (Washington, D.C.: The Carnegie Institution of Washington, 1925), p. 116.

early as 1832, fully 40 percent of the manufacturing labor force in the Northeast was comprised of females and children, a figure that began to decline soon thereafter.[17] In the American South, where women and children were, compared with men, relatively more productive in cotton growing than they were in growing the grains of the North, industrial development was far less extensive and used considerably less female and child labor. The ratio of female to male wages in southern agriculture was 0.58 among free workers in the immediate postbellum period, a figure approximately equal to that across all workers today.[18]

Early industrialization, therefore, increased the wage of females relative to males by almost 50 percent (from 0.288 to 0.440). In the briefest of periods, a mere two decades, the gender gap in manufacturing and domestic employment narrowed by 15 percentage points. The gap continued to narrow, by 12 percentage points up to about 1885, and has remained at about 0.56 ever since. The magnitude and implications of the initial advance are sufficiently important to warrant further attention.

Quantitative evidence on wages in the pre-industrial period is given in Table 5.2 and graphed in Figure 5.1, but the best evidence on the relative increase in female wages with the coming of industrialization is to be found in the narrative accounts of contemporaries.[19] Those that are most useful bear on the period prior to 1820, but of equal interest are those in the 1832 McLane Report from areas geographically isolated from industrial development.

Perhaps the best-known commentary on the relative productivity of females in the pre-industrial period and on the opportunities in manufacturing for their employment is that of Alexander Hamilton. "In general, women and children [would be] rendered more useful, and the latter more early useful, by manufacturing establishments than they would otherwise be."[20] These notions were echoed by another Secretary of the Treasury. Albert Gallatin knew in 1831, far better than Hamilton could have imagined in 1791, that "female labor employed in

[17] Goldin and Sokoloff, "Women, Children, and Industrialization," Table 1.

[18] Goldin and Sokoloff, "Relative Productivity Hypothesis."

[19] It should also be noted that certain quantitative evidence may appear to contradict the thesis that the ratio of female to male wages rose rapidly and significantly with early industrialization around the period from 1800 to 1830. Donald Adams, "The Standard of Living During American Industrialization: Evidence from the Brandywine Region, 1800–1860," *Journal of Economic History* 42 (Dec. 1982), pp. 903–917, demonstrates that the ratio of female to male wages was, on average, 0.433 as early as 1815–1824. But the Brandywine area industrialized very early.

[20] Alexander Hamilton, "Report on Manufactures," reprinted in F. W. Taussig, ed., *State Papers and Speeches on the Tariff* (Cambridge, Mass.: Harvard University Press, 1892), p. 9.

the cotton and woollen [sic] manufactures appears from the rate of their wages to be more productive than applied to the ordinary occupation of women."[21] Henry Carey, whose essay on wage rates appeared in 1835, noted that

> agricultural labor has not varied materially in these forty years [1793 to 1833] in its money price . . . the wages of men having been very steadily about nine dollars per month [with board] . . . [but] the wages of females have greatly advanced being nearly double what they were forty years since.[22]

The observations of the commentators who lived through the transitionary times of the early nineteenth century support the fragile quantitative evidence that the relative wage of females rose considerably over this period. But these learned individuals may not have been entirely uninterested in the impact industrial development would have on particular groups, including female laborers.

Additional evidence is readily available from the rather ordinary individuals surveyed by the McLane Report of 1832. This extraordinary source contains information on the period of transition and on relative wages in areas yet untouched by industrial development. One Aaron Tufts, in Dudley, Massachusetts, noted in his schedule that, "Comparatively nothing is done in the household manufactory: a female can now earn more cloth in a day than she could make in the household way in a week."[23] The Oaksville Manufactory of Cotton Goods, in Otsego, New York, noted that "The wages of females are 75 cents per week . . . but if those now engaged in the factories were thrown out of employ, wages would probably be reduced below that sum."[24] One fairly typical McLane Report respondent referred to the factories as affording the employment of "females who had little else to do."[25] Thus the commentary of the exceptional individuals who lived through these transitionary decades is corroborated by the many respondents to the McLane Report, all of whom support the quantitative evidence on the increase in the ratio of female to male wages from around 1800 to 1830.

Relative wages continued to rise in the manufacturing sector across

[21] Albert Gallatin, "Free Trade Memorial," reprinted in ibid., p. 129.
[22] Henry C. Carey, *Essay on the Rate of Wages: With an Examination of the Causes of the Differences in the Condition of the Labouring Populations Throughout the World* (Philadelphia, 1835), p. 26.
[23] U.S., Congress, House, *Documents Relative to the Statistics of Manufactures in the U.S.*, vol. 1, serial set no. 222 (Washington, D.C., 1833), p. 69. The two volumes of this report (serial set nos. 222 and 223) are commonly referred to as the *McLane Report*, after the then Secretary of the Treasury, Louis McLane.
[24] Ibid., vol. 2, p. 20.
[25] Ibid., vol. 1, p. 819.

most of the nineteenth century.[26] But the ratio stabilized sometime before 1900 and has remained constant ever since, as can be seen in Table 5.2 and the accompanying Figure 5.1. The only indications of an increase in the ratio are the data prepared by M. Ada Beney for the National Industrial Conference Board report, particularly those from the Depression, and the data for the immediate post-World War II years. The Beney data appear to produce a somewhat inflated ratio in comparison with the Brissenden data, which are consistent with those from the 1890 Census of Manufacturing, the Dewey Report from Long, and the First Annual Report of Commissioner of Labor, also from Long, for 1885.[27] The years of overlap between the Beney and Brissenden data, the early 1920s, suggest that the Beney ratios are inflated by about 10 percent.[28]

The immediate and widespread employment of women in manufacturing establishments during the early to mid-nineteenth century does not imply that their occupations or industries were identical to those of men. Occupations were almost always segregated by sex, and when they were not, incentive pay, generally piece-rate payment, was often used.[29] Furthermore, particular industries hired virtually no females. The segregation of occupations by sex within manufacturing and the absence of females in a wide range of industries suggest that part of the proximate cause of the gender gap in manufacturing might be the allocation of female labor across industries.

In 1890, over 50 percent of all male manufacturing workers were employed in twenty-two industries in which there were virtually no female employees. More than 80 percent of all female manufacturing workers were employed in just twenty-four industries. These twenty-four industries were almost uniformly those considered "female-

[26] Data from Adams, "Standard of Living during American Industrialization," for the early-industrializing Brandywine region indicate that the ratio of female to male wages rose from 0.433 during 1815–1824 to 0.536 in the 1850s.

[27] M. Ada Beney, *Wages, Hours, and Employment in the United States, 1914–1936* (New York: National Industrial Conference Board, 1936); Paul F. Brissenden, *Earnings of Factory Workers, 1899 to 1927: An Analysis of Pay-roll Statistics*, Census Monographs, no. 10 (Washington, D.C.: Government Printing Office 1929); Clarence D. Long, *Wages and Earnings in the United States, 1860–1890* (Princeton: Princeton University Press for the NBER 1960).

[28] The reasons for the inflated ratio in the Beney data probably concern the industries surveyed. Although the Brissenden data are consistent with the somewhat earlier ratios, they are virtually stable from 1899 to 1925. The Beney ratio rises in the immediate post-World War I period and then declines somewhat, a pattern that is consistent with the general rise in the unskilled to skilled wage ratio in that period.

[29] Claudia Goldin, "Monitoring and Occupational Segregation by Sex: An Historical Analysis," *Journal of Labor Economics* 4 (Jan. 1986), 1–27 presents evidence on the use and role of piece-rate payment in manufacturing.

intensive," containing labor forces that were over 30 percent female while the national average was 18 percent.

Were the male-intensive industries those having the highest wages? Was sex segregation by industry a factor responsible for the gender gap within manufacturing, and was this factor of increasing importance over time? Regression equations were estimated to explain both the male and female wage by the percentage of the industry labor force that was either adult male or adult female. Two data sets were used: for 1890, forty-eight major industries from the Census of Manufacturing were selected; for 1899 and 1925, Brissenden's data on full-time wages across forty industries were used.[30]

In all three years the percentage male was not a significant explanatory variable in the male-wage equation, and the percentage female was not a significant explanatory variable in the female-wage equation. Indeed, in 1890 the male-intensive industries had a slightly lower average male wage than did the female-intensive industries, and in 1899 the coefficient on the female wage was positive (not negative) but of marginal statistical significance. Women earned less than men in manufacturing not because males constituted 99 percent of the labor force in iron and steel, agricultural implements, shipbuilding, and masonry. Women earned proportionately less even in the industries in which they were very numerous, such as boots and shoes, cotton, woolens, boxes, and clothing.

One other aspect of the manufacturing data in Table 5.2 should be noted. Beney's data for 1914–1935 indicate that the ratio of hourly wages in manufacturing was more than 10 percent higher than that for weekly or annual earnings because women in manufacturing worked a smaller number of hours per week. O'Neill reports similar findings for more recent data.[31] There is little indication, however, that hours worked differed for the earliest years being considered. Thus the increase in the ratio, corrected for hours worked, of female to male wages in manufacturing is somewhat understated by the uncorrected figures in Table 5.2.

Earnings Ratios in the Aggregate and for Six Sectors, 1890–1983

Data from the manufacturing sector have provided nearly two centuries of information on the gender gap, indicating that it narrowed to about the turn of this century and then remained constant. But the manufac-

---

[30] Brissenden, *Earnings of Factory Workers*.
[31] June O'Neill, "The Trend in Sex Differentials in Wages," Xerox (Washington, D.C.: Urban Institute, March 1983) p. 9; June O'Neill, "Statement on the Consultation on Comparable Worth" (Paper presented to the U.S. Commission in Civil Rights, Consultation on Comparable Worth, June 1984).

TABLE 5.3
Full-Time Earnings (Current $) and Occupational Distributions of the Female and Male Labor Forces, 1890, 1930, and 1970, Entire United States

	1890 Male $	%	1890 Female $	%	1930 Male $	%	1930 Female $	%	1970 Male $	%	1970 Female $	%
Professional	1391	10.2%	366	9.6%	3713	13.6%	1428	16.5%	12250	24.9%	8700	18.9%
Clerical	943	2.8	459	4.0	1566	5.5	1105	20.9	8750	7.6	6000	34.5
Sales	766	4.6	456	4.3	1580	6.1	959	6.8	10150	6.8	4450	7.4
Manual	587	37.6	314	27.7	1532	45.2	881	19.8	8891	48.1	4950	17.9
Craft, supervisory		(12.6)		(1.4)		(16.2)		(1.0)		(21.3)		(1.8)
Operative, laborer		(25.0)		(26.3)		(29.0)		(18.8)		(26.8)		(16.1)
Service	445	3.1	236	35.5	1220	4.8	730	27.5	7100	8.2	3965	20.5
Farm	445	41.7	236	19.0	1220	24.8	730	8.4	7050	4.5	4151	0.8

NOTE: Sources and notes for this table will be found in the Appendix, pp. 168–170.

The Gender Gap

TABLE 5.4
Ratios of Female to Male Earnings, 1890, 1930, and 1970

	1890	1930	1970

A: THE RATIO OF FEMALE TO MALE EARNINGS WITHIN EACH OCCUPATION

	1890	1930	1970
Professional	0.263	0.385	0.710
Clerical	0.487	0.706	0.686
Sales	0.595	0.607	0.438
Manual	0.535	0.575	0.557
Service	0.530	0.598	0.558
Farm	0.530	0.598	0.589

B: MALE AND FEMALE EARNINGS IN CURRENT DOLLARS (θ = OCCUPATIONAL SHARE)

	Male	Female	Male	Female	Male	Female
$\sum \theta_i w_i$	624	275	1741	968	9581	5776
$\sum \theta_i w_{1890}$	624	289	683	325	809	368
$\sum \theta_i w_{1930}$	1618	864	1741	968	2043	1035
$\sum \theta_i w_{1970}$	8306	4834	8874	5411	9581	5776

C: RATIOS OF FEMALE TO MALE EARNINGS (θ VARIES ACROSS THE COLUMNS)

	1890	1930	1970
(1) $[w_{fi}/w_{mi}]$	0.463	0.556	0.603
(2) $[w_f/w_m]_{1890}$	0.463	0.489	0.455
(3) $[w_f/w_m]_{1930}$	0.534	0.556	0.507
(4) $[w_f/w_m]_{1970}$	0.571	0.610	0.603

SOURCE: Table 5.3.

turing sector hired only one-third of all female employees at any time over the last century. It becomes necessary, therefore, to construct earnings data for a wider range of occupations. These constructed earnings data cannot extend to the early nineteenth century, as could those in manufacturing, but they do indicate that the gender gap across all sectors narrowed from 1890 to about 1940, despite the apparent constancy in the gap in the manufacturing sector alone.

Full-time earnings for females and males are given in Table 5.3 for six major occupational groupings for three benchmark years, 1890, 1930, and 1970. In Table 5.4 average earnings are constructed by weighting by the occupational distributions, and ratios of female to male earnings for each and across all occupations are presented.

The ratio of female to male full-time earnings increased from 0.463 to 0.603 over the period 1890 to 1970 (Table 5.4, Part C), or by 30 percent.[32] The latter figure is unadjusted for differences in average hours of work per day for men and women and increases to 0.657 when

---

[32] The 1970 ratio of 0.603 is a weighted average of the median earnings of various occupational groups. The ratio of the actual medians (for weekly, as opposed to year-round employment—see Table 5.2 for distinction) is 0.623 for 1970 and 0.617 for 1973, the date to which the data in Table 5.3 pertain.

TABLE 5.5
Partitioning Change in the Ratio of the Log of Female to Male Earnings, 1890 and 1970 Weights

	1890 Weights	1970 Weights
1. $\sum \theta_f(R^1 - R^0)$	+0.1452	+0.3018
2. $\sum R(\theta_f^1 - \theta_f^0)$	−0.0880	+0.0687
3. $\sum a(W_m^1 - W_m^0)$	+0.0071	−0.0981
4. $\sum W_m(a^1 - a^0)$	+0.0679	−0.0373
5. $\sum (R^1 - R^0)(\theta_f^1 - \theta_f^0)$	+0.1567	−0.1567
6. $\sum (W_m^1 - W_m^0)(a^1 - a^0)$	−0.1052	+0.1052
TOTAL CHANGE	+0.1836	+0.1836

SOURCE: Table 5.3.
NOTE: Average earnings are geometrically weighted averages of the six occupations, where $w = \sum w_i^{\theta_i}$, for males and females. A geometrically weighted average enables a partitioning of the various factors accounting for change in the ratio of female to male earnings. $W = \log(w)$; $R = (W_f - W_m)$; $a = (\theta_f - \theta_m)$; $1 = 1970$; $0 = 1890$. Note that the total change in the ratio when earnings are a geometrically weighted average is considerably less than when average earnings are the arithmetic mean. The geometrically weighted results are: $(w_f/w_m) = 0.487$, but 0.463 for the arithmetic mean in 1980; the results for 1970 are 0.586 for the geometric weights, but 0.603 for the arithmetic means. Therefore the geometrically weighted averages understate the total increase. Columns may not add up due to rounding error.

the implied earnings per hour are used.[33] The rise in relative earnings from 1890 to 1970 would understate the true rise, corrected for hours, if the hours worked by males and females per week differed more in 1970 than in the two earlier years. One late-nineteenth century study that distinguished between male and female hours indicated that the ratio of male to female weekly hours was 1.076 in 1895–96.[34] The ratio was 1.0893 in 1980. The adjustment for hours worked would increase the 1890 ratio to 0.498 (to be compared with the hours-constant ratio of 0.657 in 1970), yielding an increase of 32 percent, rather than the uncorrected figure of 30 percent. While an increase of about one-third over the eighty-year period considered may not seem to be very large, the finding overturns the notion that the gender gap was stable for a period extending into the distant past. Furthermore the gender gap closed until about 1940 and, with some ups and down, remained virtually stable up to about 1980. Thus the narrowing from 1890 by about one-third extended over only a forty-to-fifty-year period.

Data for the black population, computed in a similar manner, indicate that the earnings gap decreased to about the same degree that it

[33] June O'Neill and Rachel Braun, "Women in the Labor Market: A Survey of Issues and Policies in the United States," Xerox, Urban Institute, (Washington, D.C.: Nov. 1981), p. 60, gives a value of 1.0893 for the ratio of weekly work hours of males to females in 1980.

[34] *Report of the Commissioner of Labor, 1895/96.*

did for the entire population, from 0.484 to 0.587. But the 1970 figure of 0.587, derived by applying the aggregate earnings ratios to occupational distributions for blacks, clearly understates the figure of 0.739 computed using the actual black earnings data.[35] Black males in 1970, it appears, earned less relative to their female counterparts than did white males. It is not yet possible to extend this finding to the 1930 and 1890 data. Because over 60 percent of the black male population and 44 percent of the black female population worked in agriculture in 1890, much will depend on the ratio in that sector. Females were, compared with men, relatively more productive in cotton agriculture than they were in other agricultural employments and other tasks in general. It is possible, therefore, that the black ratio is also understated for the 1890 year.

Part A gives the ratios of female to male earnings within each occupation and indicates a rise over time, particularly, for most of the occupational groups in the period from 1890 to 1930. An exception would be the manufacturing sector, as discussed above.[36] The increase was greatest in the professional and clerical categories, for which increases in education appear to have affected both the ratio of female to male earnings and the numbers employed in these sectors.

Part B constructs aggregate earnings for each year using the earnings and occupational weights of that year. Average earnings data using the earnings of a particular year but the occupational weights of another are also given. Part C uses these data to construct a matrix of female to male earnings ratios in which the occupational structure varies across the rows and the earnings data vary down the columns.

It is generally presumed that the occupational distribution between men and women is a prime determinant of the gender gap, and that changes in the occupational distribution, therefore, provide the primary way of altering relative earnings between men and women. There are two ways of assessing such a proposition. The first concerns whether changes over time in the occupational distribution have affected the gender gap more than have other factors, such as changes in relative wages within occupations. The second is to test whether the occupational distribution is important in determining the gender gap at a particular date. If women are relegated to lower-paying occupations, then

[35] Note that none of the black population wage and occupation data is given in Table 5.3. The computed figure of 0.739 is derived from data on median weekly earnings of black males and females in 1973 in the same manner as the aggregate data in Table 5.3 were. The actual ratio of median earnings for black females and black males is 0.72 for 1973. The same figure is obtained for 1970.

[36] The 1930 figure for male manual workers must be corrected because it is based on the Beney manufacturing ratio of 0.57, which is slightly too high.

giving them the male occupational distribution should substantially increase relative earnings.

The matrix of Part C has been constructed to examine the first part of the proposition on the importance of the occupational distribution. The first row presents the actual ratio of female to male earnings over the three years. The next three rows hold the occupational distribution constant down each column and hold relative wages within occupational groups constant across each row. The ratio of female to male earnings increases going down the columns more than it does going across the rows. Thus the increase over time in relative earnings of females was due far more to changes in relative earnings within occupations than it was to changes in the distribution of occupations between men and women. The narrowing of skill premia from 1890 to 1930, with the increase in schooling levels, raised relative earnings for women more than did any other single factor.[37] This finding is particularly noteworthy, because it is generally presumed that the occupational distribution is the primary determinant of relative wages.

The ratio of female to male earnings rose from 0.463 to 0.556 over the first forty-year period. If the earnings figures by occupation had remained at their 1890 levels but the structure of occupations had changed, the ratio would have increased from 0.463 to 0.489 (row 2, Part C). Part of the remaining seven-tenths of the difference in relative earnings was due to changes in the structure of earnings, both between the sexes and across all occupations. Similar findings result from holding the structure of earnings at the 1930 and 1970 levels (rows 3 and 4, Part C).

Across the last period, 1930–1970, the male labor force moved, relatively, into the high-paying positions, out of the farm sector and into professional activities. The share of the male labor force in the professional category increased from 14 to 25 percent; that for females increased from 17 to only 19 percent, but the proportion of female employment in the clerical sector continued to expand. As in the previous forty years, the ratio of female to male earnings rose during the 1930 to 1970 period, from 0.556 to 0.603. But had the earnings figures remained at their 1930 levels, this ratio would have declined, from 0.556 to 0.507. Alternatively, had the 1970 earnings prevailed, the ratio would have been 0.610 in 1930 but would have declined to 0.603 by 1970. Thus the relative shift of both males and females across sectors from 1930 to 1970 reduced the relative earnings of women. That the

---

[37] Goldin, "Female Earnings Functions," presents evidence on the role of educational advances in increasing the supply of clerical workers during the first few decades of this century.

aggregate ratio increased at all was due to the increase in the ratio of female to male earnings for professionals and to the reduction of skill differentials for men.[38] Over the last ten years (not covered in these tables, but see Table 5.2) the average earnings of women relative to those of men have risen precisely because women have progressively shifted into the professional sector, a move previously accomplished by males from 1950 to 1970.

Sidney Webb, in 1891, had cited a difficulty, common today as well, in comparing wage-rates for men and women. It was "the impossibility of discovering any but a very few instances in which [they] do precisely similar work, in the same place and at the same epoch."[39] The data in Tables 5.3 and 5.4 are too highly aggregated to demonstrate the degree to which occupations have been segregated by sex, although it is clear that women have always been relatively more numerous in clerical occupations and men have always been more frequently found in craft and supervisory positions. The aggregate level of segregation by sex across about 350 occupations has remained constant throughout the twentieth century, and fully two-thirds of all women or all men would have had to change occupations to eliminate all distinctions by gender.[40] How has this degree of occupational segregation affected earnings of females relative to males?

If women had the occupational distribution of the male labor force, would their average earnings have been substantially greater? The answer is no. If females in 1890 had the male occupational distribution given in the table for 1890, the ratio of female to male earnings would have been 0.473, but it was actually 0.463; if females in 1970 had the male occupational distribution for 1970, the ratio would have been 0.629, but it was 0.603. While these findings hold for the limited number of occupations in Tables 5.3 and 5.4, there is reason to believe that they would hold as well for far more numerous occupational classifications.[41]

---

[38] Jeffrey Williamson and Peter Lindert, *American Inequality: A Macroeconomic History* (New York: Academic Press, 1980), presents evidence on skill differentials. See also the earlier work by Paul G. Keat, "Long Run Changes in Occupational Wage Structure, 1900–1956," *Journal of Political Economy* 68 (Dec. 1960), pp. 584–600.

[39] Webb, "On Alleged Differences in Wages," p. 638.

[40] Edward Gross, "Plus ça Change . . .? The Sexual Structure of Occupations Over Time," *Social Problems* 16 (Fall 1968), pp. 198–208.

[41] Polachek, "Women in the Economy," finds a similar result for recent data and notes that the occupational classification would have to be considerably finer to overturn the conclusion that changes in occupational structure matter less than changes in relative wages within occupations. Polachek uses 195 occupations and finds that occupational segregation explains 17 to 21 percent of the 1970 earnings gap. Following Polachek's definition of a narrowing of the earnings gap and using the data in Table 5.3 gives only 5.7 percent for 1970. This result suggests that while increasing the number

While the occupational distribution mattered far less than did earnings within occupations in determining the overall earnings ratio of females to males, the structure of occupations did experience considerable change. Over the first half of the eighty-year period under consideration, the female labor force shifted, relatively, out of service and manufacturing jobs and into the clerical sector and, to a lesser degree, professional activities. The percentage of the total female labor force in the clerical sector rose from 4 to 21, and that in professional occupations increased from just under 10 to just over 16; in manual jobs it dropped from 28 to 20 percent, and the percentage working on farms was more than halved. The male labor force experienced somewhat similar shifts, although not nearly as extensive. In general, male laborers moved out of farm activities and into all other sectors.

The matrix of Table 5.4, Part C, provides a convenient way of tracking the impact of changes in earnings and occupational structure on the earnings of females relative to males, but it is not a complete partitioning of the two factors. A full partitioning must use a geometrically weighted average of earnings by occupation for each of the three benchmark years. The use of the geometric mean can be defended on the grounds that the underlying structure of earnings is a function of the log of earnings.[42] But it is used here out of necessity, even though the geometric means are not entirely good substitutes for their arithmetic counterparts. The implied ratio of female to male earnings using the geometric means is 0.487 in 1890, rising to 0.586 in 1970, while the arithmetic means are 0.463 to 0.603.

Six terms result from the partitioning in Table 5.5; the two columns present the results using both 1890 and 1970 weights. (Note that occupation subscripts (i), over which the terms are summed, have been dropped.) The first term is the change in the ratio of female to male earnings within occupations weighted by the female share in the occupation, $\sum \theta_f (R^1 - R^0)$. The third term is the change in male earnings by occupation, weighted by the ratio of the female to male share of

---

of occupations does not overturn the conclusion of the exercise, the use of only 6 occupations is rather limiting. Treiman and Hartmann, *Women, Work, and Wages*, Table 9, presents evidence pertaining to 12, 222, and 479 occupations. Occupational segregation explains only 11 to 19 percent of the differential for 222 occupations. Although the authors claim that occupational segregation explains 35 to 39 percent of the differential for 479 occupations, there is an error in the table bringing one of the figures down to 19 percent. Furthermore, it is unclear that 479 occupations is an appropriate number. On the issue of the optimal number of occupations to use see Polachek, "Women in the Economy."

[42] See the justification for this assumption in Jacob Mincer, *Schooling, Experience, and Earnings* (New York: NBER, distributed through Columbia University Press, 1974).

employment by occupation, $\sum a(W_m^1 - W_m^0)$. The change in male earnings by occupation captures changes in skill differentials within the male labor force. Thus the first and third terms measure the portion due to relative wage changes, both within and across the male and female labor forces.

The second term is the change in the structure of female occupations weighted by the ratio of female to male earnings for each occupation, $\sum R (\theta_f^1 - \theta_f^0)$, and the fourth is the change in the ratio of the female to male employment share weighted by male earnings, $\sum W_m(a^1 - a^0)$. Thus the second and fourth terms capture changes in the structure of occupations, both within and across the male and female labor forces. The final two terms are interactions, which are added when using the 1890 weights and subtracted when using the 1970 weights.

This partitioning of the change is the earnings of females relative to males reinforces the results given in the matrix of Part C. Over the entire period from 1890 to 1970, the change in relative earnings (terms 1 and 3) encompassed 83 to 111 percent of the entire change (e.g. (0.1452 + 0.0071)/.1836 = 83 percent) depending on the weights used. However the change in structure (terms 1 and 4) added only −11 to 17 percent, with the interaction terms providing the remainder.

The largest of the first four terms, the first, demonstrates that the rise in the earnings of females relative to males within occupations greatly increased the overall ratio. The effect is greater given the structure of female occupations in 1970 than it is for the 1890 structure, as would be expected if female employment increased more in sectors experiencing greater increases in female to male earnings. The second, third, and fourth terms, while relatively small, change signs depending on the year chosen for the weights. The second term weights the change in the structure of female occupations by the ratio of female to male earnings. Females moved, relatively, into their more highly paying pursuits; thus the 1970 weights yield a positive effect and the 1890 weights a negative one. The same logic holds for the fourth term, which weights the occupational shift of females relative to males by male earnings. Females moved into those occupations that were high paying within the male earnings distribution. The third term, negative for the 1970 structure while small but positive for 1890, indicates that for the 1970 weights, male earnings increased relatively more in occupations that contained more males. In this manner it serves to diminish the effect of the first term.

The complete partitioning and the matrix demonstrate that the change in relative earnings within occupations was of greater importance in altering the overall earnings ratio than was the change in occupational structure between males and females. Both methods yield proximate determinants, or mechanical features, of the gender gap. But what are the determinants of the magnitude of the earnings gap, and why has it changed over time? Why have women earned substantially less than men in the past and why, despite a decrease in the gap, do they continue to do so?

III. Explaining the Gender Gap at Various Dates and Over Time

The magnitude of the earnings gap at various points in time is addressed first, followed by an examination of the variables accounting for its change and stability over time. There is no attempt in the second section, however, to assess the relative impact of each of the four factors—education, strength, job experience, and life-cycle labor force expectations—that are discussed.

Explaining the Gap at Various Dates

The degree to which human capital measures can explain differences in the earnings of males and females is a matter of continuing debate. It is generally agreed that around 30 to 50 percent of the gap can be explained by gender-based differences in job experience, education and hours of work, among other relevant factors. There are, however, outliers; several studies in which age is used as a proxy for job experience explain considerably less, and one study that uses a measure of expected life-cycle human capital explains almost the entire gap.[43] Differences in male and female hours of work are on the order of 10 percent, and this one factor increases the female wage in 1980 from 60 to 66 percent of the male wage.

There are many who claim that the difference between male and female earnings that remains unexplained, on average 60 percent or 24 percentage points, is due to discrimination against women, and that the figure is substantial. Others cite omitted factors under the control of individuals, such as work intensity, which could close the entire gap. Still others, moving in the opposite direction, note that the factors

[43] See Polachek, "Differences in Expected Post-School Investment," in which a measure of life-cycle human capital is developed and estimated. Inclusion of this variable in the earnings equation explains over 90 percent of the gap between married male and female workers. Polachek's method and estimation have been the subject of debate and criticism; see his "Women in the Economy" for an answer to his critics. Treiman and Hartmann, *Women, Work, and Wages,* Table 10, and O'Neill and Braun, "Women and the Labor Market," summarize studies accounting for sex differences in earnings.

The Gender Gap

used to explain the gap are themselves endogenous, possibly rooted in discrimination against women. Whether women accumulate less experience on the job than men do because their wages and occupations are worse than those of men, or whether they receive lower wages and less rewarding occupations because they have less job experience is a most elusive problem in this debate.

Has our ability to explain the gap in earnings increased over time as the gap has narrowed, or has it decreased? While evidence with which to answer this question is rather scarce, it appears that the explanatory power of the earnings equations used to explain the gender gap has remained constant over time. As women have accumulated more experience in the labor force and as education has played a more important role in wage determination for both groups, the proportion of the gap explained by these two factors has been reduced. However, the absolute value of the unexplained portion or residual has remained roughly constant. A reconciliation between these two statements can be found in the narrowing gap over this century.

The method used to explain earnings differentials involves estimating earnings equations for males and females separately. Separate rather than pooled earnings equations are used because most studies have found that the coefficients in the two equations are statistically different. The difference in the earnings of the two groups (or, as in most estimations, the difference in the log of earnings) is then due to differences in the coefficients and differences in the mean values of the independent variables, each multiplied by the appropriate weight. Either male or female weights can be used.[44]

To evaluate how much of the difference between nineteenth century male and female earnings can be explained by human capital variables involves estimating earnings equations for both males and females. While many studies have estimated such equations with nineteenth century data, there is only one that has done so for both sexes using the

---

[44] This technique is generally attributed to Ronald Oaxaca, "Male-Female Wage Differentials in Urban Labor Markets," *International Economic Review* 14 (Oct. 1973), 693–709. If the male earnings equation is given by
$$\ln w_m = \beta_0 + \Sigma \beta_i X_i + \mu,$$
and the female earnings equation is given by
$$\ln w_f = \alpha_0 + \Sigma \alpha_i Z_i + \varepsilon,$$
where the Xs and the Zs measure an individual's human capital and certain attributes of the job, such as hours of work, then the difference in the two equations can be expressed either as
$$\ln \bar{w}_m - \ln \bar{w}_f = (\beta_0 - \alpha_0) + \Sigma \beta_i (\bar{X}_i - \bar{Z}_i) + \Sigma \bar{Z}_i (\beta_i - \alpha_i)$$
using the male weights, or, using the female weights, as
$$\ln \bar{w}_m - \ln \bar{w}_f = (\beta_0 - \alpha_0) + \Sigma \alpha_i (\bar{X}_i - \bar{Z}_i) + \Sigma \bar{X}_i (\beta_i - \alpha_i).$$

same data set.[45] That study is for male and female workers in California manufacturing industries in 1892.

The difference in the log of male and female earnings in this sample is 0.767, of which 0.466 to 0.492 can be accounted for by differences in the mean values of the independent variables (depending on whether the male or female weights are used). The remaining 0.302 or 0.275 is explained by differences in the coefficients, including the constant terms. Therefore, if one defines "discrimination" as a difference in earnings that cannot be explained by a difference in the means of the independent variables, discrimination accounts for from 36 to 39 percent of the difference in the log of earnings in this sample.

Contemporary studies find that discrimination, computed in this manner, accounts for about 58 percent of the difference in the log of earnings, or roughly 0.30 out of 0.51.[46] Thus the absolute value of the unexplained portion of the gap has remained roughly constant, while the explained proportion has declined over the past century. Studies that measure discrimination in one occupation or sector, however, find that discrimination accounts for a greater percentage of the difference. The earliest such study, with which to compare the manufacturing data, is of clerical workers in 1940.[47] The difference in the log of earnings in this sample is 0.44, of which only 0.15 to 0.20 can be explained by differences in the means of the independent variables, depending on the weights used. Therefore, from 55 to 67 percent of the difference in the log of earnings is due to discrimination, that is, it is unexplained.

[45] Goldin, "Female Earnings Functions," contains estimates for female manufacturing workers in 1888 and 1907; Joan Underhill Hannon, "The Immigrant Worker in the Promised Land: Human Capital and Ethnic Discrimination in the Michigan Labor Market, 1888–1890" (Ph.D., diss., University of Wisconsin, 1977), contains estimates for various ethnic groups of males in Michigan industries. Barry Eichengreen, "Experience and the Male-Female Earnings Gap in the 1890s," *Journal of Economic History* 44 (Sept. 1984), 822–834, estimates equations for both males and females in manufacturing in California in 1892. The ratio of female to male earnings in his sample, 0.464, is considerably lower than that in all U.S. manufacturing industries at that time (see Table 5.2). The coefficients from this sample differ from those in the Goldin and Hannon studies but not by an amount that would reverse the findings. Eichengreen adds a "schooling" variable to his equation that is defined as the age at which work began minus 6. Because many of these individuals did not attend school for that period of time (the derived years of attendance are far too high), this variable probably measures, in part, the returns to maturity. See Claudia Goldin, "The Work and Wages of Single Women, 1870 to 1920," *Journal of Economic History* 40 (March 1980), 81–88, for a discussion of this factor in determining manufacturing earnings.

[46] Oaxaca, "Male-Female Wage Differentials," concludes that discrimination, defined in this manner, accounts for 58.4 percent of the difference in the log of earnings.

[47] Goldin, "Monitoring and Occupational Segregation," contains earnings functions for male and female clerical workers in 1940 from the original schedules of a Woman's Bureau Bulletin.

The Gender Gap

However, comparisons with this data set also indicate that the unexplained portion in absolute value has remained roughly constant at 0.275 and that the increase in the proportion relegated to discrimination owes to the higher ratio of female to male earnings.

Explaining the Gap Over Time

A framework was outlined earlier in which the demand for various occupations in the economy evolves over time, and the returns to characteristics, some of which can be produced and acquired, are thereby altered. Certain characteristics feature prominently in the determination of wages and change in value and quantity over time: education, strength (or other gender-specific physical differences), job experience, and expected life-cycle labor force participation have been singled out for analysis. Earnings can change because of changes in the characteristics embodied in individuals or because of changes in the returns to these characteristics. Both aspects will be explored for each of the variables.

Educational Advances. The narrowing of the gender gap from 1890 to 1940 was achieved primarily through a reduction in the earnings differentials of male and female workers within occupational groups, particularly in the clerical trades and the professional sector. Educational advances by both males and females served to reduce the skill differential in general, and that between males and females in particular.

Economists have for some time recognized that the overall skill differential in the economy has declined since around 1940, and they have sought the reasons for these changes in the reduction in immigration and the increase in educational attainment.[48] Earnings ratios for male workers in 1970 contain a considerably smaller skill premium than do those for the other two years in Table 5.3.[49] This reduction in the skill premium increased the ratio of female to male earnings in 1970, given the relative increase of males in skilled occupations over the twentieth century.

But the change in the skill differential was only one factor altering the structure of earnings across this century. Another was the increase in the earnings of females relative to males within the occupations women began to dominate early in this century. Relative earnings in the clerical sector rose markedly between 1890 and 1930, while the

---

[48] See, for example, Keat, "Long Run Changes in Occupational Wage Structure," and Williamson and Lindert, *American Inequality*.

[49] The ratio of professional to manual workers' earnings for males is 3.01 in 1890, 2.55 in 1930, and 1.38 in 1970, from Table 5.3.

percentage of women in this sector expanded greatly. The combination of these two factors served to increase the ratio of female to male earnings to a considerable degree. The reason for this increase in relative earnings is to be found in the rapid increase in high school graduates and those obtaining commercial degrees in the period just following World War I.

There were two important periods of rapid increase in educational attainment. The first was the increase in high school attendance and graduation among the cohorts born around 1900 and leaving high school from about 1915 to 1930. The second was the increase in college graduates beginning with the cohorts born around 1945 and leaving college around 1967.

The first large increase in educational attainment, that of completing high school, enabled young women to enter a new set of occupations, those in the clerical field, rather than those in manufacturing or sales.[50] At the same time, clerical occupations attracted women who were not yet in the labor force and thus led to an expansion in the labor force participation rate. Jobs in manufacturing paid less than those in clerical work, particularly at entry level. But manufacturing positions offered the opportunity for advancement in wages with time on the job, particularly in craft positions. These positions were rarely occupied by females, in part because the limited number of years women stayed on the job in manufacturing made such investments in training too costly for them, their families, and their employers. The clerical labor force enabled females to gain entry to an occupation in which formal education substituted rather well for on-the-job training and was even a prerequisite for job entry. The nineteenth century male amanuensis was rapidly replaced by the female clerical worker, and the ratio of female to male wages in clerical work rose substantially.

Rewards for Differences in Size and Strength. What was the premium paid to men during the nineteenth century for their larger size and strength? Can it be demonstrated that this premium declined over time with technological advance?

Within the agricultural sector, the ratio of female to male earnings (and of young boys relative to adult males) have been shown to be highly dependent on the crop. In the early nineteenth century the ratio of female (or boy) to adult male wages was very low in the northeastern United States, but it was considerably higher in the cotton-

---

[50] Goldin, "Historical Evolution of Female Occupations," discusses the rise of the clerical sector and the role of educational advances; Goldin, "Changing Economic Role of Women," and Smith and Ward, *Women's Work and Wages* for the educational data.

growing regions of the South.[51] The introduction of the factory system and its machinery almost doubled the ratio of female to male earnings in the Northeast. But relative earnings within manufacturing were still much below one.

The extensive use of piece-rate wages for females in manufacturing at the turn of this century enables an estimate of the wage premium for strength or other physical differences correlated with gender. The premium can be measured only for jobs in which both men and women were employed and, given extensive occupational segregation, this condition makes for a rather short list. Males may have been temporarily placed in such jobs until a "male" position became available; alternatively, those employed in these jobs may have been less productive than the average male. Therefore the difference between the wages of males and females working on piece rates for a particular job may understate the difference that would have existed across all occupations, had men and women been found in all jobs.

Data on piece-rate earnings in 1895 indicate that males earned on average 30 percent more than females did (that is, the wage ratio was 0.77) when the piece rate was identical for both and when both worked at the same job in the same factory and were in the same age group.[52] Because piece rates are paid on actual physical product, any difference in earnings for full-time workers occupying the same position in the same firm must reflect a difference in strength, dexterity, determination, or the quality of the complementary inputs. The average ratio of female to male earnings for time-rate work in the factories sampled was about 0.60 in the 1895 report. The ratio for piece-rate work was 0.77. Thus the difference in physical product accounts for 23 percentage points and the residual is 17 percentage points, out of a possible 40 percentage points. If the basis ratio in manufacturing for this period was 0.77, rather than 1.00, the gender gap would narrow to 0.78 (=0.60/0.77) from 0.60.

Thus the premium paid to men for gender-specific abilities, of which strength may have been a factor, was at least 58 percent of the actual difference of 40 percent. It was at least this amount because time-rate jobs, in which there were few women, paid more, and men may have been preferred to women in such jobs because of gender-specific skills. Comparable data for other periods of time and other occupations are

---

[51] Similarly, in Southeast Asia, areas with a comparative advantage in tree crops, for example, have a much lower relative wage for females and children than do areas that cultivate rice.

[52] All printing and cigar factories were sampled from *Report of the Commissioner of Labor,* 1895/96. The figure of 0.77 is an equal firm-weighted mean.

not presently available. But it is clear that as desk jobs have replaced manual labor, the returns to such gender-specific differences as strength must have decreased, and the piece-rate data give one measure.

Life-Cycle Labor Force Experience. It is clear from the data in Table 5.1 that for most of American history the vast majority of women have not participated in the labor market on a par with men, and that the participation rate of white married women was low until the 1950s. Despite the low degree of labor market participation of married women, those in the labor force could have remained in for substantial periods of time, if their labor market turnover was low. Because labor force participation expanded over time for this group, new entrants must have joined the existing workers. These new entrants would have had very little prior labor force experience, and their entry would have tended to decrease the average level of experience of the currently working population of women.

Direct information on life-cycle labor force participation for adult women would inform the relative earnings data in two ways. The absolute level of labor market experience is important in evaluating differences between average male and female earnings, as is frequently the case in earnings functions, that is, regression equations of earnings on individual characteristics. Changes over time in the earnings ratio ought to be related to changes in the experience levels.

Data on life-cycle labor force participation and the average labor market experience of working women are scarce even for the post-World War II period, with the exceptions of certain panel surveys that begin in 1967. Two separate studies have constructed estimates of these variables for the period from 1930 to 1980.[53] The findings indicate that average years of labor market experience for currently working women have barely increased over this period, despite the rather large increases in labor force participation so evident from the data in Table 5.1. Years of job experience for the currently working population of married women increased from 9.06 in 1930, to 9.78 in 1940, to 10.52 in 1950.[54] The labor market experience of working women aged forty remained roughly constant at 13.5 years from 1940 to 1980, while the work experience of the entire population of women aged forty rose by over 4 years.[55]

[53] Claudia Goldin, "Life-Cycle Labor Force Participation of Women: Historical Evidence and Implications," National Bureau of Economic Research Working Paper (Cambridge, Mass.: Dec. 1983) no. 1251, produces estimates of life-cycle labor force experience for 1920–1950, and Smith and Ward, *Women's Work and Wages* constructs estimates for 1940–1980.

[54] Goldin, "Life-Cycle Labor Force Participation," p. 26.

[55] Smith and Ward, *Women's Work and Wages.*

The Gender Gap 165

The apparent paradox afforded by these two disparate trends, that for working women and that for the entire population of women, is easily resolved. Adult women in the labor force have had a strong tendency to remain in the labor force for substantial periods of time, that is, their turnover has not been very high. But those just entering the labor force have had relatively low experience levels. The average work experience of the entire population of working women increased greatly over the last fifty years, but the average work experience of those currently working did not, as new entrants continually brought down the average. For similar reasons, the educational attainment of the working population of women did not increase along with that of the entire population, until recently.

These data cut in two different ways in the explanation of the relative earnings data and the changes in these ratios. In terms of the absolute level, the tendency for women to remain in the labor force should have led to high wages and good jobs. But the stability of average years of experience should have lessened the relative gains in the ratio of female to male earnings. Because earnings are only observed for individuals in the labor market, the experience level and educational attainment of the working, and not the entire, population are the relevant variables.

The findings with respect to changes over time in life-cycle work experience are consistent with those concerning changes over time in the ratio of female to male earnings. But the findings with respect to the average length of employment at any point in time may not be. Many studies report that no life-cycle labor force participation could justify the choice of occupations by women on the basis of financial considerations alone. What then could account for such differences?

Expected Life-Cycle Labor Force Participation. The labor force participation among cohorts of white married women has increased within marriage (at least until age fifty-five) for every cohort of women born in the United States since about 1890. Each successive decade brought an expanded participation of married women in the market economy. The actual cohort labor force participation rates have been substantially different from the cross-section ones.

The differences between the true cohort participation profiles and those of the cross sections are not merely of academic interest. They are of critical importance in understanding how older generations socialize the younger, how the younger form their own expectations about their future labor market participation, and how society and employers do the same. The vast differences between the true cohort profiles and those in the cross sections imply that young women of no generation in America could have predicted solely from the experiences of their elders what their own work histories would turn out to be.

In 1930, for example, a cohort of twenty-year-old daughters born in 1910 would have been off by a factor of about 4 in predicting their own participation rates in twenty-five years had they simply used the experiences of their forty-five-year-old mothers born in 1885 as a guide. But they were far more informed than this simple extrapolation would suggest. They knew, for example, that their years of schooling were higher than their mothers', and they may have been aware that the jobs they held when unmarried were different from their mothers'. Knowledge of these differences would have narrowed the gap between the simple extrapolation and the actual value of the daughters' labor force participation. However, there is empirical evidence that many cohorts have vastly underestimated their own future labor force participation and therefore may have underinvested in job-related skills.

In 1968 the National Longitudinal Survey asked its young female sample, who were then fourteen to twenty-four years old, what percentage believed they would be in the labor force at age thirty-five.[56] The response was 29 percent for whites and 59 percent for blacks. More than half of these young women are now aged thirty-five, and their labor force participation rate already exceeds 60 percent if they are married and runs even higher if they are not. The figures they had reported when young were more in line with their mothers' labor force participation rates than with their own.

Although the expectations of young women in 1968 were much below their eventual labor force participation, a similar question asked of young women in 1973 indicates a rapid convergence of expected and actual participation rates. Of the women who were nineteen to twenty-nine years old in 1973, 60.3 percent of the whites believed they would be in the labor force at age thirty-five and 73.8 percent of the blacks did.[57] In 1968, young women expected a labor force participation when they were thirty-five years old that was more in line with that of their mothers when they had been thirty-five. By 1973 these young women were forming their expectations more on the basis of current conditions in the labor market for their cohort.

These data suggest that during periods of rapid labor market change it may be difficult to forecast one's future labor force participation. Individuals extrapolate from the world around them, and in doing so,

[56] Steven Sandell and David Shapiro, "Work Expectations, Human Capital Accumulation, and the Wages of Young Women," *Journal of Human Resources* 15 (Summer 1980), pp. 335–353.

[57] It should be noted that the extreme shift in the response might be related to a change in the question asked in the survey.

they may underestimate their need for formal and on-the-job training. The result may be that the actual returns for job experience for women are less than those for men, and that wage ratios are less than 1 even when job experience is equal.

IV. Summary Remarks

Is the scenario described at the beginning of this paper an accurate depiction of the historical record? Have technological advances, economic progress, education, and increased female labor force participation served to raise the average earnings of females relative to males?

The answer is somewhat mixed. Relative earnings across all occupations have increased through most of this century and advanced within manufacturing across the nineteenth century. Certain occupations that rewarded intellect more than strength witnessed increased earnings for women relative to men, but others that required a long labor force commitment have not done so until very recently. Earnings ratios have been surprisingly stable during the last century for occupational groups requiring little skill and education.

Increased female labor force participation during this century has served to stabilize, and not increase, the accumulated years of labor force experience and educational attainment of the average female worker.[58] But regression equations estimated by most discrimination studies find that females do not advance across jobs as men do, with years on the job or with the firm, and that this relative lack of job advancement accounts for a large percentage of the difference in the wages between males and females. Job investments seems to be "too low" for women.

The rapid expansion of the female labor force throughout this century may have made the future highly unpredictable for many cohorts, and surveys of young women indicate that this explanation is a plausible one for many cohorts in the past. One should not underestimate the extent of the social revolution that has occurred in the labor market and the difficulties in forecasting the future in times of rapid change. Current cohorts, however, seem to have revised their expectations in light of past change and may provide a true test of the ideals of the competitive marketplace.

The stability of the gender gap over the last thirty-five years has

---

[58] If the increase in the labor force participation of women is, in part, due to a shifting out of their labor force supply function over time, then a relative wage decline would be expected. This point was raised by Moses Abramovitz in his discussion at the conference at Wesleyan. Estimates in Smith and Ward, *Women's Wages and Work,* of the selectivity effect indicate that it is rather small compared with the other factors.

raised questions about the meaning of the increased labor market participation of women over that period. But the historical record indicates that the greatest narrowing within the industrial and agricultural sectors took place during the period of early industrialization, and that the gender gap across all occupations narrowed up until about 1930 or 1940. The presence of change during the period from 1800 to 1940 did not indicate social advancement, just as the absence of change in the period after 1940 does not mean that other factors were stable. It was precisely because the participation rate of women increased so markedly in the post-World War II period that the gender gap failed to narrow, and the relative increase in female earnings over the last five years finally reflects gains made by women in education and labor market experience.

Appendix: Sources and Notes for Table 5.3

OCCUPATIONAL DISTRIBUTION

Historical Statistics, series D 182–232, pp. 139–140. The 1900 occupational distribution was used for 1890. The professional category includes professional, technical, and kindred workers, and managers, officials and proprietors (lines 218 + 219).

EARNINGS

All earnings are annual, full-time, and in current dollars.

1890, Male, Professional: Weighted average of professional (34 percent) and managerial (66 percent) workers. Professional earnings for six categories, representing over 75 percent of all professionals, were obtained from Stanley Lebergott, *Manpower in Economic Growth: The American Record Since 1800* (New York: McGraw-Hill, 1964); p. 500, gives $1,662 for 1st- to 3rd-class postal workers (government officials); *Historical Statistics,* series D 793, p. 168, gives $731 for ministers (clergy); a value of $460 for male teachers was derived from *Historical Statistics,* series D 763, p. 167, given the assumptions that the ratio of female to male teacher salaries was 0.8 and earnings of the 5 percent who were college teachers was $1,505. The figures for physicians ($2,540), lawyers ($2,691), engineers ($2,108), and college teachers ($1,505) were derived from *Historical Statistics,* series D 913–920, p. 176 for 1929, extrapolated back to 1900 on federal employee earnings, *Historical Statistics,* series D 764, p. 167. Managerial earnings were derived from U.S. Census Office, *Report on Manufacturing Industries in the United States at the Eleventh Census: 1890,* Part 2, "Statistics of Cities" (Washington, D.C., 1895), Table 6, using the category "officers or firm members actively engaged in the industry or in supervision." A figure of $1,264 was converted into a 1900 figure of $1,285, based on nonfarm money (when employed) earnings, *Historical Statistics,* series D 735, p. 165. The final estimate of $1,391 ($1,414, for 1900) was constructed by weighting the actual

The Gender Gap

occupational distribution, and it is consistent with the notion that the ratio of full-time earnings in manufacturing jobs to those in professional occupations must have been smaller in 1890 than it was in 1930; Jeffrey Williamson and Peter Lindert, *American Inequality: A Macroeconomic History* (New York: Academic Press, 1980).

Clerical: U.S. Census Office, *Report on Manufacturing, 1890,* Part 2, p. 10, yields data for urban clerical workers excluding salaried personnel.

Sales: Data for dry goods salesmen in U.S. Commissioner of Labor, *Eleventh Annual Report of the Commissioner of Labor, 1895/96: Work and Wages of Men, Women and Children* (Washington, D.C., 1897), for 11 states yield a mean of $13.58/week or $706/year for 1895, and conversion to 1890 based on nonfarm money (when employed) earnings gives $766.

Manual: Paul F. Brissenden, *Earnings of Factory Workers, 1899 to 1927: An Analysis of Pay-roll Statistics* (Washington, D.C.: Government Printing Office, 1929), p. 94; full-time manufacturing earnings are used. Although these are given for 1899, the accompanying actual figures are identical to those for 1890. See also Elyce Rotella, *From Home to Office: U.S. Women at Work, 1870–1930* (Ann Arbor: UMI Research Press, 1981), pp. 197–212, Appendix B on the 1890 figures. The implied ratio of full-time to actual earnings is 1.18.

Service and Farm: Lebergott, *Manpower;* common laborer's wage × 310 days. The figure for service is almost identical to that in Lucy Maynard Salmon, *Domestic Service* (New York: Arno Press, 1972; orig. pub. 1897), p. 96, of $6.93/week, given 52 weeks and $100/year board. Conversion was made to 1890 based on full-time annual earnings. The farm figure poses problems because no data exist for owner-operator farmers in 1890, and those for more recent periods indicate lower earnings for operators than for farm laborers. Farm wage laborers received less than the wage for common laborers, but owner operators earned far more. The ratio of female to male farm workers for yearly contracts in 1909 was 0.578, and those for seasonal contracts (with board) was 0.538, see George Holmes, *Wages of Farm Labor,* U.S. Department of Agriculture Bureau of Statistics, Bulletin 99 (Washington, D.C. 1912). Therefore the relationship between male and female earnings on farms does not differ significantly from that given by the rate for farm wage laborers.

1890, Female, Professional: Historical Statistics, series D 760, 763, p. 167, for 1900.

Clerical: Rotella, *From Home to Office,* pp. 197–212, Appendix B.

Sales: See source for male earnings. The 1895 figure is $421.

Manufacturing: U.S. Census Office, *Census of Manufacturing: 1890,* Part 1.

Service: Historical Statistics, series D 758, p. 167, for 1900. Salmon, *Domestic Service,* gives an average of $3.23/week or $268/year, including $100 board. Lebergott, *Manpower,* p. 542, gives an estimate of $3.14/week in 1900.

1930, Male, Professional: A weighted average of the earnings of lawyers, physicians, engineers, and dentists from Milton Friedman and Simon Kuznets,

Income from Independent Professional Practice (New York: NBER, 1945), and of semiprofessionals, clergy, professors, and teachers from *Historical Statistics,* series D 793, D 792, D 913, yields $4099. Earnings of proprietors, managers, and officials from *U.S. Sixteenth Census, 1940, Population,* vol. 3, "The Labor Force," Part 1, United States Summary (Washington, D.C., 1943), p. 121, for males who worked 12 months in 1939, adjusted to 1929 dollars, $3,500.

Clerical: Rotella, *From Home to Office,* pp. 197–212, Appendix B.

Sales: U.S. *Sixteenth Census, 1940,* Vol. 3, p. 121, for males who worked 12 months in 1939, adjusted to 1929 dollars.

Manual: The weekly full-time wage from Beney, *Wages, Hours, and Employment in the United States, 1914–1936* (New York: National Industrial Conference Board, 1936), for 50 weeks; also in *Historical Statistics,* series D 835, p. 172. The Beney data imply a ratio of female to male earnings for manufacturing workers of 0.575 in 1929, which might be too high in light of Brissenden's ratios for the 1920s which are lower than Beney's for the same period.

Service and Farm: Unskilled manufacturing laborers, *Historical Statistics,* series D 841, p. 172 × 50 weeks.

1930, Females, Professional: A weighted average of professors, teachers, nurses, and attendants from *Historical Statistics,* series D 763, p. 167, and Harriet A. Byrne, "The Age Factor as it Relates to Women in Business and the Professions," Bulletin of the Women's Bureau, no. 117 (Washington, D.C.: Department of Labor, 1934).

Clerical and Manual: The weekly full-time wage from Beney, *Wages, Hours, and Employment,* for 50 weeks; Rotella, *From Home to Office,* pp. 197–212, Appendix B, gives $868. Ethel Erickson, "The Employment of Women in Offices," Bulletin of the Women's Bureau, no. 120 (Washington, D.C.: Department of Labor, 1934), gives median clerical earnings for 1931 of between $1,044 and $1,308.

Sales: U.S. Sixteenth Census, 1940, vol. 3, p. 125; see 1930, Males above.

Service: Historical Statistics, series D 758, p. 167, for 1929.

1970, Male and Female, All Sectors: U.S. Department of Labor, Bureau of Labor Statistics, *Labor Force Statistics Derived from the Current Population Survey: A Databook,* vol. I, Bulletin 2096 (Washington, D.C., 1982), p. 732, Table C-23, for median, full-time, weekly earnings for each sex-occupational group. The manufacturing group for males and the service group for females are weighted averages of subgroups. Earnings for the farm sector are those of nonfarm laborers. Annual wages are weekly × 50 weeks.

6
Demographic Aspects of the Urban Population, 1800–1840

THOMAS WEISS

It has been twenty years since Stanley Lebergott published his long term series on the American work force. Those figures have been scrutinized closely and have escaped largely intact.[1] The estimates have become part of the furniture of U.S. economic history, being among other things one of the key building blocks in virtually all long-term analyses of nineteenth century American economic performance.

Clearly an important step in labor force measurement is to move on to the task of disaggregation. The variety of experience that characterized nineteenth century America cannot be fully captured until localized measures of labor's contribution are developed. Indeed, quite apart from a descriptive knowledge of the nature and quantity of the labor input in the regional, state, and urban economies, a deeper understanding of structural transformation at the national level requires insight into the determinative interaction of the constituent parts of the labor force.

The economically active share of the population (labor force/population) is determined by the relative size of the various demographic components and the participation rate of each component. The greater our knowledge of those determining influences, the better able we are to understand variations in the aggregate statistic. As yet, the diversity

---

I am grateful to Stan Engerman, Claudia Goldin, Hannah Hiles, and Gavin Wright for their comments on an earlier version. Financial assistance for the research was provided by the National Science Foundation, Grant No. SES 8308569.

[1] Paul David, "The growth of Real Product in the United States before 1840," *Journal of Economic History* 27, no. 2 (1967), pp. 151–197, was the first to examine Lebergott's estimates and discovered some "minor inconsistencies" in the implementation of the described procedures. David's chief revision, a reduction in the size of the 1800 work force, has been adopted by Lebergott in *The Americans* (New York: Norton, 1984), p. 66. A more recent evaluation, Thomas Weiss, "Revised Estimates of the United States Workforce, 1800 to 1860," *Studies in Income and Wealth* (New York: National Bureau of Economic Research, forthcoming, concluded that the cumulative effect of a number of possible estimation errors was small; the aggregate figures are quite reliable.

of labor force participation across states and regions, as well as in urban and rural areas within regions, has not been fully depicted, and it has certainly not been understood.[2] For the late nineteenth century, we know there were substantial differences between the aggregate urban and rural participation rates. In 1870, the participation rate for the entire United States was 32.4 percent, while for the urban population it was 36.2, and the rural rate was 31.9 percent.[3] Four decades later, in 1910, the economically active population share had risen to 40.7 percent, while the urban-rural differential had widened, with shares of 45.4 and 33.6 percent respectively. Further, we know there was regional variation, but we know little about differences across states or individual cities. Our ignorance of the antebellum period is even greater, and consequently we cannot form a good picture of the long-term changes that occurred at the subnational levels.

Behind these subnational figures we find that labor force participation rates varied across demographic groups, and the age-sex composition of the population differed by urban and rural location, as well as by state and region. A sense of the differences in age-sex specific participation rates by urban-rural residence can be obtained from the Public Use Sample for 1900. Some of this evidence is presented in Table 6.1. There one can see the well-known differences by sex and age that prevailed in the aggregate. More to the point, these data show the differences between urban and rural participation behavior for almost all groups, and that these differences varied noticeably by region. For example, for the nation, the urban and rural participation rates for females aged 10–15 years were nearly identical, 10.2 versus 10.9 percent, but in New England, the urban rate was five times the rural, although it was well below the national figure. In the South, the rural rate exceeded the urban one for this age-sex group. Even among males 16 years and over—the bulk of the labor force—there were differences, ranging from 88.1 percent for rural males in New England to 94.2 percent for urban males in the Mountain region.

Clearly, a full understanding of our labor force history and record of industrialization requires that we know more about the participation of the various demographic groups and their differences over space and time. This paper represents a first modest step toward that goal. Specifically, for the period 1800–1840 the paper sets out a consistent and ac-

---

[2] Claudia Goldin, 1977, has examined the variation among females for a small set of cities.

[3] Thomas Weiss, "The Industrial Distribution of the Urban and Rural Workforces: Estimates for the United States, 1870–1910," *Journal of Economic History* 32, no. 4 (1972), p. 929.

TABLE 6.1
Male and Female Participation Rates by Age, Residence, and Region, 1900
(Percentages)

	Urban males		Rural males		Total males	
Region	10–15	16+	10–15	16+	10–15	16+
New England	15.2%	90.1%	07.6%	88.1%	13.7%	89.7%
Mid-Atlantic	17.4	92.0	18.7	90.0	17.9	91.3
East N. Central	12.3	90.8	17.4	88.8	15.3	89.8
West N. Central	12.4	89.4	21.2	88.5	19.2	88.8
S. Atlantic	20.9	90.0	46.3	91.5	43.0	91.3
East S. Central	16.0	92.8	51.1	92.6	47.9	92.6
West S. Central	31.6	91.0	43.4	91.5	40.0	91.4
Mountain	00.0	94.2	15.6	91.6	12.4	92.4
Pacific	14.8	92.4	11.6	92.3	12.8	92.4
East	16.9	91.5	17.4	89.7	17.0	90.9
N. Central	12.3	90.4	19.3	88.7	17.0	89.4
South	23.2	91.2	47.3	91.9	45.0	92.0
West	10.4	92.9	13.5	92.0	12.6	92.4
UNITED STATES	15.9%	91.2%	31.7%	90.3%	26.2%	90.7%

	Urban females		Rural females		Total females	
	10–15	16+	10–15	16+	10–15	16+
New England	06.1%	33.2%	01.2%	17.4%	05.0%	30.3%
Mid-Atlantic	13.1	29.4	04.6	14.6	09.9	24.4
East N. Central	09.4	23.5	06.0	12.9	07.4	18.0
West N. Central	07.7	24.3	02.7	11.3	03.1	15.4
S. Atlantic	20.4	33.9	20.6	24.7	20.6	26.3
East S. Central	06.5	32.8	19.3	21.2	17.6	23.2
West S. Central	10.5	26.0	11.9	15.4	11.7	17.6
Mountain	00.0	25.6	01.6	13.4	01.1	17.5
Pacific	00.0	23.9	07.1	14.1	04.7	19.1
East	11.2	30.5	04.1	15.0	08.8	25.9
N. Central	09.0	23.7	04.4	12.2	05.6	17.0
South	12.9	31.1	17.9	21.2	17.3	23.0
West	00.0	24.4	05.2	13.8	03.4	18.5
UNITED STATES	10.2%	28.1%	10.9%	16.5%	10.4%	21.7%

SOURCE: Public Use Sample, 1900.
NOTE: The regions are those defined by the Department of Commerce. For a listing of the states that compose each region see *Historical Statistics* (1975, p. 5). The groupings are based on geographic location and not some characteristic economic function.

curate series on the urban and rural population of each state, with the requisite age and sex detail needed for the eventual construction of an equally detailed labor force estimate.[4]

[4] Surprisingly, it is somewhat easier to construct the necessary demographic data for the earliest census years. The published census reported details of the population for almost all cities in the years 1800–1840, but not for 1850 and 1860. In the latter two

To produce these series, the census data had to be revised and enhanced in several ways. In each year, the census failed to report certain age and sex details for a few cities, so estimation was required. There was inconsistency over time in the details that were reported, so some revisions were necessary. For free whites it was necessary to revise the age groupings, since those reported for 1800–1820 differed from those for 1830 and 1940. This involved only a minor adjustment for those fifteen years of age.[5] For free blacks and slaves, more extensive revisions and estimations were required. In 1830 and 1840, the census reported these groups by sex and age, but the age breakdowns were quite different from the desired ones; 10 to 23 years and 24 and over, as opposed to 10 to 15 years and 16 and over. In 1820, different age breakdowns were used, while in 1800 and 1810 the data were more highly aggregated, providing no age or sex details. Thus, construction of this complete and consistent series required the use of other evidence and a number of estimating procedures. These estimation procedures are described in the appendix at the end of this chapter.

I

The rise in the urban share of the U.S. population over the antebellum years is well known, and the wide divergence across states in the levels and rates of urbanization, while not known with as much precision, could be easily derived from readily available evidence. This evidence is summarized in Table 6.2. Beyond these summary and highly aggregated statistics, we have little information. Goldin's work on slavery provided valuable data on a major component of the South's urban population but contained no evidence for the North and went back to only 1820.[6] Curry's work on free blacks covered the period 1800–1850, included all regions, and did complement Goldin's study, but the evidence was confined to a subset of cities, albeit the more important ones.[7] Neither of these studies gave much attention to the free white population, presuming perhaps that we are all familiar with that his-

---

years, the census only reported some data for selected cities. Those censuses did report demographic detail by township, so it should eventually be possible to construct the urban figures for those years, but this has not been accomplished as yet. Consequently, the present paper focuses on the years 1800–1840.

[5] The censuses of 1830 and 1840 reported the age group 10–14 years, which is consistent with the 1850 and 1860 data but differs from the earlier census years, in which the breakdown was 10–15 years.

[6] Claudia Goldin, *Urban Slavery in the American South, 1820–1860* (Chicago: University of Chicago Press, 1976).

[7] Leonard Curry, *The Free Black in Urban America, 1800–1850* (Chicago: University of Chicago Press, 1981).

TABLE 6.2
Urbanization by State and Region, 1800–1840
(Percentages)

	1800	1810	1820	1830	1840
STATE					
Alabama	—	—	—	1.0%	2.2%
Arkansas	—	—	—	—	—
Connecticut	5.2%	6.1%	7.6%	9.4	12.5
Delaware	—	—	—	—	10.2
D.C.	75.0	86.7	91.3	90.0	91.2
Florida	—	—	—	—	—
Georgia	3.1	2.1	2.4	2.7	3.6
Illinois	—	—	—	—	2.1
Indiana	—	—	—	—	1.6
Kentucky	—	1.0	1.6	2.3	4.0
Louisiana	—	22.0	17.6	21.3	29.8
Maine	2.6	3.1	3.0	3.3	7.8
Maryland	7.9	12.3	16.2	20.4	24.3
Massachusetts	15.4	21.3	22.7	31.1	37.8
Michigan	—	—	—	—	4.2
Mississippi	—	—	—	2.2	1.8
Missouri	—	—	—	3.6	4.2
New Hampshire	2.7	3.3	2.9	4.8	10.2
New Jersey	—	2.4	2.5	5.6	10.7
New York	12.7	12.6	11.7	15.0	19.4
N. Carolina	—	—	2.0	1.4	1.7
Ohio	—	1.2	1.7	3.9	5.5
Pennsylvania	11.3	12.8	13.0	15.3	17.9
Rhode Island	20.2	23.3	22.9	30.9	44.0
S. Carolina	5.5	6.0	5.0	5.9	5.7
Tennessee	—	—	—	0.9	0.8
Vermont	—	—	—	—	—
Virginia	2.4	3.3	3.3	4.1	6.3
REGION					
Northeast	9.3	10.9	11.0	14.2	18.5
North Central	—	1.0	1.2	2.6	3.8
South	3.0	4.1	4.6	5.3	6.7
UNITED STATES	6.1%	7.3%	7.2%	8.8%	10.8%

SOURCE: *Historical Statistics,* 1975, Series 172:194, and 195:209.

tory. Recently, Berlin and Gutman made comparisons between free and slave workers, but only for the urban South in 1860.[8] The only treatment of urban whites is that of Adna Weber, published nearly a century ago.[9]

[8] Ira Berlin and Herbert Gutman, "Natives and Immigrants, Free Men and Slaves: Urban Workingmen in the Antebellum American South," *American Historical Review* 88, no. 5 (1984), pp. 1175–1200.
[9] Adna Weber, *The Growth of Cities in the Nineteenth Century* (1899; reprint ed., Ithaca, N.Y.: Cornell University Press, 1963).

II. Urban Slavery

The diversity of experience across states and cities regarding the importance of slaves is quite remarkable. Goldin stressed this diversity in her analysis of urban slavery, focusing on the differential changes in urban slave populations across cities and over time.[10] Over the period from 1820 to 1860, eight of the ten cities in her study showed "a cyclical pattern in the percentage rate of change in slave populations,"[11] and she developed a model that explained this variation. Here, let me emphasize the variation in the importance of slaves in urban populations across cities and states. While most individual cities and states showed a decline in the slave portion of their urban population, in any year there were wide differences in the levels of these shares.[12] The data on slave shares are presented in Tables 6.3 and 6.4.

TABLE 6.3
Comparison of Urban and State Populations, by State, Selected Dates
(Slaves as a percentage of the population)

	1800 Urban	1800 Total	1820 Urban	1820 Total	1840 Urban	1840 Total
SOUTH ATLANTIC						
D.C.	23.9%	—	16.8%	19.3%	8.1%	10.7%
Delaware	—	9.6%	—	6.2	0.2	3.3
Georgia	46.0	36.4	40.9	43.9	42.0	40.6
Maryland	10.7	30.8	7.2	26.4	3.9	19.1
N. Carolina	—	27.9	48.1	32.1	43.3	32.6
S. Carolina	48.1	42.3	51.1	51.4	49.8	55.0
Virginia	34.9	39.2	32.5	39.9	28.4	36.2
SUB REGION	29.7	35.0	24.8	37.7	20.5	36.3
EAST SOUTH CENTRAL						
Alabama	—	—	—	32.7	30.5	42.9
Kentucky	—	18.3	28.8	22.5	20.7	23.4
Mississippi	—	39.4	—	43.5	33.8	52.0
Tennessee	—	12.9	—	19.0	30.5	22.1
SUB REGION	—	17.1	28.8	23.7	25.6	31.6
WEST SOUTH CENTRAL						
Louisiana	—	—	27.1	45.0	22.8	47.8
Arkansas	—	—	—	11.4	—	20.4
SUB REGION	—	—	27.1	42.2	22.8	41.9

SOURCES: U.S. Census, 1800, 1820, and 1840.

---

[10] Goldin, *Urban Slavery*, p. 7, Table 13, p. 52. [11] Ibid., p. 51.

[12] The slave share of the population in southern cities declined steadily from 29.7% in 1800 to 21.2% in 1840. The share fell sharply after 1840, declining to 15.3% in

TABLE 6.4
Slave Shares of City Populations, Southern Cities, 1800–1840
(Ranked by 1840 population within each group)

City	1800	1810	1820	1830	1840
I. CITIES IN EXISTENCE BY 1820					
Baltimore, Md.	.107	.100	.069	.051	.031
New Orleans, La.	.000	.346	.271	.314	.229
Charleston, S.C.	.481	.472	.511	.507	.501
Washington, D.C.	.239	.175	.147	.124	.073
Louisville, Ky.	.000	.000	.257	.233	.162
Richmond, Va.	.400	.385	.364	.395	.373
Savannah, Ga.	.460	.421	.409	.414	.419
Petersburg, Va.	.422	.384	.363	.342	.327
Norfolk, Va.	.393	.416	.385	.383	.340
Alexandria, Va.	.176	.206	.175	.153	.127
Georgetown, Va.	.000	.235	.207	.139	.107
Lexington, Ky.	.000	.349	.311	.343	.355
Wilmington, N.C.	.000	.000	.544	.531	.519
Frederickton, Md.	.000	.000	.120	.120	.075
Fayetteville, N.C.	.000	.000	.379	.373	.330
New Bern, N.C.	.000	.000	.524	.465	.429
II. CITIES IN EXISTENCE AFTER 1820					
St. Louis, Mo.	.000	.000	.000	.144	.093
Mobile, Ala.	.000	.000	.000	.368	.305
Wilmington, Del.	.000	.000	.000	.000	.002
Wheeling, Va.	.000	.000	.000	.000	.014
Nashville, Tenn.	.000	.000	.000	.325	.305
Portsmouth, Va.	.000	.000	.000	.000	.292
Augusta, Ga.	.000	.000	.000	.491	.467
Lynchburg, Va.	.000	.000	.000	.379	.428
Columbia, S.C.	.000	.000	.000	.454	.474
Fredericksburg, Va.	.000	.000	.000	.340	.309
Macon, Ga.	.000	.000	.000	.000	.409
Hagerstown, Md.	.000	.000	.000	.109	.093
Natchez, Miss.	.000	.000	.000	.426	.333
Winchester, Va.	.000	.000	.000	.000	.151
Lafayette, La.	.000	.000	.000	.000	.191
Columbus, Ga.	.000	.000	.000	.000	.343
Vicksburg, Miss.	.000	.000	.000	.000	.343
Annapolis, Md.	.000	.000	.000	.220	.179
Maysville, Ky.	.000	.000	.000	.000	.178
REGION	.297	.279	.253	.255	.212

SOURCES: U.S. Census, 1800, 1810, 1820, 1830, and 1840.

1850 and 11.1% in 1860. The largest cities showed the same pattern for the most part, and this decline is the issue confronted by Goldin in *Urban Slavery*. The newer, and smaller, cities must have behaved differently, because in the aggregate the urban slave population increased between 1840 and 1860 by 35.7%, whereas the slave population in Goldin's sample cities showed no change over the twenty years.

In 1800, while the urban population of Maryland (actually Baltimore) was composed of only 10.7 percent slaves, that of South Carolina was 48.1 percent slave. While slaves increased their share of the urban population in South Carolina between 1800 and 1840, all other states showed a decline. Still, in 1840 slaves made up substantial, but noticeably different proportions of the urban population in most states. If we exclude the northernmost states of Delaware and Maryland, as well as the District of Columbia, the slave share of the urban population exceeded 20 percent in every state, and ranged over 40 percent in three states. If we again exclude Delaware, the District of Columbia, and Maryland, there was a clear difference by subregion. The older South Atlantic area had much higher slave shares than the newer regions.

If we look at individual cities (Table 6.4), the range of experience was even wider. In 1800, of course, the paucity of cities made the urban totals for each state synonymous with the single existing city, except for Virginia. But by 1840, there were thirty-five cities in the slave South, and they had slave shares ranging from 0.2 percent of the city's population in Wilmington, Delaware to 51.9 percent in Wilmington, North Carolina.

While some of the variations across cities can perhaps be readily explained, it seems unlikely that a single factor, such as city size, can account for the wide diversity. For example, in 1840 there was a slight decline in the slave share as city size increased.[13] But this decline was very small, only 0.143 percentage points for each 10,000 increase in population, and the result is not statistically significant. That result is also heavily influenced by the inclusion of Baltimore and New Orleans. If we exclude these two unusually large cities, there was no statistical relationship between slave shares and city size. The correlation coefficient is .0001.

Wealth, industrial structure, and location, of course, account for some variation. As noted above, the urban areas in the border states had the smallest slave shares, and this is evident in the individual city shares in Delaware and Maryland, as well as in the District of Columbia and St. Louis. Older cities, such as Charleston and Savannah, in which wealthy planters comprised a substantial share of the white poulation, had some of the highest slave shares, as did cities such as Richmond and Petersburg, which emphasized manufacturing. It would also appear that slaves must have been of use in shipping and port ac-

---

[13] The following regression equation was estimated for 1840:
$$S = .286 - .00000143 \text{Pop} \qquad r^2 = .0483$$
$$(t = 1.29)$$
where S is the slave share and Pop is the city population.

tivities, as evidenced by their relative importance in Norfolk, Wilmington, Portsmouth, and Annapolis. But questions still remain. Why did the border cities of Louisville and Lexington have such high shares, and why was there such great disparity between two cities so close geographically? Why should the slave share in border cities be lower than that in the rural areas of those states? How general an explanation is the wealth variable? Goldin's model did well in explaining changes in urban slave populations, and some of the same factors surely influenced the level of slave importance. An important part of her explanation was the greater elasticity of demand for urban slaves than for rural ones, and this was due in part to the greater availability of free-labor substitutes in cities. But this is what needs to be explained in the present case. Why did cities have different amounts of free labor available? The situation in border state cities might be readily explained by accessibility to immigrants, but why would there be noticeable differences within the deeper South?

It should also be noted that in several states slaves made up a larger share of the urban population than they did of the state's total population. This was true for Georgia and South Carolina in 1800, North Carolina and Kentucky in 1820, and Georgia, North Carolina, and Tennessee in 1840. Why would slaves be overrepresented in the urban populations of some states? It may be that slaves made up a disproportionate share of populations in some cities simply because their presence there deterred nonslaveholding whites from locating in those cities, although this begs the question of why the slaves were there at all. Whatever the explanation, one implication of these incidents of high slave proportions is that the upper bound to urban slavery may not have been reached by most cities. There would appear to have been room for more urban slavery in the South, and as a corollary, slavery may not have hindered urbanization or industrialization.

The increased coverage of urban slavery does not alter the picture revealed in Goldin's study (see Table 6.5). The extension to 1800 does make for a more complete series, and since her sample cities were of greater importance in the earlier years, there can be virtually no difference between her sample cities and the entire urban population. In 1800, 90 percent of southern urban slaves lived in Goldin's sample cities, in contrast to only 49 percent in 1860. The extension in time does add two more observations that conform to the fluctuating pattern of urban slave growth, with a rapid increase from 1800 to 1810—61 percent in the sample cities—followed by a gain of only 14 percent over the next decade. The greater number of cities does produce a different percentage increase in each decade, except the 1840s, and

TABLE 6.5
Comparison of All Southern Urban Slavery with Goldin's
Selected City Data, 1800–1860

Year	Slave population Goldin's cities	Slave population All southern cities	Decadal % change Goldin's cities	Decadal % change All southern cities	Slave share of urban population Goldin's cities	Slave share of urban population All southern cities
1800	20,764	23,126	—	—	29.9%	29.7%
1810	33,509	39,841	+61%	72%	27.7	27.9
1820	38,063[a]	51,539	+14	29	23.8	25.3
1830	51,679	78,056	36	53	22.7	25.5
1840	67,775	102,440	31	30	19.4	21.2
1850	76,946	124,000	14	21	13.6	15.3
1860	68,013	139,000	−12	+12	8.3	11.1

SOURCES: Goldin, *Urban Slavery*, Table 13, and p. 12; U.S. Census, 1800–1840.
NOTE: Goldin's cities include Baltimore, Charleston, Louisville, Mobile, New Orleans, Norfolk, Richmond, St. Louis, Savannah, Washington, D.C.
[a] I deducted her estimates for Mobile and St. Louis.

shows a noticeable difference in the 1850s, a 12-percent increase versus a 12-percent decline in Goldin's sample cities.[14] But the pattern of fluctuating rates of change is still evident, and the decline in the slave share of the urban population is also unmistakable, albeit not quite as pronounced as in the ten sample cities.

Among slaves, females exceeded males for all southern cities in about the same proportion as prevailed in Goldin's set (see Table 6.6). In 1820, for example, the demographic sex ratio of males to females averaged 0.78 for Goldin's cities, only slightly below the 0.80 figure for all cities. The average ratio for her set of cities declined to 0.76 in 1840, while that for all cities rose slightly to 0.81. Still, this is a difference of only 5 percent. In her set of cities the male/female ratio ranged from 1.11 in Richmond to 0.57 in Baltimore, with Richmond being the only city with an excess of male slaves. For the full set of thirty-five cities, there were seven with an excess of males: Richmond, Petersburg, and Lynchburg, Virginia; Augusta, Georgia; Wilmington, North Carolina; Lexington, Kentucky; and Lafayette, Louisiana. And the ratio for the full set ranged from 1.21 in Lafayette to 0.47 in Wheeling, Virginia. These exceptions to the pattern of excess females must surely reflect unusually strong opportunities for employment of male slaves in

[14] Goldin expanded her sample to twenty-two cities in 1850 and 1860 and concluded that the additional data mitigated the decline evident in the smaller sample, showing a decrease of only 4%, but felt that the smaller sample showed no serious bias (*Urban Slavery*, p. 54). The greater increase evident in the total urban slave population reflects to some extent an increase in the number of places that qualified as urban in 1860, as opposed to 1850.

TABLE 6.6
Male Shares of the Urban and Total Population, by Region, 1800–1840

	1800 Urban	1800 Total	1810 Urban	1810 Total	1820 Urban	1820 Total	1830 Urban	1830 Total	1840 Urban	1840 Total
New England										
Whites	.479	.499	.486	.498	.478	.485	.475	.494	.484	.497
Blacks	n.a.	n.a.	n.a.	n.a.	.444	.481	.440	.472	.499	.499
Mid-Atlantic										
Whites	.496	.515	.501	.513	.484	.509	.488	.509	.481	.506
Blacks	n.a.	n.a.	n.a.	n.a.	.419	.482	.431	.488	.428	.482
West										
Whites	—	.544	.552	.526	.534	.525	.526	.520	.528	.523
Blacks	—	n.a.	n.a.	n.a.	.506	.524	.478	.505	.478	.512
S. Atlantic										
Free whites	.545	.511	.524	.508	.503	.506	.494	.505	.491	.503
Free blacks	n.a.	n.a.	n.a.	n.a.	.422	.482	.415	.475	.413	.477
Slaves	n.a.	n.a.	n.a.	n.a.	.456	.513	.454	.504	.453	.500
East S. Central										
Free whites	—	.520	.572	.521	.575	.517	.594	.518	.572	.518
Free blacks	—	n.a.	n.a.	n.a.	.452	.545	.493	.529	.476	.511
Slaves	—	n.a.	n.a.	n.a.	.483	.505	.510	.500	.456	.501
West S. Central (La.)										
Free whites	—	—	.566	.550	.609	.562	.597	.556	.589	.560
Free blacks	—	—	n.a.	n.a.	.390	.432	.408	.434	.440	.453
Slaves	—	—	n.a.	n.a.	.368	.529	.397	.527	.421	.513
United States										
Whites	.499	.510	.502	.509	.490	.508	.492	.508	.494	.511
Blacks	n.a.	n.a.	n.a.	n.a.	.420	.483	.425	.485	.434	.483
Slaves	n.a.	n.a.	n.a.	n.a.	.444	.512	.450	.504	.446	.501

SOURCES: U.S. Census, 1800, 1810, 1820, 1830 and 1840; Appendix at the end of this chapter. The census did not report males and females separately for blacks or slaves in 1800 or 1810.

manufacturing or perhaps as dock workers. More generally, females predominate because of their greater employment as domestic servants in cities and the male slaves' comparative advantage as farm labor.

III. Urban Free Persons

While Goldin and Curry made some comparisons with the free white populations, they focused almost exclusively on either slaves or free blacks. Moreover, their sample sizes were fairly small. Much earlier, Weber provided some data and hypotheses about urban demography, but his U.S. evidence was confined primarily to "great cities" in the latter half of the nineteenth century. Thus, the present compilation pertaining to all cities provides a more comprehensive picture than was previously available, particularly for smaller-sized cities. The advantage of the larger set of cities is immediately obvious, for the sample cities of these other researchers showed different pictures. Weber noted an excess of females,[15] while Curry found an excess of females among free blacks, but a shortage among whites.[16]

Table 6.6 summarizes data on the male share of the urban and state populations by region, for each racial group, at each census date from 1800 to 1840. Clearly, Weber's finding for great cities in the late nineteenth century prevailed as well in earlier decades. With the exception of 1810, the U.S. cities in the aggregate showed a slight excess of females among free whites, and larger excesses among free blacks and slaves. This situation contrasted with that of the population as a whole, as well as with that of the rural population, where males were a slight majority in each year for whites and slaves, but a minority for free blacks.[17] Moreover, there appears to be virtually no trend in the male or female share of the urban population for all cities combined, in contrast with the sample of twelve cities used by Curry, in which the female share was rising for whites, but declining for free blacks.[18]

There were, however, great differences among regions. Only the Mid-Atlantic region conformed to the U.S. pattern in all regards; males comprised a majority of the region's white population in each year, but the male share of the urban population was a minority, and of course was a smaller share than prevailed for the entire region's population. This was true in each year for both whites and blacks. New En-

[15] Weber, *Growth of Cities*, p. 286.   [16] Curry, *Free Black in Urban America*, p. 9.
[17] The urban-rural difference accords with Weber's findings for 1890. He claimed that all but three of the fifteen largest cities contained a larger proportion of females than did the states in which they were situated (*Growth of Cities*, p. 286).
[18] Curry, *Free Black in Urban America*, p. 9.

TABLE 6.7
Male Shares of the Free White Urban Population, by State, 1800–1840

	1800	1810	1820	1830	1840
Connecticut	.485	.497	.472	.492	.481
Maine	.499	.478	.465	.467	.483
Massachusetts	.477	.487	.481	.475	.491
New Hampshire	.491	.479	.482	.442	.441
Rhode Island	.472	.477	.473	.480	.475
New Jersey	—	.507	.486	.497	.478
New York	.496	.501	.490	.492	.487
Pennsylvania	.497	.494	.475	.483	.473
Illinois	—	—	—	—	.591
Indiana	—	—	—	—	.518
Michigan	—	—	—	—	.538
Ohio	—	.552	.534	.526	.521
D.C.	.536	.500	.499	.496	.477
Delaware	—	—	—	—	.466
Georgia	.587	.527	.545	.549	.545
Maryland	.540	.526	.499	.487	.480
N. Carolina	—	—	.512	.498	.486
S. Carolina	.520	.507	.500	.502	.528
Virginia	.565	.541	.509	.494	.497
Alabama	—	—	—	.664	.684
Kentucky	—	.572	.575	.594	.534
Mississippi	—	—	—	.586	.610
Tennessee	—	—	—	.564	.530
Louisiana	—	.566	.609	.597	.589
Missouri	—	—	—	..578	.613
U.S. URBAN	.499	.502	.490	.493	.494
U.S. TOTAL	.509	.509	.508	.509	.511

SOURCE: U.S. Census, 1800–1840.

gland was very similar, but the males made up a minority of the region's population as a whole. More striking differences occurred in the other regions. In the West, for example, males were a majority among whites in both the urban and rural areas, but among urban blacks they were a minority in 1830 and 1840. In the South Atlantic States, white males went, sometime after 1820, from being in the majority in urban areas to being the minority. In the South Central regions white males were always a substantial majority in cities, and the urban male shares exceeded the rural ones in all years.

Almost without exception the urban population of each state conformed to the regional pattern of excess males or females. This can be seen in Table 6.7 where the male shares are reported for the free white urban population of each state. There are, of course, a handful of exceptions. In the North, the only exception is in 1810, when the region had a slight excess of males in cities, but Pennsylvania had more females. Most of the exceptions occurred in the South Atlantic region. There, all states showed an exess of males in 1800 and 1810, but thereafter the urban population of most states shifted to a female majority. As the process was underway, some states shifted sooner than others; so they would appear as deviations in 1820, while the laggards would appear so in later years. As of 1840, an excess of males occurred in only Georgia and South Carolina.

Even though regions showed certain patterns, there were still wide differences across states in each region. In New England, where each state showed an excess of females in each year, the shares ranged noticeably, with the smallest variation occurring in 1820. In all years, the Mid-Atlantic region had the smallest variation, while outside the Northeast the shares ranged widely, and in many cases were far below the national norm. At the extreme was Alabama, where the urban white population was approximately two-thirds male in 1830 and 1840. As would be expected, the variation across individual cities was more striking. In 1820, for example, the male shares ranged from 63.3 percent in Louisville to 43.3 percent in Norwich, Connecticut, while in 1840 the range went from 71.9 percent in Joliet, Illinois, to 35 percent in Lowell, Massachusetts. In 1820 only 17 cities out of 61 had an excess of males, while in 1840 there was an excess of males in 54 cities out of 131. In both 1820 and 1840 the 10 cities with the largest share of females were located almost exclusively in New England.

To a great extent, these patterns and trends in the sex composition conform to the well-established ones of males predominating in frontier areas, with the female share increasing as settlement proceeds.[19] Thus the older, more established regions and states tended to have smaller male shares in both urban and rural areas than did the newly developing regions such as the West and the West South Central region. The variations in the urban male shares within regions must reflect primarily that the sex composition of the immigration to certain cities and states differed in response to the different types of employment opportunities that existed. This situation shows up most clearly in certain

---

[19] The variation across cities was not systematically related to city size. The following regression results were obtained for 1840:

$$F/M(U.S.) = 1.009 + .00000021 * \text{Size 1} \ (r^2 = .0014)$$
$$F/M(NE) = 1.071 + .00000011 * \text{Size 1} \ (r^2 = .0008)$$
$$F/M(NON) = 0.971 - .0000002 * \text{Size 1} \ (r^2 = .0005)$$

where F/M is the female/male ratio, U.S. is all cities ($n = 132$), NE means cities in the Northeast ($n = 81$), NON means cities outside the Northeast ($n = 51$), and Size means the population of the city. None of the coefficients is significant.

New England cities where textile manufacturing provided a substantial number of jobs for females. Weber has also argued that higher female/male sex ratios in cities resulted in part from higher male mortality rates resulting from males' greater employment in dangerous occupations.[20] It would seem that this argument could be extended to explain variations in the sex ratio across cities, because the industrial composition of employment varied over cities and the mortality rate varied by industry.[21]

IV. Implications

These sex differences across cities, states, and regions are especially relevant because of their impact on the work-force estimates. This is also the case for the age composition of urban and rural areas and for its variations across states and regions. In particular, differences in the share of males 16 years and older would have the most decided impact, and in some states this could be reinforced by a lower share of those aged 0–9 years. Taken in conjunction with the state differences in the extent of urbanization, these factors could produce noticeable differences in state participation rates and, perhaps more so, differences in the industrial or occupational distributions of the work force. And while urban-rural differences might be mitigated by a greater participation of males 10–15 years old in rural areas (see Table 6.1), the nature and quality of the work forces would still be different.

In each year, the urban population of the northeastern regions had a smaller share of males 16 and over than did other regions, and the differences were not small: as many as 7 percentage points in 1800, 12 in 1810, 18 in 1820, and 13 in 1830 and 1840 (see Table 6.8). At the same time, outside the Northeast there were substantial differences between the urban and rural areas in each region: 4–8 percentage points in the West and the South Atlantic region and 10–15 percentage points in the South Central regions. If age-sex specific participation rates were everywhere the same, then the variation in the importance of this demographic group would translate directly into noticeable differences in aggregate participation rates.

Of course, other things were not equal. Not only did the composition of the population vary across states, regions, and urban areas, but the evidence in Table 6.1 suggests that age-sex participation rates also differed by location. Table 6.9 presents some hypothetical aggregate par-

[20] Weber, *Growth of Cities*, pp. 289–300.
[21] Paul Uselding, "In Dispraise of the Muckrakers: United States Occupational Mortality, 1890–1910," *Research in Economic History* 1 (1976), pp. 334–371.

TABLE 6.8
Males 16 years and over as a Share of the Urban and Total Population, by Region, Free Whites, 1800–1840

	Urban population					Total population				
	1800	1810	1820	1830	1840	1800	1810	1820	1830	1840
New England	.264	.276	.279	.300	.312	.255	.259	.268	.283	.297
Mid-Atlantic	.292	.287	.275	.295	.296	.258	.255	.263	.275	.284
West	—	.297	.315	.321	.358	.258	.244	.253	.251	.270
S. Atlantic	.331	.318	.300	.299	.300	.250	.250	.255	.260	.261
East S. Central	—	.332	.380	.416	.401	.235	.238	.240	.250	.253
West S. Central	—	.401	.463	.427	.428	—	.295	.330	.318	.328
UNITED STATES	.289	.288	.287	.303	.313	.253	.253	.259	.267	.276

SOURCES: U.S. Census, 1800–1840. See Appendix at end of the chapter, for details of compilation and estimation.

TABLE 6.9
Hypothetical Variations in the White Work Forces, by Region, in 1820
(Percentages)

Region	Hypothetical participation rates			Composition of the work force						
				Urban			Rural			
	Urban	Rural	Total	Males 16+	Males 10–15	Females 16+	Males 16+	Males 10–15	Females 16+
New England	36.7%	29.1%	29.8%	68.6%	2.8%	28.6%	80.7%	2.1%	17.1%
Mid-Atlantic	35.2	28.6	29.3	71.9	3.1	25.0	82.2	5.2	12.6
S. Atlantic	38.1	32.9	33.2	70.9	3.5	25.5	70.3	10.9	18.8
East S. Central	43.4	31.0	31.1	81.3	2.3	16.4	71.3	13.5	15.1
West S. Central	48.9	34.3	36.6	86.2	2.3	11.5	81.5	9.1	9.4
West[a]	35.0	26.6	26.7	81.5	2.2	16.3	84.4	5.2	10.4
UNITED STATES	36.6%	29.8%	30.2%	71.4%	3.0%	25.5%	77.7%	7.1%	15.1%

NOTE: For each population component, a hypothetical number of workers was estimated as the product of the age-sex-location–specific participation rate for 1900 (Table 6.1) times the urban or rural population in that group (Tables 6.10–6.14 at the end of the chapter, and R-1–R-4 available from the author). The participation rates are the sums of those estimates divided by the appropriate total white population in the region.
[a] Missouri is included with the West.

ticipation rates, which take into account these variations in demographic composition and participation rates. Since there is nothing to be gained by spawning a series of preliminary labor force figures, this calculation has been confined to just one year, 1820.

As can be seen, differences in the shares of the population composed of males 16 and over show up as substantial variation in the participation rates, and it is the variation due to this group that accounts for most of the differences in the aggregate rates. In every region, there is some difference between the urban and rural rates, but a few regions show smaller discrepancies than prevailed for the nation. The South Central regions and the West, however, show much larger differences. In the South Central regions, urban males 10–15 years old had a lower participation rate than their rural counterparts, so this group helped narrow the urban-rural divergence, whereas in the West this group had little effect.

The table also reveals considerable differences across regions in the participation rates for the total white population, and even larger variation for the urban white population. Here, while large discrepancies arose from the adult male component's differences in population shares, the female component tended to narrow the regional differences in all cases. The teen-age group had little effect, especially in the urban areas, but did contribute some slight additional divergence in the total participation rates.

The variation in the composition of the work force is also striking, with notable differences between the urban and rural work forces in certain regions, as well as sharp differences across regions. For example, adult males were a much larger component in the rural areas of New England and the Mid-Atlantic region than in the urban areas, because of the greater labor force participation of urban females (see Table 6.1); while in the two South Central regions, adult males were a relatively larger part of the urban work force than of the rural one, because of their greater importance in the urban population (see Table 6.8). Across regions, adult males were a larger share of the urban work force in the South Central regions than they were in the Northeastern areas, primarily because of differences in the composition of the urban populations. A comparison of the rural shares reveals a different pattern: adult males were more important in the Northeast than in the east South Central region, but there was little difference between the Northeast and the west South Central region. The east South Central diverges because its rural population contained a smaller share of adult males and because its rural families were more economically active than those in other regions.

These differences hold important implications for productivity comparisons, if one assumes that a young or female worker was not the equivalent of an adult male worker. For example, if a youth or a female were equivalent to only 60 percent of a male worker, the full-time equivalence of the Mid-Atlantic rural work force would be 92.9 percent of the reported number, while for the South Atlantic region it would be only 88.1 percent. Applying these figures to the rural participation rates narrows the regional differences, producing full-time equivalent participation rates of 29.0 for the South Atlantic region and 26.7 for the Middle Atlantic area. As can be readily seen from the evidence in Table 6.9, this full-time equivalent adjustment would widen the disparities in urban participation rates.

To be sure, these preceding exercises are not meant to indicate what actually prevailed in 1820. Rather, they are intended to suggest that variations in the composition of population across regions and urban-rural locations, and over time, had real consequences, giving rise to differences in participation rates and the composition of the work force. In turn, such variations in the quantity and nature of the labor inputs have an impact on the measurement of regional and sectoral productivity levels and changes, and on the progress of regional and national income per capita. While the actual variations and consequences were not precisely the ones depicted here, the exercise suggests that the actual differences were of noticeable importance. Of course, there are other factors, such as variations in hours of work or days worked per year, that might alter even more our estimates of differences and changes in productivity and income per capita. Whatever their impact, this paper makes clear that the effect of locational differences in population and participation were surely nonnegligible.

Appendix: Urban Population Estimates, 1800–1840

The set of estimates of the urban population is presented in Tables 6.10–6.19. These contain the major breakdowns by race, sex, and age. Additional age details are available for the white population, and the details presented here are available for the rural population as well.

Any measurement of the urban population is, of course, influenced by the definition of *urban*. The definition used by the census has varied over time and today includes population living in urban-fringe areas and in unincorporated places of 2,500 or more. The inclusions add a certain amount of arbitrariness to the totals, since boundaries that do not legally exist must somehow be established to mark off the population to be included in these places. The justification for doing so is that these places are essentially urban, being as large and as densely settled as those areas that qualified as urban because of their size and the fact of incorporation.

The inclusion of these places is a recent phenomenon, having started with the 1950 census. The historical series, however, has not been reconstructed to conform to this new definition. Instead, the historical series is based on the 1940 census definition of urban. According to that definition, the urban population includes all persons living in incorporated places of 2,500 or more and in areas (usually minor civil divisions) classified as urban under special rules relating to population size and density. The historical series, then, includes the population living in urban places, while the newer concept includes some people living outside urban areas.

Depending on the use to which the data are put, the inclusion or exclusion of the inhabitants of these locations and the arbitrariness of the boundaries could be of some consequence. Clearly, the "true" urban total may be misstated, and the distinctions between urban and rural economic behavior may be blurred.

However, it is not likely that the differences in these classifications would significantly affect nineteeth century data. In 1950, the effect of the change in definition was an increase in the urban population of 8.9 percent. This is not a negligible number, but its size reflects that the change was made because these places were becoming more urbanized and more important.[22] In earlier years the problems of definition must have been less serious, especially in the nineteenth century. An indication that the growth of these places was a recent phenomenon is that their share of the urban population increased from 8.2 percent in 1950 to 10.7 in 1970. Moreover, the revision of the definition is made up of two items: the population living within the artificially designated fringe of urbanized areas (i.e., the settled areas around cities of 50,000 or more) and the population living in unincorporated places of 2,500 or more and located outside the urban fringes. It is the former component that had been increasing in importance, and that would have been negligible in the nineteenth century. This component is also the larger of the two revisions, accounting for 78.8 percent of the addition to the urban total.[23] The other part, the population residing in larger, densely settled, but unincorporated areas, was a problem in all years, but it was of lesser importance. In 1950, the population in unincorporated places of 2,500 inhabitants or more outside fringe areas accounted for only 2.1 percent of the urban population. Since some of these places, especially those in New England, were classified as urban under special rules of the 1940 Census, the possible understatement in earlier years would be still smaller.

In summary, the historical data are not exactly comparable with the urban counts since 1950, but they are measured consistently over the period 1790–

---

[22] U.S. Bureau of the Census, "The Development of the Urban-Rural Classification in the United States: 1874 to 1949," *Current Population Reports*, series P-23, no. 1 (1949), p. 12.

[23] *Census of 1950*, vol. 1, p. 17.

1940. And they have the advantage of being less influenced by the establishment of artificial boundaries to mark off urban areas. Whatever classification and measurement problems beset the historical series, they would be of minor importance at the national and regional levels but could be of concern in certain states, especially in New England. These problems, of course, have been recognized for decades and are not a product of the present estimation. The present estimates are simply providing additional detail for the urban population time series produced by the census.

As noted at the beginning of this essay, construction of this detailed set of population figures involved dealing with a number of data deficiencies. First, it required the estimation of the race, age, and sex composition of the population for certain cities in each year. Second, it was necessary to reconcile a few discrepancies between the census totals and my summary figures for certain cities and states and to correct some census arithmetical errors. Third, a certain consistency in the age breakdowns over time was desired, which necessitated the estimation of the number of fifteen-year-old free whites in 1830 and 1840 and of the age and sex composition of free blacks and slaves. Each of these adjustments is explained in a subsequent section.

Estimation of Omitted Details for Selected Cities

In the aggregate, the proportion of the urban population for which it was necessary to estimate some missing components was small in each year, as can be seen in Table 6.20.

Census Year 1800. The published census of 1800 did not distinguish between the urban and rural components of the population in two urbanized areas: the District of Columbia and Hartford, Connecticut. While we know what the total urban population was in each of these places, we do not know the breakdown by free versus slave, or by age for the free whites. These breakdowns were derived as follows. First, an estimate was made of the number of urban slaves and deducted from the urban population. Then the sex and age distribution of all free persons was estimated.

The evidence of 1810 and 1820 was used to determine the share of slaves residing in the urban area. For the District of Columbia, that share declined from 0.748 in 1820 to 0.731 in 1810, the latter being 97.7 percent of the former. The assumed 1800 share (0.716) was derived as 98 percent of the 1810 figure. There were only eighteen slaves in Hartford County, and it was assumed they were all in the city of Hartford, as was true in 1810.

For the District of Columbia, the age distribution of the urban free population was assumed to be the same as the population for the entire county in 1800. The two distributions were extremely close in 1810, and the urban population comprised 78 percent of the county in 1800. For Hartford, the age distribution was assumed to be the same as that for the other Connecticut cities for which details were reported in 1800 (New Haven and New London). This approach had the merit of assuring that the Hartford estimates would not unduly alter any urban-rural differences that existed for the state.

TABLE 6.10
Urban Population by State, 1800–1840: Free White Males, All Ages and Ages 0–9

	Free white males, all ages					Free white males, ages 0–9				
	1800	1810	1820	1830	1840	1800	1810	1820	1830	1840
Alabama	—	—	—	1094	5655	—	—	—	198	736
Connecticut	5889	7492	9189	13044	17728	1780	2082	2598	2901	4168
Delaware	—	—	—	—	3146	—	—	—	—	776
D.C.	2361	4573	7261	9629	10486	699	1430	2037	2874	2933
Georgia	1526	1313	2106	3848	7340	307	273	390	867	1719
Illinois	—	—	—	—	5572	—	—	—	—	1221
Indiana	—	—	—	—	5265	—	—	—	—	1635
Kentucky	—	1562	3682	6796	12518	—	429	844	1424	3191
Louisiana	—	3586	8266	11962	36527	—	697	1477	2347	7650
Maine	1850	3327	3887	5736	18679	578	1038	1227	1476	5248
Maryland	11294	19065	25436	33831	43185	3035	5315	7374	9159	12054
Massachusetts	30217	47433	55279	88447	134618	8949	13995	15064	21856	31656
Michigan	—	—	—	—	4796	—	—	—	—	1191
Mississippi	—	—	—	895	2575	—	—	—	172	460
Missouri	—	—	—	2404	8830	—	—	—	629	1428
New Hampshire	2553	3256	3453	5913	12504	829	1052	992	1599	3275
New Jersey	—	2629	3167	8342	17910	—	733	938	2116	5394
New York	33180	65449	72265	129661	218902	10077	20790	21001	35039	58928
N. Carolina	—	—	2901	2340	2953	—	—	757	593	759
Ohio	—	1358	4919	18297	41330	—	434	1454	5029	10346
Pennsylvania	30941	46571	59114	91462	135165	8316	13469	18390	25994	38595
Rhode Island	6178	7867	8293	13688	21686	1695	2081	2218	3412	5537
S. Carolina	4590	5863	5323	7306	8004	1189	1292	1408	1889	1789
Tennessee	—	—	—	2006	2337	—	—	—	477	618
Virginia	6792	9128	10071	12938	23761	1464	2142	2755	3408	6428
TOTAL	137371	230472	284612	469639	801472	38918	67252	80924	129117	207735

NOTE: As of 1840 there was no urban population in Arkansas, Florida, Iowa, Vermont, or Wisconsin.

TABLE 6.11
Urban Population by State, 1800–1840: Free White Males, Ages 10–15 and 16 and Over

	Free white males, ages 10–15					Free white males, ages 16 and over				
	1800	1810	1820	1830	1840	1800	1810	1820	1830	1840
Alabama	—	—	—	51	218	—	—	—	845	4701
Connecticut	908	1091	1266	1536	1812	3201	4319	5325	8607	11748
Delaware	—	—	—	—	346	—	—	—	—	2024
D.C.	273	611	977	1140	1450	1389	2532	4247	5615	6103
Georgia	224	173	154	335	723	995	867	1562	2646	4898
Illinois	—	—	—	—	424	—	—	—	—	3927
Indiana	—	—	—	—	623	—	—	—	—	3007
Kentucky	—	227	403	627	1271	—	906	2435	4745	8056
Louisiana	—	347	495	1048	2322	—	2542	6294	8567	26555
Maine	299	452	587	735	2318	973	1837	2073	3525	11113
Maryland	1849	2439	3296	4516	5259	6410	11311	14766	20156	25872
Massachusetts	4422	6552	7765	10733	14322	16846	26886	32450	55858	88640
Michigan	—	—	—	—	437	—	—	—	—	3168
Mississippi	—	—	—	73	142	—	—	—	650	1973
Missouri	—	—	—	288	553	—	—	—	1487	6849
New Hampshire	536	465	513	794	1582	1188	1739	1948	3520	7647
New Jersey	—	410	474	1122	2175	—	1486	1755	5104	10341
New York	4253	8771	9295	15059	22713	18850	35888	41969	79563	137261
N. Carolina	—	—	350	292	393	—	—	1794	1455	1801
Ohio	—	194	569	2246	4174	—	730	2896	11022	26810
Pennsylvania	3792	6115	7829	11564	15337	18833	26987	32895	53904	81233
Rhode Island	942	1179	1129	1692	2720	3541	4607	4946	8584	13429
S. Carolina	614	832	649	909	874	2787	3739	3266	4508	5341
Tennessee	—	—	—	217	281	—	—	—	1312	1438
Virginia	768	1250	1281	1676	2905	4560	5736	6035	7854	14428
TOTAL	18880	31108	37032	56656	85372	79573	132112	166656	289524	508365

NOTE: As of 1840 there was no urban population in Arkansas, Florida, Iowa, Vermont, or Wisconsin.

TABLE 6.12
Urban Population by State, 1800–1840: Free White Females, All Ages and Ages 0–9

	Free white females, all ages					Free white females, ages 0–9				
	1800	1810	1820	1830	1840	1800	1810	1820	1830	1840
Alabama	—	—	—	553	2607	—	—	—	165	727
Connecticut	6248	7581	10289	13478	19097	1797	2050	2576	2983	4166
Delaware	—	—	—	—	3605	—	—	—	—	815
D.C.	2047	4566	7285	9795	11481	705	1484	2151	2715	2915
Georgia	1072	1177	1760	3161	6125	204	302	422	841	1661
Illinois	—	—	—	—	3860	—	—	—	—	1215
Indiana	—	—	—	—	4893	—	—	—	—	1529
Kentucky	—	1170	2727	4647	10922	—	416	896	1296	3128
Louisiana	—	2745	5318	8082	25462	—	726	1500	2522	8122
Maine	1854	3640	4470	6549	19999	520	1124	1162	1523	5094
Maryland	9606	17147	25577	35634	46868	2675	5183	7192	9206	12057
Massachusetts	33067	50001	59652	97680	139817	8580	13799	15172	22171	31374
Michigan	—	—	—	—	4113	—	—	—	—	1175
Mississippi	—	—	—	632	1646	—	—	—	194	486
Missouri	—	—	—	1754	5577	—	—	—	614	1394
New Hampshire	2648	3547	3714	7456	15866	840	1005	935	1571	3222
New Jersey	—	2561	3356	8452	19560	—	716	933	2134	5303
New York	33697	63819	75133	133933	230947	10137	20190	20983	35116	59897
N. Carolina	—	—	2766	2359	3122	—	—	742	621	787
Ohio	—	1100	4290	16463	38001	—	380	1362	4814	10340
Pennsylvania	31363	47755	65215	97921	150493	8651	13347	17870	25445	38647
Rhode Island	6904	8610	9226	14801	23939	1663	2156	2164	3418	5352
S. Carolina	4230	5705	5330	7240	7162	1214	1683	1359	1758	1789
Tennessee	—	—	—	1548	2069	—	—	—	455	638
Virginia	5228	7753	9704	13263	24054	1450	2132	2628	3385	6294
TOTAL	137964	228877	295792	485401	821285	38436	66693	80047	128375	208127

NOTE: As of 1840 there was no urban population in Arkansas, Florida, Iowa, Vermont, or Wisconsin.

194

TABLE 6.13
Urban Population by State, 1800–1840: Free White Females, Ages 10–15 and 16 and Over

	Free white females, ages 10–15					Free white females, ages 16 and over				
	1800	1810	1820	1830	1840	1800	1810	1820	1830	1840
Alabama	856	—	—	54	287	—	—	—	334	1593
Connecticut	—	1045	1477	1806	2237	3595	4486	6236	8689	12694
Delaware	—	—	—	—	450	—	—	—	—	2340
D.C.	272	655	1064	1312	1700	1070	2429	4070	5768	6866
Georgia	176	174	197	381	861	692	701	1141	1939	3603
Illinois	—	—	—	—	418	—	—	—	—	2227
Indiana	—	—	—	—	675	—	—	—	—	2689
Kentucky	—	190	442	645	1288	—	564	1389	2706	6506
Louisiana	—	416	885	1152	2558	—	1603	2933	4408	14782
Maine	271	480	718	796	2593	1063	2036	2590	4230	12312
Maryland	1621	2410	3705	4858	5744	5310	9554	14660	21570	29067
Massachusetts	4399	6684	8452	11721	15305	20088	29518	36028	63788	93138
Michigan	—	—	—	—	560	—	—	—	—	2378
Mississippi	—	—	—	92	176	—	—	—	346	984
Missouri	—	—	—	269	683	—	—	—	871	3500
New Hampshire	511	510	576	923	1831	1297	2032	2203	4962	10813
New Jersey	—	366	490	1119	2292	—	1479	1933	5199	11965
New York	4233	9440	10863	16979	25360	19327	34189	43287	81838	145690
N. Carolina	—	—	414	273	385	—	—	1610	1465	1950
Ohio	—	188	686	2505	4637	—	532	2242	9144	23024
Pennsylvania	4077	6523	9289	12795	17613	18635	27885	38056	59681	94233
Rhode Island	868	1218	1196	1793	2737	4373	5236	5866	9590	15850
S. Carolina	599	953	825	922	774	2417	3069	3146	4560	4599
Tennessee	—	—	—	286	285	—	—	—	807	1146
Virginia	687	1193	1426	1923	2973	3091	4428	5650	7955	14787
TOTAL	18570	32445	42705	62591	94418	80958	129739	173040	299863	518740

NOTE: As of 1840 there was no urban population in Arkansas, Florida, Iowa, Vermont, or Wisconsin.

TABLE 6.14
Urban Population by State, 1800–1840: Free Black Males, All Ages and Ages 0–9

	Free black males, all ages					Free black males, ages 0–9				
	1800	1810	1820	1830	1840	1800	1810	1820	1830	1840
Alabama	—	—	—	172	250	—	—	—	65	85
Connecticut	185	349	569	547	877	38	72	121	130	214
Delaware	—	—	—	—	691	—	—	—	—	203
D.C.	131	598	1125	1841	2538	44	202	385	620	852
Georgia	74	217	224	288	356	24	69	77	93	105
Illinois	—	—	—	—	97	—	—	—	—	29
Indiana	—	—	—	—	257	—	—	—	—	64
Kentucky	—	41	95	234	535	—	11	20	66	147
Louisiana	—	2044	2432	4722	8506	—	777	1040	1712	2972
Maine	58	100	98	157	360	13	22	20	43	66
Maryland	1150	2353	4483	6664	7876	347	711	1348	2016	2382
Massachusetts	1045	1339	1372	1672	2836	189	242	248	346	438
Michigan	—	—	—	—	106	—	—	—	—	22
Mississippi	—	—	—	41	111	—	—	—	15	45
Missouri	—	—	—	37	267	—	—	—	10	61
New Hampshire	64	60	82	45	71	15	14	17	12	16
New Jersey	—	170	330	651	920	—	49	86	195	272
New York	1664	3897	5168	7517	9272	379	889	1127	1751	2133
N. Carolina	—	—	342	393	634	—	—	117	121	212
Ohio	—	40	219	908	1969	—	10	52	250	458
Pennsylvania	2550	4033	5183	7158	9420	643	1016	1345	1786	2329
Rhode Island	460	589	596	725	831	85	108	114	140	209
S. Carolina	376	581	623	845	645	158	244	235	384	274
Tennessee	—	—	—	97	193	—	—	—	32	55
Virginia	739	1560	1735	2909	3744	241	509	580	948	1182
TOTAL	8497	17973	24675	37623	53362	2176	4946	6933	10735	14825

NOTE: As of 1840 there was no urban population in Arkansas, Florida, Iowa, Vermont, or Wisconsin. The 1820 figures include the "other free" persons reported in the census, distributed by sex and age in the same proportion as the free blacks. There was a total of 1,421 other free persons.

196

TABLE 6.15
Urban Population by State, 1800–1840: Free Black Males, Ages 10–15 and 16 and Over

	Free black males, ages 10–15					Free black males, ages 16 and over				
	1800	1810	1820	1830	1840	1800	1810	1820	1830	1840
Alabama	—	—	—	32	43	—	—	—	75	122
Connecticut	22	42	66	79	86	124	235	383	338	577
Delaware	—	—	—	—	104	—	—	—	—	384
D.C.	27	121	231	370	514	60	275	510	851	1172
Georgia	12	36	37	42	64	38	112	110	153	187
Illinois	—	—	—	—	17	—	—	—	—	51
Indiana	—	—	—	—	54	—	—	—	—	139
Kentucky	—	6	14	35	90	—	24	60	133	298
Louisiana	—	384	462	892	1580	—	883	930	2118	3954
Maine	7	11	12	16	40	39	66	65	98	254
Maryland	212	433	836	1255	1396	591	1210	2299	3393	4098
Massachusetts	105	134	138	192	242	752	963	986	1134	2156
Michigan	—	—	—	—	15	—	—	—	—	69
Mississippi	—	—	—	8	15	—	—	—	18	51
Missouri	—	—	—	7	34	—	—	—	20	172
New Hampshire	11	10	18	5	13	38	36	47	28	42
New Jersey	—	20	39	89	92	—	101	205	367	556
New York	195	456	612	878	1084	1090	2553	3429	4888	6055
N. Carolina	—	—	62	81	129	—	—	163	191	293
Ohio	—	7	38	180	348	—	23	128	478	1163
Pennsylvania	352	557	722	968	1321	1556	2460	3115	4404	5770
Rhode Island	50	64	60	73	102	326	417	422	512	520
S. Carolina	68	105	116	142	123	150	232	271	319	248
Tennessee	—	—	—	16	34	—	—	—	49	104
Virginia	148	312	352	635	674	350	740	803	1326	1888
TOTAL	1207	2699	3816	5996	8215	5114	10329	13926	20892	30322

NOTE: As of 1840 there was no urban population in Arkansas, Florida, Iowa, Vermont, or Wisconsin.

197

TABLE 6.16
Urban Population by State, 1800–1840: Free Black Females, All Ages and Ages 0–9

	Free black females, all ages					Free black females, ages 0–9				
	1800	1810	1820	1830	1840	1800	1810	1820	1830	1840
Alabama	—	—	—	200	291	—	—	—	66	84
Connecticut	253	478	749	775	1207	50	94	144	149	249
Delaware	—	—	—	—	910	—	—	—	—	203
D.C.	180	820	1465	2496	3673	44	202	378	580	912
Georgia	107	313	358	401	481	27	80	98	103	113
Illinois	—	—	—	—	72	—	—	—	—	20
Indiana	—	—	—	—	301	—	—	—	—	83
Kentucky	—	44	115	219	573	—	10	25	48	146
Louisiana	—	2906	3805	6840	10843	—	834	1118	1836	3217
Maine	60	102	126	156	304	14	25	31	39	69
Maryland	1621	3318	6108	9313	11559	360	736	1449	1994	2471
Massachusetts	1021	1307	1499	1857	2182	203	260	286	372	451
Michigan	—	—	—	—	87	—	—	—	—	15
Mississippi	—	—	—	34	118	—	—	—	14	31
Missouri	—	—	—	65	264	—	—	—	15	54
New Hampshire	74	71	78	60	90	12	11	17	8	12
New Jersey	—	214	406	841	1143	—	52	99	209	270
New York	2261	5294	7458	10013	12142	411	964	1300	1843	2275
N. Carolina	—	—	483	514	835	—	—	121	103	206
Ohio	—	42	214	990	2193	—	11	58	269	546
Pennsylvania	3423	5413	6946	9375	12898	688	1088	1443	1896	2482
Rhode Island	708	906	950	1150	1203	107	137	151	159	186
S. Carolina	575	891	852	1351	1062	176	273	283	450	267
Tennessee	—	—	—	107	216	—	—	—	25	49
Virginia	1017	2146	2432	4152	4886	239	504	595	992	1078
TOTAL	11299	24263	34044	52909	69533	2332	5280	7597	11170	15489

NOTE: As of 1840 there was no urban population in Arkansas, Florida, Iowa, Vermont, or Wisconsin. The 1820 figures include the "other free" persons reported in the census, distributed by sex and age in the same proportion as the free blacks. There was a total of 1,421 other free persons.

TABLE 6.17
Urban Population by State, 1800–1840: Free Black Females, Ages 10–15 and 16 and Over

	Free black females, ages 10–15					Free black females, ages 16 and over				
	1800	1810	1820	1830	1840	1800	1810	1820	1830	1840
Alabama	—	—	—	38	40	—	—	—	96	167
Connecticut	30	57	80	109	138	173	326	525	517	820
Delaware	—	—	—	—	174	—	—	—	—	533
D.C.	26	119	204	362	559	109	499	883	1554	2202
Georgia	16	47	52	61	74	64	186	208	237	294
Illinois	—	—	—	—	15	—	—	—	—	37
Indiana	—	—	—	—	71	—	—	—	—	147
Kentucky	—	6	14	33	76	—	28	76	138	351
Louisiana	—	421	518	1026	1629	—	1650	2168	3978	5997
Maine	7	12	18	15	39	38	65	77	102	196
Maryland	266	544	852	1677	1984	995	1037	3806	5642	7104
Massachusetts	110	141	166	198	235	707	906	1048	1287	1496
Michigan	—	—	—	—	12	—	—	—	—	60
Mississippi	—	—	—	5	12	—	—	—	15	75
Missouri	—	—	—	12	36	—	—	—	38	174
New Hampshire	12	12	14	12	11	50	48	48	40	67
New Jersey	—	27	39	135	146	—	135	267	497	727
New York	269	630	822	1240	1502	1580	3701	5336	6930	8365
N. Carolina	—	—	74	91	133	—	—	287	320	496
Ohio	—	8	41	205	437	—	23	115	516	1210
Pennsylvania	452	714	827	1333	1745	2283	3610	4676	6146	8671
Rhode Island	66	84	85	111	113	535	685	714	880	904
S. Carolina	98	151	139	224	194	301	467	430	677	601
Tennessee	—	—	—	19	41	—	—	—	63	126
Virginia	178	376	430	748	821	600	1266	1408	2412	2979
TOTAL	1531	3351	4376	7653	10237	7437	15631	22072	32086	43807

NOTE: As of 1840 there was no urban population in Arkansas, Florida, Iowa, Vermont, or Wisconsin.

TABLE 6.18
Urban Population by State, 1800–1840: Male Slaves, Ages 0–9 and 10 and Over

	Male slaves, ages 0–9					Male slaves, ages 10 and over				
	1800	1810	1820	1830	1840	1800	1810	1820	1830	1840
Alabama	—	—	—	132	375	—	—	—	479	1526
Connecticut	21	6	—	—	—	53	15	4	—	38
Delaware	—	—	—	—	—	—	—	—	—	5
D.C.	196	343	469	452	280	469	821	1121	1083	696
Georgia	319	296	394	726	1215	752	697	931	2202	3584
Illinois	—	—	—	—	—	—	—	—	—	4
Indiana	—	—	—	—	—	—	—	—	—	—
Kentucky	—	216	378	557	755	—	522	912	1858	2093
Louisiana	—	615	743	1515	2484	—	1742	1966	4237	7646
Maine	—	—	—	—	—	—	—	—	—	—
Maryland	368	604	620	642	464	909	1494	1533	1641	1218
Massachusetts	—	—	—	—	—	—	—	—	—	—
Michigan	—	—	—	—	—	—	—	—	—	—
Mississippi	—	—	—	131	242	—	—	—	393	793
Missouri	—	—	—	113	173	—	—	—	236	523
New Hampshire	—	—	—	—	—	—	—	—	—	—
New Jersey	—	49	4	—	—	—	125	87	13	6
New York	486	343	112	—	—	1207	853	278	—	—
N. Carolina	—	—	738	662	786	—	—	2017	1608	1959
Ohio	—	—	—	—	—	—	—	—	—	—
Pennsylvania	10	2	—	2	—	27	4	1	18	1
Rhode Island	11	1	2	—	—	28	2	6	3	1
S. Carolina	1181	1523	1666	2303	2127	2851	3675	4029	5244	5199
Tennessee	—	—	—	232	260	—	—	—	629	644
Virginia	914	1392	1429	2121	2555	2561	3900	3996	5924	8091
TOTAL	3506	5390	6555	9588	11716	8857	13849	16881	25568	34028

NOTE: As of 1840 there was no urban population in Arkansas, Florida, Iowa, Vermont, or Wisconsin.

200

TABLE 6.19
Urban Population by State, 1800–1840: Female Slaves, Ages 0–9 and 10 and Over

	Female slaves, ages 0–9					Female slaves, ages 10 and over				
	1800	1810	1820	1830	1840	1800	1810	1820	1830	1840
Alabama	—	—	—	160	439	—	—	—	404	1529
Connecticut	18	5	—	1	—	56	15	4	3	5
Delaware	—	—	—	—	1	—	—	—	—	9
D.C.	204	357	468	462	326	615	1077	1413	1509	1196
Georgia	356	330	481	800	1284	939	871	1269	2588	4273
Illinois	—	—	—	—	—	—	—	—	—	2
Indiana	—	—	—	—	—	—	—	—	—	—
Kentucky	—	204	421	505	873	—	566	961	1551	2679
Louisiana	—	623	803	1783	2849	—	2980	3843	6941	11083
Maine	—	—	—	—	—	—	—	—	—	—
Maryland	398	654	671	677	544	1169	1920	1970	2639	2199
Massachusetts	—	—	—	—	—	—	—	—	1	—
Michigan	—	—	—	—	—	—	—	—	—	—
Mississippi	—	—	—	157	299	—	—	—	506	935
Missouri	—	—	—	113	200	—	—	—	253	635
New Hampshire	—	—	—	—	—	—	—	—	1	—
New Jersey	—	56	5	—	—	—	173	102	34	9
New York	469	332	122	10	1	1766	1248	460	7	2
N. Carolina	—	—	855	736	856	—	—	2399	1845	2164
Ohio	—	—	—	—	—	—	—	—	—	—
Pennsylvania	12	2	1	3	—	37	6	6	25	—
Rhode Island	16	1	1	—	—	48	3	12	5	2
S. Carolina	1230	1586	1704	2492	2363	3791	4887	5253	6818	7039
Tennessee	—	—	—	215	321	—	—	—	732	889
Virginia	1038	1581	1621	2241	2691	2865	4362	4465	6809	9071
TOTAL	3742	5731	7152	10355	13047	11287	18110	22158	33386	43721

NOTE: As of 1840 there was no urban population in Arkansas, Florida, Iowa, Vermont, or Wisconsin.

TABLE 6.20
Estimated Share of Urban Population,
1800–1840

Census year	Population of cities subject to some estimation[a]	Total urban population	Share of total subjected to some estimation
1800	9726	322521	3.02%
1810	25115	524918	4.78
1820	12502	691869	1.81
1830	25904	1121749	2.31
1840	12572	1848164	0.68

[a] These figures refer to the population estimates that were made to provide the same detail for all cities as the census provided for most. These numbers do not include subsequent estimates of additional details, such as the age and sex composition of free blacks.

At the state level in 1800, the race/sex and age breakdowns were not provided for 766 "souls" added to the count for Indiana. It was assumed that none of these were slaves, and that the others were distributed by race, sex, and age in the same proportions as prevailed for the 4,740 free persons in the state for whom a breakdown was provided.

Census Year 1810. In 1810, the details of the urban population were not reported separately for four cities in New York: Albany, Brooklyn, Hudson, and Schenectady. For each city the total population was known. The distribution was made as follows.

First, the number of slaves and the number of free black persons in each of the four cities was estimated as a share of the total population; the shares were equal to each city's mean for 1800 and 1820. (For Brooklyn, there were no figures for 1800, so the 1820 figures alone were used).

The age distribution of the free white population was assumed equal to that for the other urban areas in the Middle Atlantic region. The region was used instead of the state because the only urban data for the state were those for New York City. The data for 1800 and 1820 indicate that the distributions for these cities and for the region were very similar.

Census Year 1820. In 1820, the age and sex distributions of the population components were not reported for four cities in North Carolina: Fayetteville, New Bern, Raleigh and Wilmington. Fortunately, the census did report the total population, as well as the number of free whites, slaves, and free blacks for each of the four cities. The age and sex distribution for each component was assumed to be the same as that for the urban population in seven southern cities: Savannah, Georgia; Lexington, Kentucky; Charleston, South Carolina; and Alexandria, Norfolk, Petersburg, and Richmond, Virginia. This sample of cities used all available cities south of D.C., except Louisville and New Orleans. Examination of the data for 1830 indicated that the sex distributions of these two cities were markedly different from the distribution for the North Carolina cities and other southern cities in 1830.

Census Year 1830. In 1830, the distribution of the total population by race, sex, and age was not reported for four cities: Augusta and Savannah, Georgia; St. Louis, Missouri; and Wilmington, North Carolina. For a fifth city, Middletown, Connecticut, the detailed figures pertained to a larger township and had to be adjusted to the city basis.

The details of Augusta were estimated using the 1830 county data and the 1840 city shares of the county population. Between 1830 and 1840, there was virtually no change in the total population of either Augusta or Richmond County. The city population declined by 300 (4.5 percent), while the county's increased by 288 (2.5 percent). It seemed reasonable to assume that any changes in the distribution of the city's population must have paralleled that reported for the county. Thus, for 1830, each component of Augusta's population was estimated as a fraction of the county total for that component. The fraction was set equal to that which prevailed in 1840. Each component was then adjusted to assure that the sum of the components would equal the known totals. This was done in successive steps so that the adjusted figures for each racial group became the "known totals" for use in adjusting the estimates for each age/sex component.

The details for Savannah were reported in 1820 and 1840, so the 1830 figures for each race/sex category were estimated as equal to the mean of that component's share in those two census years. The age distributions were not comparable over time. For free whites aged 0-9 years, figures were available in both years and indicate little change between 1820 and 1840, so it was assumed that the 1830 age distributions would be similar to those for 1840.

An examination of the 1840 data indicated that St. Louis was atypical of other midwestern cities and of cities in nearby Kentucky. While males comprised 61.2 percent of free whites in St. Louis, in other cities in the region, with one exception, the figures were all lower, ranging from 47.9-55.4 percent. The exception, Joliet, Illinois, had a male share of 72 percent. The 1840 data for St. Louis County was much closer, showing a male share of 59.3 percent. Given the closeness of the 1840 city and county data, the 1830 urban figures were estimated on the basis of the county distributions.

The city's slave and free black populations were estimated as a fraction of the reported total city population. For each category, the fraction was set equal to that group's share of the 1830 county population adjusted by the city/county ratio of shares that prevailed for the same group in 1840. The free white total was taken as the residual 1830 city population unassigned to either the slave or free black group. The residual share was 83.5 percent of the total, slightly above the share (83.2 percent) that would have been derived if this group had been estimated using the same procedure as that for slaves and free blacks, and if the unexplained residual in that case had been distributed in proportion to the "explained" shares.

The estimated number of people in each of the three racial categories was then distributed across age and sex components in the same proportion as prevailed in the county in 1830.

For Wilmington, I first estimated the free white, slave, and free black shares of the total, using the mean of the 1820 and 1840 figures. The estimate for each racial group was then divided by sex on the basis of the mean of the 1820 and 1840 sex breakdowns. Finally, for each race/sex group the estimated count was distributed by age according to the 1840 distribution. The differences in the age classifications in the 1820 census, as compared with 1830 and 1840, precluded using the mean of the 1820 and 1840 values. The 0–9 years-of-age category was reported uniformly in both years, and those data indicated that there was little change between 1820 and 1840, and it thus seemed reasonable to assume there was no change between 1830 and 1840.

For Middletown, the total population was known, as well as the distribution for the county subdivision in which the city was located. In 1840, the distribution for the city was very similar to that for the entire township, and the city's relative importance did not change much over the decade, comprising 45.7 percent of the township in 1830 and 47.9 percent in 1840. Thus it seemed reasonable to assume that the 1830 city total was distributed across race/age/sex categories according to the distribution for the township in that year.

Census Year 1840. Three cities were listed among the principal towns in the 1840 Census Compendium, but the reported number of inhabitants differed from that reported in later census volumes. For two of these cities, Hagerstown, Maryland, and Natchez, Mississippi, it appears that the Census Bureau subsequently decided that the original figure referred to a larger township, rather than just to the city. Accordingly, the revised figures were reduced from 7,179 to 3,625 for Hagerstown, and from 4,800 to 3,612 for Natchez. The third city, Wilmington, North Carolina, was revised upward from 4,744 to 5,335, presumably on the basis of additional evidence. Given the proximity of size and location for each of these city/township pairs, the original race/age/sex distribution reported for the township was used to distribute the revised city figures.

Arithmetical Discrepancies

The published census volumes do contain a number of arithmetical errors, but they are surprisingly few in light of the technology of the time and the enormity of the task. Nonetheless, there are instances where the sum of the column or row does not equal the reported total, where digits have been transposed, and even where column entries have been transposed. In some cases, it appears that the errors were caught in later census compilations, but some flaws seem to have persisted. Perhaps, I have not found all the errors, but those cases in which there is some discrepancy between my figures and those reported in the census are summarized in Table 6.21. I offer an explanation of each discrepancy in the table notes.

TABLE 6.21
Discrepancies Between Present Estimates and Census Figures

Census year	Cities and states with discrepancies	Present estimate	Original census figures	Current census figures
I. 1800	Cities			
	Liberties, Pa.	10728	10718	10718
	New Bedford, Mass.	4411	4361	4361
	New York, N.Y.	60489	60515	60515
	Portland, Maine[a]	3822	3704	3704
	States[b]			
	Georgia	163879	162686	162686
	Kentucky	220955	220959	220955
	New York	586203	586050	589051
	Pennsylvania[c]	602365	602545	602365
II. 1810	Cities			
	Boston, Mass.[d]	33250	33250	33787
	Petersburg, Va.	5666	5668	5668
	Trenton, N.J.	3000	3002	3002
	States—No discrepancies			
III. 1820	Cities—No discrepancies			
	States			
	Arkansas	14246	14273	14273
	Pennsylvania	1049458	1049398	1049458
	Tennessee	422613	422813	422823
IV. 1830	Cities			
	Alexandria, Va.	8221	8241	8241
	New York, N.Y.[e]	197112	197112	202589
	States—No discrepancies			
V. 1840	Cities			
	Vicksburg, Miss.[f]	3104	3104	0
	Zanesville, Ohio	4768	4766	4766
	States—No discrepancies			

SOURCES: The current census figures for cities come from a Bureau of the Census worksheet, "Urban Places Ranked by Population Size," provided by the bureau. The state figures come from U.S. Census of 1950, "Population," vol. 2, Table 1, for each state. The present estimates and original census figures come from the census volumes of the respective years.

NOTE: Unless otherwise noted, the difference is taken to be only a minor error of addition.

[a] The census reported 3,704 in the column headed "Slaves," with a footnote indicating that the column should be titled "Whole Numbers of Persons." In fact, the figure of 3,704 is the number of free whites. There were 118 other free persons, bringing the total to 3,822. The correct figure was included in the summation for the state total.

[b] The U.S. Summary Table in the original census volume (page a) contains a number of flaws in the components of certain states, but these did not affect the state totals. The most notable errors occur in New York and Virginia. In New York the number of free white males under 10 years of age should read 83,161, not 33,161. For Virginia, in the District of Columbia free white males under 10 should read 689, not 889; and those 16–25 should read 483, not 83; and in the Western District the figures for free white females aged 26–44 and for those 45 and over should be reversed.

[c] The original census simply transposed two digits in the total for the Eastern District. The correct figure is 327,799.

[d] Presumably, the current census figure reflects corrections determined appropriate by the census bureau at a later date; but the age and sex details are not known. I chose to use the figure for which the details were known.

[e] The current figure may reflect the inclusion of population that was residing in certain county subdivisions that were consolidated with New York City around 1900. When the details can be ascertained, I shall adjust my figures accordingly.

[f] Vicksburg, Miss., was reported in the 1840 Compendium, but has been excluded from later census counts. The reported population of 3,104 falls in line with that for 1850 (3,678), so it was decided to include Vicksburg in the urban list for 1840. While its addition has virtually no impact on the U.S. urban total, it nearly doubles the Mississippi urban figure. The town of Vicksburg was incorporated in 1825, and historians have estimated that its population exceeded 2,500 persons by 1835 [Virginia Harrell, *Vicksburg and the River* (Jackson: University Press of Mississippi, 1982), p. 14; Peter Walker, *Vicksburg: A People at War, 1860–1865* (Chapel Hill: University of North Carolina Press, 1960), p. 6].

Estimation of Free Whites Aged 15 Years

The estimation of those aged 15 years seems a trivial matter to pursue, but it is necessary in order to have a consistent population series, especially one with age groupings suited to labor force estimation. Specifically, what is desired is a population series for those aged 10–15 years, and those aged 16 years and above. Unfortunately, the census did not report the population in these categories in all years. For the years 1830–1870, the census reported the age groups 10–14 years and 15 years and above, but for the years 1800 to 1820 it reported those aged 10–15 years, and 16 and over.

Fortunately, sufficient age detail was available for the years 1880–1920, as was sample data for 1860, to permit a reorganization of the census data into the desired categories. All that was necessary was some way of estimating the number of 15-year-olds in any year so that their number could be shifted from one age group to the other. While it may be impossible to estimate precisely the number of 15-year-olds, small discrepancies would be of little consequence, since the number of 15-year-olds was only some fraction of those aged 10–15, and an even smaller fraction of those aged 15 and over. A number of age relationships were examined with the data for 1880–1920, and it was decided that the ratio of those aged 15 to those aged 10–14 would give very reliable estimates.

This ratio was preferred over alternatives for two reasons. First, given the proximity of their ages there should be a closer movement between those aged 15 and those aged 10–14 than there would be for other groups. Consequently, if there were any trend in the importance of those aged 15 years it would be picked up in the movement of the extrapolating variable. And there was clearly a trend in the extrapolating variable as those aged 10–14 decreased from 12.5 percent of the population in 1830 to 10.3 percent in 1900; and apparently one existed for 15-year-olds as well, as the share of 10–15-year-olds declined from 16.0 percent in 1800 to 12.3 percent in 1900. Second, the census is known to have been deficient in the counts of youths in several postbellum years. The most notable problem was the overcount of teen-age workers in 1910, but there are known problems in 1870 and 1890 as well. However, any census problems affecting 15-year-olds is likely to affect the 10–14 group equally, so that the calculated ratio would still be close to the true one.

For each state, the mean of the ratios for the census years 1880, 1890, 1900, and 1920 was derived. This mean could be compared to a sample mean of the ratio of 15-year-olds to those aged 10–14 for the census year 1860. The comparisons are presented in Table 6.22 and indicate that the postbellum mean is close to the antebellum figures in most states. The difference between the two figures is significant in only three cases for males and three for females, out of a total of sixteen possible cases for each. And most of the significant differences occurred in states whose population was unimportant in the early nineteenth century. The exception is New Jersey, but in this case it seems clear that the 1860 sample must be flawed. Given the closeness of the

TABLE 6.22
Comparison of Population Mean and Sample Data, Selected States
(Ratio of 15-year-olds to 10–14-year-olds)

	Males			Females		
State	Four-year mean	1860 sample	Difference	Four-year mean	1860 sample	Difference
Connecticut	.186	.164	.022	.195	.155	.040
Illinois	.179	.168	.011	.183	.182	.001
Indiana	.181	.179	.002	.184	.189	−.005
Iowa	.182	.177	.005	.183	.187	−.004
Kansas	.175	.206	−.031	.177	.219	−.042[b]
Maryland	.180	.195	−.015	.182	.160	.022
Michigan	.178	.188	−.010	.181	.183	−.002
Minnesota	.180	.191	−.011	.179	.184	−.005
Missouri	.176	.144	.032[a]	.181	.159	.022[b]
Ohio	.162	.182	−.020	.166	.173	−.007
New Hampshire	.189	.151	.038	.197	.247	−.050[a]
New Jersey	.180	.088	.092[a]	.184	.147	.037
New York	.182	.188	−.006	.189	.201	−.012
Pennsylvania	.178	.192	−.014	.183	.179	.004
Vermont	.189	.257	−.068	.195	.205	−.010
Wisconsin	.183	.117	.066[a]	.186	.163	.023
UNITED STATES	.176	.177	.001	.180	.185	.005

SOURCES: Fred Bateman and James Foust, "Agricultural and Demographic Records for Rural Households in the North, 1860," MS Census Sample.

NOTE: The four-year mean is that for 1880, 1890, 1900, and 1920.

[a] The difference is significant at the .05 level.

[b] The difference is significant at the .10 level.

mean ratios and the wide variety of states in the comparison, the postbellum ratios were accepted as being good estimates for the antebellum period, as well as for 1870.

For each state, the mean postbellum ratio was used to estimate the number of 15-year-olds in each year from 1830 to 1870. This was done separately for white males and white females. The resulting figures were then subtracted from the census totals of those 15 and over in the years 1830–1870 and added to the 10–14-year-old group.

These state totals were then divided between the urban and rural population in the same proportion as the 10–14-year-olds were distributed in each state. The resulting figures for the urban population are those reported in the tables at the beginning of this appendix.

Estimation of Age and Sex Composition of Free Blacks and Slaves

Constructing a series for free blacks and slaves that contained the age details desired for estimation of the labor force required more extensive estimation than was needed for the white series. Four adjustments to the census statistics were required. The reported data for 1830 and 1840 required revision in the age groupings for free blacks; while the 1820 data required a different revision in the age groupings for both free blacks and slaves. For 1800 and 1810, it was necessary to estimate first the sex breakdown and then the age composition. For each group, the estimates were first constructed at the state level, and then an urban-rural breakdown was derived.

Free Blacks. The state figures for free blacks were estimated as follows. The censuses of 1830 and 1840 reported the figures broken down into several age categories; 0–9, 10–23, 24–35, 36–54, and 55 and older. To get the desired groupings, it was only necessary to split the 10–23 group into those aged 10–15 and those 16–23. The 10–15 group was estimated using the ratio of those aged 10–15 to those aged 0–9 years; the ratios were derived from the census data for the period 1870 to 1910. The ratios used were those calculated for the free whites rather than for blacks. This was done because it was anticipated that the postbellum evidence for blacks would be heavily influenced by ex-slaves, and thus more representative of slaves than of antebellum free blacks. And where comparisons between free blacks, slaves, and free whites could be made, the evidence indicated that the demographic details of free blacks were closer to those of free whites than to those of slaves. Once the 10–15-year group was estimated, the 16–23 group was derived by subtraction. The latter group was then combined with the reported number of those 24 years and above to obtain the broader category of 16 and over. This procedure was followed for each sex separately.

The 1820 census reported different age categories than those used in the 1830 and 1840 censuses. In 1820, the categories were ages 0–13, 14–25, 26–44, and 45 and above. The number of free blacks aged 10–15 years and those 16 and over were estimated as fractions of the reported number of free blacks of each sex; the ratios were for the period 1830–1860.

For 1800 and 1810, it was first necessary to estimate the sex composition of the reported free black population. For most states I simply used the mean shares for the years 1820–1860. For those states for which there appeared to be a noticeable trend in the shares, the 1800 and 1810 shares were estimated using the 1820 share.

The number of free blacks of each sex was then distributed by age group using the evidence for the years 1830–1860. For most states the mean shares for the years 1830–1860 were used. In those states where the 1830 figure was noticeably different from the 1830–1860 mean values (using a criterion of 10 percent of the mean), the estimator was based on the data for the shorter period 1830–1840.

The state totals obtained were then distributed between urban and rural areas on the basis of the urban-rural distribution of the reported population. For 1830 and 1840, the 10–15-year age group was distributed on the basis of the reported 10–23 group. Since the number of those aged 0–9 was known, the 16-and-over age group was derived as a residual. For 1820, the 10–15 group was distributed using the urban-rural shares for those aged 0–13 years; while the 16-and-over group was distributed using the data for those aged 14 and above. The 0–9 age group served as the residual category.

For 1800 and 1810, the estimated number of each sex was distributed by age using the newly derived evidence on the demographic composition of the urban and rural populations for 1820-1840.

Slaves. Greater precision in the estimates of the age and sex composition of the slave population is needed because the group was of far greater numerical importance than free blacks. Fortunately, fewer details are required because little distinction has been made between sexes, or between youths and adults, as regards labor force participation. It might be desirable to consider variations in participation among the various slave components, but for the present study I intended only to derive an estimate of the number of slaves aged 10 years or more. Consequently, the reported census statistics could be used for 1830 and 1840. For earlier census years I followed procedures similar to those used to estimate the details for free blacks.

The evidence on the 0-9 share of the male and female slave population for the years 1830-1860 was examined. In some cases there was variation in the shares, but no obvious trend, so the mean share for the period was used. In other cases—the mean share for 1830 and 1840 or the 1830 share—values for years closer to the period of extrapolation were used. In six of the more important slave states there was a clear trend in the 0-9 share, so it was extended back to 1800. In a few cases, the 1820 figure for slaves aged 0-9 was derived by multiplying the postbellum ratio of those aged 0-9 to those aged 0-13 times the reported number of slaves aged 0-13 years. For a few nonsouthern states where the age data were sparse for the period 1830-1860 I used national averages.

The postbellum evidence on the age distribution of free blacks permitted the calculation of a ratio of those aged 0-9 to those aged 0-13, by sex, for each state. These ratios were used to check the reasonableness of the estimated values for 1820, a year in which the census reported the number of slaves aged 0-13 years.

The above procedures for deriving the age breakdown could be applied directly in 1820 because the census reported the number of male and female slaves separately. For 1800 and 1810, I first had to derive a sex breakdown and then estimate the age distribution. The data for each state was examined to see whether or not there was a clear trend in the sex shares. For most states there was not, and the mean for the years 1820-1860 was used to estimate the values in 1800 and 1810. For a few states, there was a noticeable trend, or else the 1820 or 1830 shares deviated noticeably from the later years, in which cases the extrapolators were based on the evidence for the years 1820-1840.

The state totals were then distributed between urban and rural areas. For 1820, the urban-rural distribution of males aged 10 and over was made on the basis of the reported distribution of those aged 14 and over. For female slaves, this procedure resulted both in a 0-9 share in urban areas that was noticeably too low in comparison with 1830 and 1840 and in a ratio of 0-9 to 0-13 that was exceptionally high. As an alternative, the urban/rural distribution of those aged 0-9 was based on the reported distribution of those aged 0-13.

For 1800 and 1810, the distribution was based on the demographic composition of the state's urban and rural population that prevailed over the period from 1820 to 1840.

Additional Appendix Tables

The data used to estimate the age and sex details of free blacks and slaves, as well as the estimates themselves, are available in a series of appendix tables. The following is a list of tables that are available from the author.

I. DATA ON FREE BLACK ESTIMATIONS

Table   Title

A. Data Used in Estimation

B:1   Male Shares of Free Black Population, by State
B:2   10–15 Shares of Free Black Males, by State
B:3   16+ Shares of Free Black Males, by State
B:4   10–15 Shares of Free Black Females, by State
B:5   16+ Shares of Free Black Females, by State
B:6   Male Shares of Urban Free Black Population
B:7   Urban Age Distribution: Free Black Males 10–15 Years Old
B:8   Urban Age Distribution: Free Black Males Aged 16 and Over
B:9   Urban Age Distribution: Free Black Females Aged 10–15 Years
B:10 Urban Age Distribution: Free Black Females Aged 16 and Over

B. Estimates of Free Blacks in Cities

B:11 Free Black Males in Cities, 1800; Estimates by Age Groups
B:12 Free Black Females in Cities, 1800; Estimates by Age Groups
B:13 Free Black Males in Cities, 1810; Estimates by Age Groups
B:14 Free Black Females in Cities, 1810; Estimates by Age Groups
B:15 Free Black Males in Cities, 1820; Revised Age Groups
B:16 Free Black Females in Cities, 1820; Revised Age Groups
B:17 Free Blacks in Cities, 1830; Revised Age Groups
B:18 Free Blacks in Cities, 1840; Revised Age Groups

C. Estimates of Free Blacks

B:20 Free Black Males by State, 1800; Estimates by Age
B:21 Free Black Females, by State, 1800; Estimates by Age
B:22 Free Black Males, by State, 1810; Estimates by Age
B:23 Free Black Females, by State, 1810; Estimates by Age
B:24 Free Black Males, by State, 1820; Revised Age Groups
B:25 Free Black Females, by State, 1820; Revised Age Groups
B:26 Free Black Males, by State, 1830; Revised Age Groups
B:27 Free Black Females, by State, 1830; Revised Age Groups
B:28 Free Black Males, by State, 1840; Revised Age Groups
B:29 Free Black Females, by State, 1840; Revised Age Groups

D. Estimates of Free Blacks in Rural Areas

B:30 Free Black Males in Rural Areas, 1800; Estimates by Age Group
B:31 Free Black Females in Rural Areas, 1800; Estimates by Age Group

Table Title

B:32   Free Black Males in Rural Areas, 1810; Estimates by Age Group
B:33   Free Black Females in Rural Areas, 1810; Estimates by Age Group
B:34   Free Black Males in Rural Areas, 1820; Revised Age Groups
B:35   Free Black Females in Rural Areas, 1820; Revised Age Groups
B:36   Free Blacks in Rural Areas, 1830; Revised Age Groups
B:37   Free Blacks in Rural Areas, 1840; Revised Age Groups

II. DATA ON SLAVE ESTIMATIONS

A. All Slaves

S:1   Male Shares of Slaves, by State, 1800 to 1860
S:2   0 to 9 Shares of Male Slaves, by State; 1800 to 1860
S:3   0 to 9 Shares of Female Slaves, by State; 1800 to 1860
S:4   Estimates of Male Slaves, by Age, by State, 1800
S:5   Estimates of Female Slaves, by Age, by State, 1800
S:6   Estimates of Male Slaves, by Age, by State, 1810
S:7   Estimates of Female Slaves, by Age, by State, 1810
S:8   Male Slaves, by State, 1820; Estimation of Revised Age Groups
S:9   Female Slaves, by State, 1820; Estimation of Revised Age Groups

B. Urban Slaves

S:10   Male Share of Urban Slaves, by State, 1800–1840
S:11   Male Slaves in Cities, 1820; Estimates of Revised Age Groups
S:12   Female Slaves in Cities, 1820; Estimates of New Age Groups for Urban Slaves
S:13   Age Shares of Urban Male Slaves, 1800 to 1840
S:14   Age Shares of Urban Female Slaves, 1800 to 1840
S:15   Estimation of Sex and Age Breakdowns of Urban Slaves, 1800
S:16   Estimation of Sex and Age Breakdowns of Urban Slaves, 1810

C. Rural Slaves

S:20   Estimates of Rural Male Slaves, by Age, by State, 1800
S:21   Estimates of Rural Female Slaves, by Age, by State, 1800
S:22   Estimates of Rural Male Slaves, by Age, by State, 1810
S:23   Estimates of Rural Female Slaves, by Age, by State, 1810
S:24   Estimates of Rural Male Slaves, by Age, by State, 1820
S:25   Estimates of Rural Female Slaves, by Age, by State, 1820

Bibliography

Bateman, Fred, and James Foust. "Agricultural and Demographic Records for Rural Households in the North, 1860." MS Census Sample.

Berlin, Ira, and Herbert Gutman. "Natives and Immigrants, Free Men and Slaves: Urban Workingmen in the Antebellum American South." *American Historical Review* 88, no. 5 (1984), 1175–1200.

Curry, Leonard. *The Free Black in Urban America, 1800–1850*. Chicago: University of Chicago Press, 1981.

David, Paul. "The Growth of Real Product in the United States Before 1840." *Journal of Economic History* 27, no. 2 (1967), 151–197.

Easterlin, Richard. "Interregional Differences in Per Capita Income, Population, and Total Income, 1840–1950." Studies in Income and Wealth, vol. 24. New York: National Bureau of Economic Research, 1960.
Gilchrist, David, ed. *The Growth of the Seaport Cities, 1790–1825.* Charlottesville: The University Press of Virginia, 1967.
Goldin, Claudia. *Urban Slavery in the American South, 1820–1860.* Chicago: University of Chicago Press, 1976.
———. "Female Labor Force Participation: the Origin of Black and White Differences, 1870 and 1880." *Journal of Economic History* 37, no. 1 (1977), 89–108.
Grabill, W. H., C. V. Kiser, and P. K. Whelpton. *The Fertility of American Women.* New York: Wiley and Sons, 1958.
Harrell, Virginia. *Vicksburg and the River.* Jackson: University Press of Mississippi, 1982.
Lebergott, Stanley. "Labor Force and Employment, 1800–1960." *Studies in Income and Wealth,* vol. 30. New York: National Bureau of Economic Research.
———. "A Reply to Harold Vatter." *Economic Development and Cultural Change* 23, no. 4 (1975), 749–750.
———. *The Americans.* New York: Norton, 1984.
Public Use Sample for 1900. MS Census Sample, 1980. Compiled under the direction of Samuel Preston and Robert Higgs. A User's Handbook was prepared by Stephen Graham, Center for Studies in Demography and Ecology, University of Washington.
U.S. Bureau of the Census. "The Development of the Urban-Rural Classification in the United States: 1874 to 1949." *Current Population Reports,* series P-23, no. 1 (1949).
———. *U.S. Census of the Population,* 1950, vols. 1 and 2.
———. "Urban Places Ranked by Population Size." Mimeo, 1977.
U.S. Census Office. *Second Census, Return of the Whole Number of Persons,* 1800; *Third Census,* 1810; *Fourth Census,* 1820; *Fifth Census,* 1830; *Sixth Census,* 1840.
U.S. Department of Commerce. *Historical Statistics of the United States.* Washington, D.C., 1975.
Uselding, Paul. "In Dispraise of the Muckrakers: United States Occupational Mortality, 1890–1910." *Research in Economic History* 1 (1976), 334–371.
Vatter, Harold. "Industrialization, Regional Change, and the Sectoral Distribution of the U.S. Labor Force, 1850–1880." *Economic Development and Cultural Change* 23, no. 4 (1975), pp. 739–748.
Wade, Richard. *The Urban Frontier.* Cambridge: Harvard University Press, 1959.
Walker, Peter. *Vicksburg: A People at War, 1860–1865.* Chapel Hill: University of North Carolina Press, 1960.
Weber, Adna. *The Growth of Cities in the Nineteenth Century.* 1899. Reprint. Ithaca: Cornell University Press, 1963.
Weiss, Thomas. "The Industrial Distribution of the Urban and Rural Workforces: Estimates for the United States, 1870–1910." *Journal of Economic History* 32, no. 4 (1972), 919–937.

———. "Revised Estimates of the United States Workforce, 1800 to 1860." *Studies in Income and Wealth,* New York: National Bureau of Economic Research, forthcoming.

Yasuba, Yasukichi. *Birth Rates of the White Population in the United States, 1800–1860.* Baltimore: Johns Hopkins University Press, 1962.

7
Investment Flows and Capital Stocks: U.S. Experience in the Nineteenth Century

ROBERT E. GALLMAN

I

There are two ways to estimate the value of the capital stock: by cumulating investment flows, following the perpetual inventory procedures developed by Raymond Goldsmith;[1] or by taking a census of the existing stock, enumerating each element and placing a value on it. With identical concepts in each case and perfectly accurate measurements, the two sets of results ought to be the same. In practice, measurements are never perfectly accurate and, historically, the concepts embedded in perpetual inventory and census-style estimates have often differed, so that one or the other had to be adjusted to permit close comparisons. Given these incongruities, the degree of consistency observed between U.S. capital stock estimates of the two types is encouraging. Census-style and perpetual inventory series have exhibited similar levels and trends, so that for many analytical purposes, it matters little which type is used. Quite the contrary obtains, however, when the focus is on short periods; in such cases the two series often trace discrepant courses.[2] In the study of business cycles or Kuznets cycles it matters a great deal which form of evidence is adopted.

Much of the research on which this paper draws was funded by the National Science Foundation, through a grant made to the National Bureau of Economic Research, and by the Kenan Foundation. Thanks are due to many people, but particularly to Steven Rosefielde, for help on a technical point; to Edward S. Howle, who, many years ago, worked on the census-style capital stock estimates described herein; to Jaikisan Desai, Michael Butler, and particularly, Colleen Callahan, who helped to complete that work and to produce the perpetual inventory estimates, also described herein; to Simon Kuznets and Raymond Goldsmith, whose work motivated the whole exercise; to the two discussants, Lance Davis and Gary Yohe, for useful comments; and to Karin Gleiter, who kindly read the paper in draft, more than once, and offered helpful suggestions.

[1] See, e.g., Raymond Goldsmith, *A Study of Saving in the United States,* vol. 1 (Princeton: Princeton University Press, 1955).

[2] See John Kendrick's remarks on pp. 24–25 of *Measuring the Nation's Wealth,* Studies in Income and Wealth, vol. 29 (New York: National Bureau of Economic Research, 1964). (This volume is a congressional document that the National Bureau of

The two types of estimates differ in other respects. Censuses of wealth have been taken only intermittently, so that capital series assembled from them are discontinuous, whereas perpetual inventory series are continuous, a great advantage for many purposes. Economists have worked out several useful concepts of value and systems of measuring capital consumption. Perpetual inventory procedures can be readily adapted to generate, from a given set of flow data and prices, a variety of capital stock estimates, reflecting different systems of valuation, average service lives, distributions of service lives, and systems of capital consumption. It is much more difficult—in many cases, virtually impossible—to manipulate census-style capital data to achieve similar ends.

On the other hand, census-style estimates are likely to be the more comprehensive of the two. Perpetual inventory estimates depend upon measurements of investment flows. Such measurements are almost bound to be more complete with respect to repetitive, market-bound events than with respect to their opposite. Should a farmer build a log cabin or a split-rail fence, his activity would probably not be captured in any official record of investment on which perpetual inventory estimates are based. But his log cabin and his fence would almost certainly turn up in a census of wealth, should one be taken. In modern economies, homemade cabins and fences are so few as to be negligible sources of discrepancies between aggregate stock and flow capital series. That would not have been the case in earlier days, when agriculture accounted for a larger fraction of activity, when the raw materials of construction lay within the reach of many farmers, when mechanization had not yet smoothed out the seasonal demands for farm labor, and when markets were so incompletely articulated that off-season work off the farm was not widely available. In those days farmers built many a fence and cabin in the off season from the materials drawn from farm woodlots.

There is ample reason, then, to attempt to build up two sets of capital stock estimates, one based on perpetual inventory and the other on census-style procedures. We can check one against the other and lay the basis for sensitivity testing with respect to analyses involving in-

---

Economic Research bound in hard covers and distributed through Columbia University Press as a volume in the Income and Wealth series.) See also Simon Kuznets, *National Product since 1869* (New York: National Bureau of Economic Research, 1946); Simon Kuznets, *Capital in the American Economy* (Princeton, N.J.: Princeton University Press, 1961); Lance E. Davis and Robert E. Gallman, "The Share of Savings and Investment in Gross National Product During the Nineteenth Century, United States of America," *Proceedings of the Fourth International Conference of Economic History* (Paris: Mouton, 1973).

vestment and the capital stock (e.g., analyses of the share of investment in income, the rate of growth of the capital stock, the sources of economic growth, the structure of the capital stock, etc.). Furthermore, in view of the existing range of analytical requirements—for gross and net series, for acquisition-cost, reproduction-cost, and market-value series—and the degree of uncertainty as to appropriate capital service lives and the pattern by which capital gives up value as it ages, there is reason to produce a wide array of perpetual inventory series, resting on a variety of assumptions with respect to these matters. Such considerations motivated the work that produced this paper.

The paper is concerned with the United States in the nineteenth century, a period that has suffered neglect with respect to the study of capital, at least as compared with the twentieth century. A new set of estimates has recently appeared, however, providing capital stock figures at decade intervals from 1840 through 1900.[3] The estimates are detailed and have been tested in a variety of ways. But many of the details depend upon census-style evidence, evidence that leaves something to be desired, in part because the capital concepts involved are not always perfectly clear. A central part of this paper is devoted to tests of these elements against perpetual inventory estimates assembled specifically for that purpose, and to a consideration of the implications of the results for the history of the U.S. capital stock. The perpetual inventory series are also used to explore the effects of choices with respect to average service life, retirement schedule, and depreciation technique on the level and rate of change of the measured capital stock.

The perpetual inventory series encompass two elements: manufactured producers' durables (i.e., tools, equipment, machines) and "other construction" (i.e., construction other than railroads, canals, farmland clearing, and construction carried out with farm materials). They are important elements, accounting for 80–90 percent of the conventionally defined nineteenth century U.S. domestic capital stock (exclusive of inventories), and 50–70 percent of the same stock, including the value of farmland clearing and first ground breaking.[4] Unfortunately, since the ultimate source for these series is production data, they do not provide the sectoral evidence available in the census-style series.

[3] Robert E. Gallman, "The U.S. Capital Stock in the Nineteenth Century," in *Long Term Factors in American Economic Growth,* Stanley L. Engerman and Robert E. Gallman, eds., Studies in Income and Wealth, vol. 51 (Chicago: University of Chicago Press, 1987). Estimates for the early part of the century have also been prepared, but have not yet been published.
[4] Ibid.

They are imperfect in other respects, as well. The basic annual investment series from which they were derived were initially assembled chiefly to establish secular levels of investment, and they have never been published in annual form, because their authors were doubtful that they could be depended upon to pick out year-to-year movements accurately.[5] For present purposes, however, that is not so serious a problem. Perpetual inventory capital stock series, after all, are cumulations of investment over several years. They are not likely to be unduly sensitive to spurious annual perturbations in the investment series, so long as errors more or less offset each other, and so long as the series are reliable indicators of, for instance, quinquennial or decennial levels. The basic annual investment series probably pass that test satisfactorily—at least their authors believed this to be so, since they published quinquennial and decennial averages.

A more serious problem is that the basic series cover only the years 1834–1859 and 1869–1909. To fill the gap of the sixties and to extend the evidence backward to the late eighteenth century (necessary if estimates of the stock of "other construction" were to be produced for the mid-nineteenth century), the basic series (hereafter, the Gallman series) had to be pieced out with evidence from Berry's monograph.[6] Since Berry's work consisted of carrying Kuznets's series backward from the late nineteenth century to 1790, and since the Gallman series are also linked to Kuznets's work (at 1909), this procedure seems reasonable enough, in a general way. But the nature of the Berry series poses some problems nonetheless. It was derived as the difference between national product and consumption (including government) and therefore has all the weaknesses of a series composed of residuals. Furthermore, components of investment are not distinguished, which means that the bases for carrying the two elements of the Gallman series, individually, across the 1860s and into the years before 1834 are by no means so strong as one could wish. Finally, the empirical bases for the Berry estimates become ever more frail as the series extends into the early nineteenth century and the late eighteenth.

[5] See Robert E. Gallman, "Gross National Product in the United States, 1834–1909," in *Output, Employment, and Productivity in the United States after 1800*, Dorothy S. Brady, ed., Studies in Income and Wealth, vol. 30 (New York: National Bureau of Economic Research, 1966), pp. 39–41, 64–71. In the period after the Civil War, the series depend importantly on Simon Kuznets' work sheets (kindly supplied by Kuznets) underlying *Capital in the American Economy*. Kuznets never published his annual series for the period 1869–1888.

[6] Thomas Senior Berry, *Revised Annual Estimation of American Gross National Product* (Richmond: The Bostwick Press, 1978).

These are important matters, but perhaps not quite so important as they seem at first. Presumably, the most doubtful elements of the pieced-together Berry-Gallman series are those that relate to the earliest period. But given the rapid rate at which the U.S. economy was growing, these elements bear only very modest weights in the determination of the capital stock estimates discussed in this paper—estimates beginning in 1840. So long as the remote Berry-Gallman figures pick up the trend level of investment, at least roughly, there is no serious problem, although it is certainly true that one should cast a more distrustful eye on the perpetual inventory estimates for 1840 than on the rest.

The rest deserve their share of suspicion, however. None of the investment flow evidence underlying the perpetual inventory estimates—the evidence of neither Berry nor Gallman nor Kuznets—can be regarded as of exceptionally high quality. The tests to be described in this paper are, then, tests of consistency between two series, both of which must be regarded with some suspicion. The consistency tests are intended as checks on both series, rather than on just one.

The required consistency tests are not easily made. As noted earlier, there are certain types of investment that appear in only one of the two series—census style or perpetual inventory. Thus, adjustments are called for before proper comparisons can be drawn. Furthermore, the conceptual content of the census-style estimates is not perfectly clear. That matter also must be clarified before proper comparisons can be made.

Section II treats the conceptual problem. Section III takes up the questions of the appropriate service life of capital, the retirement schedule, and the depreciation procedure. Section IV considers elements of the capital stock omitted from the two series, proposes appropriate adjustments, and exhibits the final comparisons. Section V pulls things together.

II

Capital stocks may be valued at acquisition cost, reproduction cost, or market value.[7] Each measure has its own special analytical uses. Acquisition cost is backward looking. A capital stock estimate so valued might be used to study savings behavior, since in this view the stock can be regarded as accumulated savings. Reproduction cost concerns the present. It conceives of the capital stock as the value of inputs re-

---

[7]Gallman, "U.S. Capital Stock," deals with this topic in greater detail.

quired to reproduce it, given current factor prices and production techniques.[8] Such measures are useful in the study of production relationships. Market value is forward looking. It is the discounted stream of anticipated returns to capital, and it would serve well, for example, in the analysis of aggregate consumption. It would be good to have all three types of measure, but commonly it is necessary to make do with one or two.

Market value is a net concept, since it takes into account only the remaining earning life of existing capital. Acquisition cost and reproduction cost can be measured both gross and net. The questions whether net measures should be produced, and if so how, are vexed; neither would be appropriately addressed here.[9] The subsequent sections rest on the assumption that both gross and net measures are legitimate, as are conventional methods of obtaining net measures.

If the economy were perpetually in equilibrium, if prices and productivity never changed, and if depreciation allowances accurately described the decline in the earning power of capital as time passed, then net acquisition cost would always equal net reproduction cost, which would always equal market value. In fact, these conditions do not obtain, and therefore the three measures are not equal.

These matters would be of no present importance, were it certain that the perpetual inventory and census-style estimates embraced the same valuation scheme. But that is not the case. The perpetual inventory series—which are expressed in constant (1860) prices—closely approximate reproduction cost, deviating from that standard only to the extent that the markets for new capital goods were out of equilibrium in 1860. The meaning of the census-style figures is less clear and may, in fact, differ from one figure or one year to the next.

---

[8] As Lance Davis pointed out at the Wesleyan Conference, a second version of net reproduction cost values each piece of capital at the price required to replace it in a given year (or the base year, in the case of constant-price estimates), replacement consisting of the substitution of an equally productive piece of capital. How "equally productive" should be defined is not perfectly clear, nor are the uses to which such a series could be put. One leading solution to the definition problem would turn the capital stock into a simple transformation of national income. See the exchanges between Edward F. Denison and Simon Kuznets, pp. 222–254 and 273–284, in *Problems of Capital Formation*, Franco Modigliani, ed., Studies in Income and Wealth, vol. 19 (Princeton, N.J.: Princeton University Press, 1957).

[9] The topic is discussed in Robert E. Gallman, "How Do I Measure Thee? Let Me Count the Ways" (and in the works cited therein, particularly vols. 2 and 14 of Studies in Income and Wealth), presented to the Conference on the Variety of Quantitative History, sponsored by the Weingart Foundation, the California Institute of Technology, and the Social Science History Association and held in Pasadena, California, in the spring of 1983.

The principal (but not exclusive) source of these data is the federal census, which collected wealth data from individuals, business firms (including farms), and tax officials. From the latter, the census requested statements of "true value," a concept that is itself ambiguous. While it is often understood by outsiders to refer to market value, tax officials seem to have something else in mind: perhaps what market value would be, were it not subject to temporary fluctuations proceeding from transitory interest rate shifts or cyclical booms and busts.

Businessmen might be thought to have provided the census takers with acquisition-cost values—book values, either gross or net. But one must remember that capital accounting was a new phenomenon in the nineteenth century. Most businessmen charged off capital as a current expense. For these people, tax records would have constituted the only 'books' from which the answers to the census taker's questions could be drawn.

The census instructions are not a great help in guiding one as to the meaning of value, and different modern analysts have interpreted them in different ways. There is strong support for the notion that the census was seeking acquisition cost—probably gross—when it approached businesses. However, the evidence indicates that net reproduction cost or market value was most often meant, at least in the latter part of the century. Consider the following definition of value, drawn from the 1890 manufacturing census questionnaire: value "should be estimated at what the works would cost in 1890, if then erected, with such an allowance for depreciation as may be suitable in the individual case." [10]

But however the matter is judged, it must be regarded as still in doubt. Thus if consistency tests are to be run between the two sets of capital estimates, and if the conceptual content of one set is uncertain, it behooves the analyst to consider first—before comparisons are drawn—the forces at work during the century driving acquisition, reproduction, and market values apart, and the strength of these forces. Only with this information in hand can the comparison of the two series be properly interpreted.

The acquisition cost and reproduction cost of a capital stock will differ if capital goods prices have changed over time, and if the changes have not offset each other. For example, if capital goods prices persistently rise, a capital stock will be smaller if measured in acquisition costs than if measured in reproduction costs. If capital goods prices persistently fall, the reverse will be true.

[10] *U.S. Eleventh Census*, 1892, vol. 6, part I, p. 10. The question addressed to farmers and householders appears to have referred to market value, or possibly to net reproduction cost.

TABLE 7.1
Construction Price Indexes, 1789–1798 through 1889–1898 (base, 1860), and Estimates of the Ratio of Acquisition Cost to Reproduction Cost for "Other Construction" Capital Stock, 1839–1899, Decennial Intervals

Dates	(1) Index of residential bldg. costs in Philadelphia	(2) Index of total U.S. construction costs	(3) Ratio of acquisition cost to reproduction cost (gross)
1789–98	96		
1794–03	110		
1799–08	110		
1804–13	118		
1809–18	121		
1814–23	126		
1819–28	108		
1824–33	102		
1829–38	101		
1834–43	97		1.06 (1839)
1839–48		95	
1844–53		95	1.05 (1849)
1849–58		97	
1854–63		110	1.00 (1859)
1859–68		117	
1864–73		125	0.93 (1869)
1869–78		107	
1874–83		109	0.96 (1879)
1879–88		118	
1884–93		109	1.00 (1889)
1889–98		100	
1894–03		104	1.05 (1899)

SOURCES:
Column 1: Donald R. Adams, Jr., "Residential Construction Industry in the Early Nineteenth Century," *Journal of Economic History* 25 (Dec. 1975), 813, Variant B, linked to Brady-Gallman index (implicit index of "new construction," Table A-3, p. 34, of Robert E. Gallman, "Gross National Product in the United States, 1834–1909," in Studies in Income and Wealth, vol. 30 (New York and London: Columbia University Press, 1966)) at census year 1839. The link was established in the following way: The Adams (calendar year) index numbers for 1839 and 1840 were averaged (1839: weight of 7; 1840: weight of 5) to approximate an index number for census year 1839 (86.3). This number was divided through the Brady-Gallman index number for census year 1839 (97.9), resulting in the ratio 1.134. The Adams index numbers, 1789–1839 (decade averages, unweighted), were multiplied by 1.134 and then rounded, to produce the values in column 1, which refer to calendar years. The Adams Variant B series was accepted in preference to Variant A, because the weighting scheme adopted by Adams in Variant B is similar to the one underlying the Brady-Gallman series.

Column 2: 1839–1848 through 1849–1858 and 1869–1878 through 1894–1903 are derived from Gallman, "Gross National Product" (see notes to column 1) and refer to all new construction.

1839–1848 through 1854–1863 are three-item averages, referring to 1839, 1844, 1849 (1839–1848), 1844, 1849, 1854 (1844–1853), etc. The years are census years, except 1863, which is a calendar year.

1869–1878 through 1894–1903 are weighted decade averages and refer to calendar years.

1859–1868 (calendar years) is based on interpolations of the Brady-Gallman estimates of 1860 and 1869, carried out on a construction-cost series derived for the purpose. The construction-cost series was computed from the David-Solar index of the common wage [Paul A. David and Peter Solar, "A Bicentenary Contribution to the History of the Cost of Living in America," in *Research in Economic History*, vol. 2 (Paul Uselding, ed.) (Greenwich, Conn.: J.A.E. Press, 1977)] and the Warren and Pearson price index of building materials. The weights used were the same as Adams' Variant B weights.

Column 3: See text. The service life adopted was 50 years, except for 1839, in which case I used 40 years. The price index numbers used to inflate the constant-price flow series are the index numbers contained in columns 1 and 2. (These inflators were required only for the years prior to 1869, since the Gallman flow series are available in both current and constant prices, 1869–1909.) Flows across each decade were inflated by decade average index numbers. The index numbers used to inflate the reproduction-cost stock series, 1869, 1879, 1889, and 1899, refer expressly to "other construction" and were derived from data underlying the data in column 2. The figures for 1879, 1889, and 1899 are weighted averages of prices for calendar years 1878 and 1879, etc., to approximate the census year. Such an adjustment was impossible for 1869, which refers to the calendar year. The index numbers used to inflate the reproduction-cost series, 1839, 1849, 1859, are the appropriate figures underlying column 2.

What happened to capital goods prices across the nineteenth century? Interestingly enough, the prices of construction goods apparently rose and fell periodically but exhibited no clear long-term trend (see Table 7.1, columns 1 and 2). Thus, one would suppose that acquisition costs and reproduction costs would differ little at the census dates, and that such differences as emerged would place acquisition costs sometimes above and sometimes below reproduction costs.

Experiments show that this is precisely what happened. The experiments were run in the following way. The constant-price annual flow series was cumulated to produce constant-price reproduction-cost estimates of the "other construction" capital stock, at decade intervals, from 1839 through 1899. The series was then inflated, using construction price index numbers relevant to the benchmark years, 1839, 1849, and so on. The annual flow series was then inflated and cumulated to form acquisition-cost estimates of the same capital stock for the same years. The ratios of the current-price acquisition-cost estimates to the current-price reproduction-cost estimates so produced are given in column 3 of Table 7.1.[11] It will be observed that the ratios are all close to a value of 1. In every case but one, acquisition cost is within 6 percent of reproduction cost. That must be regarded as very close, particularly given the fairly wide margins for error that must be allowed for all capital stock estimates in the nineteenth century. In view of these considerations, it is a matter of small importance whether the census measured "other construction" capital projects at acquisition costs or reproduction costs.

The situation with respect to manufactured producers' durables was very different, however. The prices were more variable, and they dropped throughout the century, but particularly sharply in the 1880s (see Table 7.2). Consequently, in 1890, acquisition cost (gross and net) was well above reproduction cost—a quarter to a third higher in the net variants, and two-fifths to almost three-fifths in the gross variants (Table 7.2). But by 1900 the two measures produced roughly the same values. No doubt acquisition cost was also slightly the higher in 1850 and 1880 (but not 1860), and perhaps more pronouncedly so in 1870, but not to the degree exhibited in 1890. It does matter, then, whether the census returns of the stock of manufactured producers' durables were measured in acquisition cost or reproduction cost. To interpret the census wealth estimates for 1890, one must know the concept that

---

[11] All the estimates computed were of gross stocks. We also made calculations with net stocks (straight-line depreciation) for 1869. The resulting ratio was 0.93, the same as the ratio of the gross estimates for that year.

TABLE 7.2
Manufactured Producers' Durables Price Indexes, 1839–1848 through 1899–1908 (base 1860), and Estimates of the Ratio of Acquisition Cost to Reproduction Cost, 1890, 1900, 1908

Dates	(1) Price index	(2) Ratios of acquisition cost to reproduction cost			
		Net valuation		Gross valuation	
		13 years[a]	18 years[a]	13 years[a]	18 years[a]
1839–48	114				
1844–53	109				
1849–58	108				
1869–78	88				
1874–83	72				
1879–88	55				
1884–93	42	1.24 (1890)	1.31 (1890)	1.42 (1890)	1.56 (1890)
1889–98	35				
1894–03	37	0.92 (1900)	0.96 (1900)	1.03 (1900)	1.09 (1900)
1899–08	38	0.99 (1908)	0.99 (1908)	1.02 (1908)	1.00 (1908)

SOURCES:

Column 1: See the source notes for column 2 of Table 7.1.

Column 2: See the source notes for column 3 of Table 7.1. The net valuations depend upon straight-line depreciation.

[a] Service lives.

guided the collection of the evidence in that year. It is less important, but would be useful, to have this information for other years as well.

Market price might deviate from net reproduction cost for any reason that could throw the new capital goods markets out of equilibrium, or alter the distribution of expected earnings among capital goods of differing age, or alter the appraisal of a given income stream. The first type of development is not of great interest in the context of this paper, since the measures of net reproduction cost used here are probably affected by disequilibrium in new capital goods markets and are, thus, a kind of mixture of reproduction cost and market value. The other two could be allowed for in forming net reproduction-cost estimates, service lives and depreciation systems being adjusted to reflect the changing market reality. But insofar as these decisions were left unchanged over extended periods, market value and net reproduction cost could and would diverge.

The distribution of earnings among capital goods of differing vintage might presumably change in response to technical changes, but I am unable to offer an account of precisely how earnings streams were altered and, thus, how the pattern of market capital values was affected in the nineteenth century. It is possible, however, to say something about the third (and probably most powerful) development listed above—changes in the appraisal of the income streams flowing from capital.

The market value of the capital stock represents the discounted anticipated income flowing from capital. A rise in the discount rate—the rate of interest—will tend to reduce the market value of capital, ceteris paribus, while a decline will tend to increase it. To judge the effect of changes in interest rates on the value of capital, one needs to know which interest rates are the relevant ones, the extent to which they changed, and the age distribution of the capital stock beforehand. With this information and an annuity table, one can readily compute the change in the market value of capital.

The focus here is on the capital stock values derived from census wealth data. The question is this: If the census had appraised capital at market value, how far would the interest rate changes across the nineteenth century have altered census capital stock values, relative to what they would have been, had capital been appraised at net reproduction cost, with fixed estimating parameters from one census to the next? For example, assume that in 1840 the market value of the stock of manufactured producers' durables had been equal to the net reproduction cost of this capital, the latter computed on the assumption of a thirteen-year average service life and straight-line depreciation. How far would this equality have been disturbed by the observed interest rate changes of the nineteenth century?

To answer this question, one must imagine how census appraisers (officials or respondents) went about their task. It may be safely assumed that if they attempted to place market values on capital, they were well aware of the influence of interest rates on capital values and therefore took interest rates into account. Whether they would have looked to nominal or real interest rates is by no means certain, but both possibilities were considered. Surely they would have been concerned not with the interest rate on the morning of the day on which their appraisals were made, but rather with the general level of the interest rate in the census year, and perhaps even the year or two preceding it. That is, it may be assumed that they would have left out of account what they regarded as temporary, short-run movements.

With these considerations in mind, I examined interest rate changes from one census year to the next, coming to the following conclusions concerning patterns of change:[12]

---

[12] Based on Lance E. Davis and Robert E. Gallman, "Capital Formation in the United States During the Nineteenth Century," *Cambridge Economic History of Europe*, vol. 3, part 2, (Cambridge: Cambridge University Press, 1978), p. 23, Table 7, col. B(1), and p. 26, Table 9, col. 5, and sources underlying these tables.

Investment Flows and Capital Stocks 225

Intercensal periods	Interest rates Real	Interest rates Nominal
1840–1850	fell	no change
1850–1860	rose	fell modestly
1860–1870	rose modestly	fell modestly
1870–1880	fell	fell pronouncedly
1880–1890	fell pronouncedly	fell
1890–1900	fell very pronouncedly	fell

The table suggests that the period from 1870 onward is worthy of examination. From 1870 to 1880 the nominal rate fell pronouncedly, while from 1880 to 1890 and from 1890 to 1900 it was the real rate. In the first two instances, the decline was roughly from 7 percent to 5 percent, in the third, roughly from 5 percent to 1 percent. In each case the market price of the capital stock must have gone up. Assuming that before each rise market price had been equal to net reproduction cost, how far would the former have increased above the latter as a result of the interest rate change? Answers to this question were worked out with an annuity table and the perpetual inventory estimates, using a service life of thirteen years (manufactured producers' durables) and straight-line depreciation. The ratios of market value to net reproduction cost emerging from these calculations are

$$\text{Change from } 7\% \text{ to } 5\% = 1.10$$
$$\text{Change from } 5\% \text{ to } 1\% = 1.24$$

One's first impression is that these differences are really quite small. This is particularly the case if one is concerned, as in this paper, chiefly with the probable differences between the perpetual inventory series and the measures taken from the wealth census. Some part of the effect of falling interest rates on the value of capital—that part that has to do with the pricing of new capital—is reflected in the perpetual inventory series. Thus, if the census wealth and perpetual inventory series were in all ways consistent, except in mode of valuation, and if the census-of-wealth data were expressed in market value and the perpetual inventory series in net reproduction cost of the form previously attributed to it, then the ratios in the tabulation would actually overstate the quantitative differences between the two series.

Second thoughts suggest the following qualifications, however.[13] The service life selected above—thirteen years—may not be unrepresentative of manufactured producers' durables, but it is short for im-

---

[13] This paragraph and the next are taken virtually verbatim from Gallman, "How Do I Measure Thee?" pp. 21–23.

provements. Changes in interest rates have greater effects on long-lived capital. Thus, for improvements (construction), computed changes in value would surely be greater than those recorded above. Furthermore, since the interest rate seems to have been falling from at least 1870 to 1900, it is possible that the deviation between reproduction cost and market value would continue to grow from one census date to the next, in which case the two might diverge at 1900 by as much as 50 percent ($1.10 \times 1.10 \times 1.24 = 1.50$). (This assumes that the experiences of the decades 1870–1880 and 1880–1890 were similar and ignores the qualification advanced in the previous paragraph.) Such a conclusion surely goes too far, however, since it rests on the implicit assumption that reproduction cost would remain unchanged. In fact, with the interest rate falling and investment being encouraged, one would expect some tendancy for reproduction cost to rise (relative to market price) toward a new equilibrium. This would be a factor counteracting the widening of the gap between market price and reproduction cost. Furthermore, the calculations carried out above rest on the implicit assumption that the income-earning capacity of capital remained unchanged. But, ceteris paribus, one would expect that a flood of new investment would tend to lower income and thus reduce the market value of capital.

Clearly, the calculations are less than conclusive (especially since they take into account only one element affecting market value). Nonetheless, the modest change in market value occasioned by a fall from a rate of 7 percent to one of 5 percent remains impressive and, despite all qualifications, even the effects of a change from 5 percent to 1 percent appear rather modest. In terms of the practical problems to be discussed in the next section, it seems possible to conclude that at the decennial census dates 1840–1890, reproduction cost and the market price of capital were unlikely to have been very far apart, although at the last of these dates, and perhaps the one before as well, market price probably exceeded reproduction cost. This was also almost certainly true in 1900, and the margin between these two measures was the greater at that date.

In summary, the constant-price perpetual inventory series approximate reproduction-cost series, while the series derived from the census wealth data may be valued at reproduction cost or market value or acquisition cost. The possible conceptual differences are apparently empirically unimportant so far as the antebellum period is concerned. With respect to the postbellum period, the question whether the construction elements of the capital stock are measured in acquisition costs or reproduction costs is also unimportant, from a practical point of view. Where conceptual differences *are* important, reproduction cost is a smaller value than acquisition cost (manufactured producers' durables, 1890) and market value (all capital, 1900). If the comparison of the perpetual inventory and census wealth series were to show that the latter exceeded the former in 1890 (manufactured producers' durables), it would appear to be evidence that the census valued wealth at acquisition cost. If, on the other hand, the census estimates of all types exceeded the perpetual inventory estimates in 1900, it would be evidence that the census valued wealth at market value. With this background the relevant comparisons may be examined.

III

To compute perpetual inventory estimates, one must establish service lives for the relevant types of capital, the pace at which each type of capital lost value as time passed (the depreciation schedule), and the pattern in which capital retirements took place. Since the perpetual inventory estimates under discussion were assembled to test the census wealth data, it was necessary to make allowance for casualty losses and to keep in mind, while choosing among depreciation schedules, the manner in which the census wealth data were assembled. That is, census wealth data are net of casualty losses. Comparable perpetual inventory estimates must thus also be net of casualty losses. Census wealth data represent appraisals by owners or officials. Thus comparable perpetual inventory estimates must capture the mental processes of nineteenth century appraisers.

I have not computed service lives from nineteenth century evidence, although there are surely data among census and business records by which such computations could be made. For example, the Tenth Census contains data from which the service lives of railroad rails and ties (of various specifications) have been computed.[14] Lance Davis, Teresa Hutchins, and I have assembled a set of data concerning the New Bedford whaling fleet, from which service lives and the incidence of casualty losses have been calculated.[15] There must be much more evidence of this type, particularly in business records. But I have not been able to assemble a full set of such data and have therefore accepted guid-

[14] See, for example, Robert W. Fogel, *Railroads and Economic Growth* (Baltimore, Md.: Johns Hopkins Press, 1964), p. 172.

[15] See Lance E. Davis, Robert E. Gallman, and Teresa Hutchins, "The Structure of the Capital Stock in Economic Growth and Decline: The New Bedford Whaling Fleet in the Nineteenth Century," Ch. 10 of this volume.

ance from the work of Simon Kuznets and Raymond Goldsmith, both of whom, however, have been concerned chiefly with twentieth century experience, not nineteenth.[16]

Kuznets, in his research on the late nineteenth century, adopted a service life of thirteen years for manufactured producers' durables, and fifty years for improvements, figures that include an allowance for losses by fire, but not other casualties.[17] Furthermore, the improvements include railways and waterways, unusually long-lived capital that is excluded from the perpetual inventory series discussed in this paper. Thus fifty years may be an excessive service life for this exercise. As a check, Goldsmith's service life data drawn from Bulletin F were weighted up (by sector and type of capital), with data from the census-style capital stock series.[18] These calculations yielded values of seventeen years for durables and fifty-two years for improvements (exclusive of railroads and canals); neither figure includes an allowance for casualty losses.

Perpetual inventory producers' durables series based on both thirteen- and seventeen-year service lives were computed, but while the latter presumably consists of upper-limit estimates in each year, the former may not constitute lower limits, in view of Kuznets's evidence. With respect to improvements, series using forty- and fifty-year service lives were computed. It is possible that these two values do describe limits within which the appropriate service life lies, although one can be surer with respect to the upper bound than with respect to the lower.

It is possible, of course, that average service lives changed across the nineteenth century. Experiments with weighting up the Bulletin F evidence revealed no shifts in average lives occasioned by changes in weights (i.e., changes in the structure of the stock of durables and of improvements). But the information used for the purpose is not detailed (as to types of capital). In any case, there may have been shifts in durability or in the rate of obsolescence which influenced average service life by type of capital. That must be borne in mind when the two sets of capital stock estimates are compared.

Estimates were made based on three systems of capital consump-

[16] The next several sentences are adapted from Gallman, "How Do I Measure Thee?" pp. 27–28.

[17] Simon Kuznets, *National Product since 1869*, pp. 116–117 (inferred from the notes to col. 1, lines 1–10 and col. 4, lines 3–9, of the table). See also p. 197, where the content of capital consumption is defined.

[18] Raymond W. Goldsmith, "A Perpetual Inventory of National Wealth," in *Conference on Research in Income and Wealth*, Studies in Income and Wealth, vol. 14 (New York: National Bureau of Economic Research, n.d.), pp. 14–17, 20–24. See also Goldsmith, *A Study of Saving*, vol. 3, Table W-7, pp. 32–38.

tion: straight line, declining balances, and B.L.S. concave. The first would presumably come closest to replicating census values, if census enumerators or their respondents in fact estimated the cost of reconstructing each piece of capital, then chose a service life, and then computed the depreciation to be deducted from the value of each piece of capital. But it is possible that estimators did not go through all of these steps, at least consciously. It is also possible that they used rules of thumb that in fact reflected a different depreciation scheme. Certainly it would not be surprising if they believed that capital lost value with particular rapidity—or, for that matter, with particular slowness—in the first years of life, adopting in this way attitudes that are embodied in declining-balances and B.L.S. concave procedures. Therefore, while it was expected that the best results would come from the first technique, computations were carried out for all of them.

Two separate retirement schedules were made. The first rests on the assumption that all pieces of capital of a given type were retired at the same age. For example, in the case of the lower-bound durables estimates, it was assumed that all durables lasted exactly thirteen years. The second set of estimates computed makes provision for both early and late retirements.[19] While it is based on twentieth century experience, rather than nineteenth, it is more realistic than the assumption of a uniform retirement age. But it poses a problem: it is a formidable consumer of data. Thus, to produce estimates of the value of improvements for years before 1889, based on a fifty-year service life, requires data running deep into the eighteenth century, data that do not appear to exist. Estimates can be produced, of course, if zeros are entered for missing values, and such computations were made. Given the rapid pace at which investment grew in the nineteenth century and given the nature of the retirement distribution, these estimates are unlikely to deviate very far from true values in the years with which this paper is concerned. But they are biased downward, and the bias is more serious the earlier the date to which the estimate refers. For this reason, one set of estimates was computed resting upon these procedures and another depending upon the assumption of a common retirement age. The two sets of estimates, in fact, differ little with respect to level, and even less with respect to trend. Consequently, it matters little, for present purposes, which set of estimates is employed.

[19] R. Winfrey, *Statistical Analysis of Industrial Property Retirements,* Iowa Engineering Experiment Station, Bulletin 125, 1935, as reported in Allan H. Young and John C. Musgrave, "Estimation of Capital Stock in the United States," in *The Measurement of Capital,* Dan Usher, ed., Studies in Income and Wealth, vol. 45 (Chicago and London: University of Chicago Press, 1980).

TABLE 7.3
Ratios of Gross and Net Perpetual Inventory Capital Stock Estimates to Census-Style Capital Stock Estimates, 1840–1900

PANEL A: GROSS ESTIMATES

	Improvements 40 yrs.[a]	Improvements 50 yrs.[a]	Producers' durables 13 yrs.[a]	Producers' durables 17 yrs.[a]
1840	1.19	1.21	1.21	1.35
1850	1.45	1.48	1.31	1.54
1860	1.40	1.43	1.26	1.43
1870	1.67	1.73	1.89	2.17
1880	2.20	2.31	2.10	2.42
1890	1.89	1.99	1.51	1.73
1900	2.15	2.31	1.47	1.76
MEAN	1.71	1.78	1.54	1.77

PANEL B: NET ESTIMATES

	Improvements, Straight line 40 yrs.[a]	Improvements, Straight line 50 yrs.[a]	Improvements, Declining balance, 50 yrs.[a]	Producers' durables, straight line 13 yrs.[a]	Producers' durables, straight line 17 yrs.[a]
1840	0.90	0.96	0.79	0.77	0.89
1850	1.05	1.14	0.92	0.83	0.98
1860	1.02	1.10	0.89	0.77	0.91
1870	1.12	1.24	0.97	1.31	1.49
1880	1.49	1.65	1.29	1.22	1.48
1890	1.28	1.42	1.12	.95	1.11
1900	1.46	1.61	1.25	.86	1.04
MEAN	1.19	1.30	1.03	.96	1.12

SOURCES: See text.
[a] Service lives.

Table 7.3 contains the results of a first effort to compare the perpetual inventory and census-style capital stock estimates. Each entry expresses one of the former estimates as a ratio of one of the latter. Both gross and net perpetual inventory estimates were prepared; the gross figures represent each of the four average service lives deemed relevant: 40 years and 50 years, in the case of improvements; 13 years and 17 years, in the case of producers' durables (machinery and equipment). Only the net calculations that produced the closest fits to the census-style estimates figure in the ratios computed for the table. At least one estimate for each service life is included. A common age of retirement was assumed in the case of improvements, to avoid the computational problem discussed above. In the cases of the producers' durables, that assumption was unnecessary.

Investment Flows and Capital Stocks 231

Even a casual study of the table reveals several important points. All of the ratios in the columns headed "gross" are greater than one—several substantially so—while this is not true of the ratios in the columns headed "net." *Gross* and *net* refer to the perpetual inventory estimates. Since the net values correspond the more closely to the values derived from census data, the results are consistent with the notion that the census returns are expressed in net values. Of course, they are also consistent with the idea that the census data are gross, but that they or the perpetual inventory data are subject to serious measurement error that seriously understates the former or seriously overstates the latter. These possibilities cannot be excluded, of course, but they seem less probable. It is likely that the census data are truly net.

Assuming that this judgment is correct, what do the net ratios reveal about the degree of consistency between the two sets of series? Since the two sets of series do not contain precisely the same components (see above), the data underlying Table 7.3 need to be adjusted before a final answer to this question can be given. But a preliminary answer can be offered, if an appropriate standard of consistency can be established.

Suppose that the margin for measurement error in each series were as low as 10 percent—it may very well be higher—and that none of the series were biased, so that in any given year a positive error was as likely as a negative error. The maximum relative deviation between the two series in two successive years would then appear if a set of errors of the following type were to emerge:

	Year 1	Year 2
Perpetual inventory	−10%	+10%
Census style	+10%	−10%

Now supposing that the two series were perfectly consistent, except for these random errors, the ratios for the two years—corresponding to those in Table 7.3—would be 0.82 in the first year and 1.22, in the second. That is, 0.82 and 1.22 are values that can occur, even if the two types of estimates are fundamentally consistent, but subject to independent measurement errors of as much as ±10 percent.

Now notice that of the thirty-five "net" ratios in Table 7.3, twenty fall within this range and another five are within 5 percentage points of the limits of this range. Is that good or bad? It seems moderately good—that is, suggests consistency—although the test is not very demanding. This is a point which will be taken up further below.

There are also some details in Table 7.3 that are worth noticing. Of the five ratios for 1900 that lie in the net columns, four exceed the value of 1, three exceed the values recorded for 1890, and the other two fall only moderately short of the 1890 values. Abstracting from the possible errors discussed in the preceding paragraph, the ratios for 1900 would have been lower than those for 1890, and it might also have been expected that they would fall well below a value of 1, *if census returns had been expressed in market values* (see section II, above). One of the two producers' durables ratios for 1890 also exceeds a value of 1, while the other is very close to 1. Had the census valued capital at acquisition cost, both of these ratios would have been well below a value of 1. The ratios for 1890 and 1900, thus, are inconsistent with the idea that the census valued capital at acquisition cost or market value, and they are consistent with the idea that capital was valued at net reproduction cost. That suggests that no differences in valuation criteria stand in the way of the comparison of the perpetual inventory and census-style capital stock estimates. It also indicates that the census-style estimates can be treated, for analytical purposes, as net reproduction-cost estimates, although one should bear in mind that for the antebellum years, and in some measure for the postbellum years as well (see section II), the three different systems of valuation are likely to have yielded very similar values.

Introducing the possibility of measurement error, of course, blurs the clear outlines of these conclusions. But they are probably not completely erased. The results of the consistency tests square with what informed students might have supposed before the fact. It is therefore reasonable to accept the view that the census-style figures are truly net and are truly valued at reproduction cost (at those few dates where the valuation concept matters); at least these conclusions can be accepted in the preliminary way in which even the strongest research results should be accepted.

IV

The census-style data include all capital in each sector covered, regardless of where it was produced and regardless of the materials used. The perpetual inventory series are more narrowly conceived. They include, under the heading "producers' durables," only the products of census establishments (adjusted for foreign trade in durables, of course); they exclude durables made by very small firms and implements produced at home or on the farm. These omissions are unimportant throughout, but they were more important at the beginning of the period under consideration than at the end. Thus the perpetual inventory series should increase faster than the census-style series, as they do (see Table 7.3).

The upward bias imparted to the rate of change of the perpetual inventory improvements series is probably more serious. The estimates include all improvements (except railroads and canals) carried out with construction materials produced by census firms (again, adjusted for foreign trade flows). The census-style capital stock estimates, however, include in addition residences, sheds, barns, and the like, produced from farm materials. For example, log cabins and barns are included in the census-style estimates, but not in the perpetual inventory estimates.[20] Since these types of capital were more important earlier in the period than later, one would expect the perpetual inventory series to exhibit higher rates of growth than the census-style series, which in fact, they do (see Table 7.3).

While the census-style data are comprehensive with respect to the industrial sectors covered, they do not cover all sectors. The principal omission consists of highways and highway bridges. Insofar as these projects were constructed from materials returned by the census, the value of such capital is included in the perpetual inventory improvements series. It seems likely that highways and bridges, so constructed, increased in relative importance over time, which is yet another reason why the perpetual inventory improvements series could be expected to exhibit higher rates of growth than the census-style series.

In summary, were it possible to remove these elements of incomparability lying between the perpetual inventory and census-style series, the ratios contained in the "net" columns of Table 7.3 would probably be closer to values of 1, although they would certainly not all achieve a value of 1.

Finally, the two sets of series from which the ratios of Table 7.3 were computed treat the losses of capital during the Civil War differently: the census-style estimates are net of such losses, while the perpetual inventory estimates are not. Removal of this inconsistency might further diminish the differences between the two sets of series.

The best estimate of Civil War destruction of capital is one prepared by Goldin and Lewis.[21] It covers only Southern losses—the implication being left that Northern losses were negligible—and its authors

---

[20] Where it was possible to identify census-style capital produced from farm materials (e.g., certain types of fences, the value of land clearing), it was of course deleted from the estimates used in the consistency tests described in this section. But the census did not distinguish buildings by the types of materials from which they were built.

[21] Claudia Goldin and Frank Lewis, "The Economic Cost of the American Civil War: Estimates and Implications," *The Journal of Economic History* 35 (June 1975), 308. The Goldin and Lewis figure is the Civil War loss, discounted back to 1861. We probably should have used the undiscounted figure (about $0.2 billion higher), but in view of the roughness of our calculations, we decided that this would represent an unjustified refinement.

regard it as an upper-bound estimate. However, if northern losses were, in fact, more than negligible, the figure may not constitute an excessive appraisal of the losses of North and South combined. The North was not the theater of much of the war, of course, although southern raiders did do some damage. Greater losses were suffered at sea. Southern cruisers appear to have injured the U.S. whaling fleet seriously and to have induced the transfer of part of the merchant marine to foreign ownership, a transfer that was not immediately reversed with the end of the war. The real value of U.S. shipping was only slightly greater in 1870 than in 1860. Presumably, the transfer of ownership of vessels simply changed the form in which U.S. capital was held, diminishing the value of shipping, and producing a compensatory change in net claims on foreigners. But since the latter are unrepresented in the series underlying Table 7.3, and the former is reflected only in the postwar census-style estimates, the transfer is a source of difference between the two sets of estimates forming the numerators and denominators of the ratios. It does not stretch the meaning of words too far to attribute this element of the difference to northern wartime "losses" of capital.

If the Goldin and Lewis estimate exaggerates Southern losses—as they believe it does—it probably does not exaggerate Southern and Northern losses, taken together, especially if the element of "loss" discussed immediately above is included. Indeed, it may even understate the true total. For present purposes, that is a matter of small importance, since this bias is offset by the fact that the Goldin and Lewis figure includes the value of certain types of destroyed capital (railroads, animal inventories) that have no bearing whatever on the ratios displayed in Table 7.3. Whether the biases precisely offset each other or not cannot be established, but in what follows it is assumed that they do.

Table 7.4 contains ratios from Table 7.3, recomputed to bring the numerators and denominators into closer conceptual conformity. Specifically, estimated Civil War losses (appropriately depreciated) were deducted from the numerators. To make the computations, it was assumed that total losses came to $1.5 billion (Goldin and Lewis's estimate of $1.487 billion, rounded) and that four-fifths of the capital destroyed ($1.2 billion) consisted of improvements, and one-fifth ($0.3 billion) of manufactured producers' durables.

It was also assumed that the improvements destroyed were distributed among vintages in the same proportions as were improvements in general, that the average service life of all improvements was forty years, and that destruction was centered on the year 1864. In the case of producers' durables, it was assumed that while the average service

TABLE 7.4
Ratios of Net Perpetual Inventory Capital Stock Estimates (adjusted for Civil War losses) to Census-Style Capital Stock Estimates, 1840–1900

PANEL A: IMPROVEMENTS

	Straight line		Declining balance, 50 yrs.	Mean of cols. 1 & 2
	40 yrs.	50 yrs.		
1840	0.90	0.96	0.79	0.93
1850	1.06	1.14	0.92	1.10
1860	1.02	1.10	0.89	1.06
1870	0.95	1.07	0.79	1.01
1880	1.42	1.57	1.22	1.50
1890	1.27	1.41	1.11	1.34
1900	1.46	1.61	1.25	1.54
MEAN	1.15	1.27	1.00	1.21

PANEL B: PRODUCERS' DURABLES

	Straight line		Mean of cols. 1 & 2
	13 yrs.	17 yrs.	
1840	0.77	0.89	0.83
1850	0.83	0.98	0.91
1860	0.77	0.91	0.84
1870	1.22	1.40	1.31
1880	1.22	1.48	1.35
1890	0.95	1.11	1.03
1900	0.86	1.05	0.96
MEAN	0.95	1.12	1.00

PANEL C: IMPROVEMENTS AND PRODUCERS' DURABLES

	Weighted means of:	
	Panel A, col. 4 & Panel B, col. 3	Panel A, col. 3 & Panel B, col. 3
1840	0.90	0.79
1850	1.05	0.92
1860	1.02	0.88
1870	1.08	0.91
1880	1.46	1.25
1890	1.23	1.08
1900	1.29	1.13
MEAN	1.15	.99

SOURCES: See text.

life of the stock, as a whole, was thirteen years, the lost property—since it must have consisted disproportionately of shipping, farm vehicles, and other long-lived equipment—had an expected average service life of twenty years, perhaps an upper bound. It was assumed that losses centered on the year 1863, a date lying between the time of the principal transfer of shipping to foreign ownership and the period of greatest military destruction in the South.[22]

The adjustments improve the results of Table 7.3 by reducing the ratios for 1870 and 1880 and bringing the mean ratios closer to values of 1. That two of the ratios for improvements fall below 1 in 1870—well below, in the case of one of them—is a little troubling. The 1870 census is widely believed to have been short, particularly in the South.[23] One would therefore expect to find the 1870 ratio to be larger than 1, even after adjustment of the perpetual inventory series for Civil War losses.

Nonetheless, given the nature of the data—and particularly that the 1870 and 1840 perpetual inventory estimates are heavily dependent on disparate series patched together—the degree of consistency attained by the two sets of series is moderately reassuring. Notice that the declining-balances improvements series (Panel A, column 3) and the mean of the two producers' durables series track the two census-style series reasonably well, particularly when one allows for the incompleteness of each set of series.

Happily, the degree of consistency improves when the level of aggregation is increased. That is as it should be, in view of the fact that several of the census-style estimates were made by distributing a total between its improvements and producers' durables components. Errors made at that level wash out with aggregation, of couse. Panel C of Table 7.4 shows that the weighted average ratios are better than the component ratios from which they were assembled, and that the combination of the declining-balances improvements estimates and the straight-line producers' durables estimates yields a fairly plausible set of ratios—the one large outlier appears in 1880. Even that value lies only barely outside the boundaries established by assuming that each series is subject to errors as large as 10 percent, and that the errors are distributed among years randomly (see above).

[22] In principle, a separate calculation of net losses should be made for each of the five primary series underlying Table 7.4, instead of the two sets of estimates I made. I did not carry through with this refinement, but doubt that it would alter Table 7.4 very far.

[23] Roger Ransom and Richard Sutch, "The Impact of the Civil War and of Emancipation on Southern Agriculture," *Explorations in Economic History* 12 (Jan. 1975), 10.

The position of 1880 as outlier calls for a little further consideration, nonetheless. Are there peculiarities surrounding the evidence for that year that account for the differences between the two sets of series? So far as the producers' durables series are concerned, we know of nothing. It is true that the prices of producers' durables fell very dramatically in the postwar years. If the weighting schemes underlying the deflation for the two series differed, that might produce contrasting results of the sort we observe. By the nature of the series, this is a difficult possibility to check. Nonetheless, the patterns of change described by the implicit deflators of the census-style stock estimates and the flow data underlying the perpetual inventory series move nicely in parallel, picking out precisely the same periods of rapid and slow decline. The 1880 problem does not appear to be rooted in deflation.

The annual construction-flow data for the years 1869–1909 that underlie the perpetual inventory improvements estimates are based on a series prepared by Simon Kuznets for *Capital in the American Economy*. To calculate the estimates here, this series was reworked, distinguishing components thereof, and altering the total construction flows, particularly for the earlier years.[24] Were these adjustments well advised? Did they influence importantly the "other" construction component relevant to the present discussion? There are two ways to approach these questions: by recomputing the perpetual inventory estimates on the basis of the Kuznets series, to see whether a better fit with the census-style estimates can be obtained; and by considering the rationale for the original adjustments to the Kuznets series.

In computing the census-style estimates, I have departed from the practice of another formidable authority, Raymond Goldsmith. In his own work with the nineteenth century capital stock, Goldsmith has assumed that nonfarm residences typically accounted for three-quarters of the value of nonfarm residential real estate. It has been assumed here that the figure was probably closer to 64 percent. Clearly, had Goldsmith's example been followed, the census-style capital stock estimate for 1880 would have been higher, and the ratios in columns 1 and 3 of Table 7.4 for 1880 would have been lower. But presumably the ratios for all the other years would also be lower, which would not be altogether a desirable result.

Table 7.5 was assembled to test the proposition that shifting to the original Kuznets flow data and adjusting the census-style estimates to reflect Goldsmith's judgment as to the relative importance of structures in nonfarm residential real estate would markedly improve the quan-

---

[24] Gallman, "Gross National Product," pp. 37–39.

TABLE 7.5
Ratios of Net Perpetual Inventory Capital Stock Estimates
(adjusted for Civil War losses) to Census-Style Capital Stock Estimates,
1840–1900, Land Improvements

	Straight line 40 yrs.	Straight line 50 yrs.	Declining balance, 50 yrs.	Mean of cols. 1 & 2
\multicolumn{5}{c}{PANEL A: GOLDSMITH ADJUSTMENT TO CENSUS-STYLE ESTIMATES}
1840	0.85	0.91	0.75	0.88
1850	0.99	1.05	0.86	1.02
1860	0.95	1.02	0.82	0.99
1870	0.89	1.00	0.74	0.95
1880	1.32	1.46	1.13	1.39
1890	1.18	1.31	1.02	1.25
1900	1.37	1.50	1.16	1.44
MEAN	1.08	1.18	.93	1.13
\multicolumn{5}{c}{PANEL B: KUZNETS ADJUSTMENT TO PERPETUAL INVENTORY SERIES}
1840	0.90	0.96	0.79	0.93
1850	1.06	1.14	0.92	1.10
1860	1.02	1.10	0.89	1.07
1870	0.88	1.00	0.72	0.95
1880	1.18	1.33	0.98	1.26
1890	1.11	1.22	0.94	1.17
1900	1.37	1.49	1.08	1.44
MEAN	1.07	1.18	.90	1.13
\multicolumn{5}{c}{PANEL C: GOLDSMITH AND KUZNETS ADJUSTMENTS}
1840	0.85	0.91	0.75	0.88
1850	0.99	1.05	0.86	1.02
1860	0.95	1.02	0.82	0.99
1870	0.82	0.93	0.68	0.88
1880	1.10	1.23	0.91	1.17
1890	1.02	1.13	1.15	1.08
1900	1.28	1.39	1.08	1.34
MEAN	1.00	1.09	.89	1.05

SOURCES: See text.

titative fit of the perpetual inventory and census-style estimates. Panel A, which incorporates only the adjustment of the census-style estimates to bring them into closer conformity with Goldsmith's views, shows that the adjustment does not altogether solve the 1880 "problem." As to the ratios for the other years, some are improvements on those appearing in Table 7.4, but others are not. The test does not provide a secure basis for choosing between the Goldsmith and Gallman judgments on this point. The differences between the Gallman estimates and the set that would be substituted for them in the event that we accepted Goldsmith's view are not very large, after all, and the test is by no means a refined one.

Investment Flows and Capital Stocks

The ratios in Panel B are only rough approximations of the ratios that would have emerged had the perpetual inventory estimates been reworked using Kuznets' data. To recompute the perpetual inventory series, it would be necessary to distribute the Kuznets flow estimates between the two components, "railroad construction" and "other construction," since Kuznets did not himself distinguish these components. For purposes of the computations underlying Panel B, it was assumed that the ratio of this total construction-flow estimate to Kuznets'[25] would be an appropriate basis for adjusting the "other construction" flow data to a basis consistent with the Kuznets' series. That assumption almost certainly resulted in too large an adjustment (for reasons to be discussed below), so that the contrasts between the relevant Table 7.4 and Table 7.5 ratios are, in fact, too great.

With that qualification in mind, it can be said that the "Kuznets adjustment" does improve the fit between the perpetual inventory and census-style series in 1880–1900, but in two of the variants it produces a much poorer fit in 1870.

Despite the seeming overall improvement occasioned by the adjustments underlying Panel B, there are good reasons why they should be rejected. There are three differences between Kuznets' annual construction-flow estimates and mine, and in each of these cases mine were the beneficiaries of work done (by other scholars) between the time when Kuznets constructed his series and when I made mine. Harold Barger worked out margin estimates for wholesale and retail trade, Dorothy Brady assembled final price index estimates, and Melville Ulmer and Albert Fishlow estimated railroad construction series. Kuznets had generated his nineteenth century construction-flow series by extrapolating his twentieth century series backward to 1869 on constant-price materials flows, and then inflating the series. There is the implicit assumption here that trade margins and value added by construction constituted a constant fraction of final product, at least in constant prices.

The assumption was clearly the best one available when the estimates were made, particularly in view of the deflators available to Kuznets. But given the data of Barger, Brady, Fishlow, and Ulmer—and particularly Brady's true price indexes—this assumption no longer has to be made.[26] I marked up the materials flows for trade margins,

---

[25] Ibid., Table A-6.

[26] Harold Barger, *Distribution's Place in the American Economy Since 1865* (New York: National Bureau of Economic Research), 1955; Dorothy S. Brady, "Price Deflation for Final Product Estimates" in Dorothy S. Brady, (ed), op cit. Albert Fishlow, "Productivity and Technological Change in the Railroad Sector, 1840–1900" in Dorothy S. Brady (ed) op cit. Melville J. Ulmer, *Capital in Transportation*.

following the work of Barger. Distinctions were made between flows into railroad construction and all others, because value added by construction is, relatively, very much more important in heavy construction—for example, railroads—than elsewhere in the sector. The materials for construction were marked up using census current-price ratios and the Fishlow work on railroads, and deflated using the final price indexes developed by Brady and Ulmer. The series thus rests on improved evidence and procedures. It does yield very much larger estimates of construction flows, especially in the 1870s. But the main explanation for this is not that my "other" construction series deviated far from what Kuznets' procedures would have yielded, but that the estimates distinguish railroad construction, where the ratio between final product flows and materials inputs is very large. The great margin of my total construction series in the 1870s over Kuznets' reflects, chiefly, the fact that railroad construction was very important in that decade.

I conclude, then, that the data underlying Table 7.4 are the best series I am able to assemble and that they are reasonably consistent. But clearly the two sets of series are far from identical. How far would the historical narrative of U.S. economic growth in the nineteenth century be affected by the choice, on the part of the narrator, of one of these sets of series rather than the other? The question is a large one and a detailed answer is best left to another occasion. However, Table 7.6 gathers together a few data that bear on the question. The estimates from which they were drawn refer to the national capital stock—land improvements and producers' durables of all kinds, as well as inventories and net claims on foreigners. One set of estimates is based chiefly on census-style data. In the other set, the perpetual inventory data underlying Table 7.4, Panel A, column 3, and Panel B, column 3, have been substituted for the census-style "other" improvements and producers' durables. Both sets of total capital stock estimates, remember, include all of the conventional components of the capital stock, as well as the value of land clearing, breaking, and fencing.

The two sets of estimates tell essentially the same story of the long-term growth of the capital stock, of shifts in its structure, and of the level and direction of change of the capital output ratio. The capital stock grew rapidly—except over the decade of the 1860s—and the pace was particularly pronounced in the 1850s, the 1870s, and the 1880s. According to the census-style series, the rate of growth was higher in the 1880s than in the 1870s, while according to the other series the reverse was true. The differences are not great, but they are

TABLE 7.6
Rates of Growth, Structure of the Capital Stock, and Capital Output Ratios,
Two Versions, 1860 Prices, 1840–1900

PANEL A: RATES OF GROWTH

	Census-style	Perpetual inventory
1840–50	4.2	4.6
1850–60	5.8	5.5
1860–70	1.6	1.8
1870–80	4.2	6.0
1880–90	6.3	5.5
1890–1900	3.8	4.2
1840–60	5.0	5.1
1860–80	4.0	3.9
1880–1900	4.8	4.8
1840–1900	4.3	4.6

PANEL B: SHARES OF "OTHER" IMPROVEMENTS AND PRODUCERS' DURABLES IN THE STOCK

	Improvements Census-style	Improvements Perpet. inv.	Durables Census-style	Durables Perpet. inv.	Improvements and durables Census-style	Improvements and durables Perpet. inv.
1840	.24	.21	.05	.05	.29	.26
1850	.27	.25	.06	.06	.33	.31
1860	.34	.32	.09	.08	.43	.40
1870	.37	.31	.11	.15	.49	.46
1880	.34	.37	.13	.15	.47	.52
1890	.37	.40	.22	.22	.60	.62
1900	.37	.43	.27	.24	.64	.66

PANEL C: CAPITAL-OUTPUT RATIOS

	Census-style	Perpetual inventory
1840	2.8	2.6
1850	2.7	2.7
1860	2.9	2.8
1870	2.8	2.9
1880	2.6	2.9
1890	3.2	3.3
1900	3.4	3.7

SOURCES: See text. For methods by which Panel C was computed, see Gallman, "U.S. Capital Stock."

great enough to affect the analysis of the business cycle and the Kuznets cycle over this period.

The structural findings drawn from the two series (Panel B) are also similar. The shares in total capital of "other" improvements and durables increase in both series, the change being particularly pronounced in the case of durables. Once again, the timing of the changes in shares is a little different from one series to the next—particularly with respect to improvements across the 1860s—but not much different.

The same kinds of results emerge from Panel C. The levels of the capital output ratios and their broad trends are similar in the two cases, the two being distinguished only by very modest differences in the timing of the changes—this time across the 1870s.

Clearly, much more needs to be done along these lines. But the data in Table 7.6 suggest that the most consistent sets of census-style and perpetual inventory estimates—estimates plausible on other grounds (e.g., average service life, system of depreciation), as well—do tell roughly the same story about the nineteenth century capital stock.

What happens, however, when less consistent perpetual inventory estimates are selected? Table 7.7 was put together as a first step toward answering this question. Panels A and C compare levels of gross and net estimates computed following a variety of plausible procedures, while Panels B and D compare rates of growth. (None of these series, incidentally, has been corrected for Civil War losses, since the adjustment is not required for present purposes.)

The table shows that while the levels of the series differ from one case to the next—declining-balances series are always very much lower than B.L.S. concave series, for example—the various durables series move in parallel, as do the various improvements series. There are differences, of course, and they emerge where one would expect to find them. Thus the durables declining-balances series shows an unusually large increase across the 1860s, as compared with the other series, because the postwar investment boom receives a much larger weight in the 1870 value in this series than in the others. Similarly, the gross series and the B.L.S. concave series exhibit especially small rates of growth across the same decade, because the poor wartime investment experience figures importantly in the 1870 value in these series.

These expected contrasts aside, the ratios in Panels A and C are quite stable, and the rates of change in Panels B and D—particularly D, which has to do with longer-lived property—are quite similar. These are fortunate findings, since they suggest that analytical results depending upon rates of change of the capital stock are unlikely to be

TABLE 7.7
Levels and Decennial Rates of Change of Stocks of "Other" Improvements
and Producers' Durables, 1860 Prices, Perpetual Inventory Estimates,
Various Versions, 1840–1900

PANEL A: PRODUCERS' DURABLES ESTIMATES EXPRESSED AS RATIOS OF NET ESTIMATES,
17-YEAR SERVICE LIFE, STRAIGHT-LINE DEPRECIATION

	17-yr. service life			13-yr. life,
	Decl. bal.	B.L.S.	Gross	Str. line
1840	.74	1.15	1.51	0.86
1850	.75	1.16	1.57	0.85
1860	.73	1.16	1.57	0.85
1870	.78	1.13	1.46	0.88
1880	.73	1.18	1.64	0.83
1890	.75	1.16	1.55	0.85
1900	.73	1.18	1.69	0.82

PANEL B: PRODUCERS' DURABLES, DECENNIAL RATES OF GROWTH (%)

	17-yr. service life				13-yr. life,
	Str. line	Decl. bal.	B.L.S.	Gross	Str. line
1840–50	96%	99%	96%	103%	93%
1850–60	130	124	132	129	129
1860–70	132	147	124	116	141
1870–80	83	71	91	105	72
1880–90	138	143	134	126	145
1890–1900	64	59	68	78	58

PANEL C: "OTHER" IMPROVEMENTS ESTIMATES EXPRESSED AS RATIOS OF NET
ESTIMATES, 50-YEAR SERVICE LIFE, STRAIGHT-LINE DEPRECIATION

	Decl. bal.	B.L.S.	Gross
1840	0.82	1.16	1.27
1850	0.81	1.18	1.30
1860	0.81	1.18	1.30
1870	0.78	1.23	1.40
1880	0.79	1.23	1.40
1890	0.79	1.22	1.40
1900	0.78	1.24	1.43

PANEL D: "OTHER" IMPROVEMENTS, DECENNIAL RATES OF GROWTH (%)

	Str. line	Decl. bal.	B.L.S.	Gross
1840–50	101%	97%	104%	106%
1850–60	116	116	116	116
1860–70	45	40	51	55
1870–80	83	84	82	83
1880–90	74	76	74	74
1890–1900	64	60	65	67

SOURCES: See text.

very sensitive to the choice of service life and depreciation scheme, the exceptions to the rule being quite obvious.

V

What has been learned from all of these data and calculations? The statistical tests suggest that nineteenth century census-style capital stock estimates reflect net values—a useful result, in view of the previous disagreements in the literature about this matter.

In most of the census years, net acquisition cost, net reproduction cost, and market value are unlikely to have differed much; in those few years in which they did, the statistical tests show that the census-style data are probably expressed in reproduction cost. The statistical finding, in this case, has support in literary evidence. Once again, the result bears on a subject on which scholars have previously disagreed.

The results of the consistency tests could, of course, reflect no more than compensatory errors. It is important not to understate the possibility that this is the case. Nonetheless, the evidence suggests that the tests do have merit, and that both sets of estimates are, in fact, net and, where the valuation scheme matters, both are valued in reproduction costs.

When plausible assumptions are made with respect to service lives and depreciation schedules, and when appropriate allowances are made for differences in coverage between the two sets of series, the levels of the two sets of series—perpetual inventory and census-style—are roughly similar. There are some suspicious results, nonetheless: the ratios of perpetual inventory to census-style estimates seem always rather low in 1840, and high in 1880, the latter being the more important deviation; the ratio for durables also seems high in 1870, and for improvements in 1890 and, especially, 1900. The 1840 results should not be surprising, since both sets of estimates are relatively weak at that date, while the deviations in 1870, 1890, and 1900 disappear when durables and improvements are aggregated, leaving 1880 as the one remaining major puzzle. No doubt more work is called for, although one cannot be confident that it will explain this particular anomaly.

Combined with other elements of the capital stock, the census-style and perpetual inventory series imply essentially the same pattern of long-term evolution of the U.S. capital stock, although moderate differences as to timing and short-period developments also emerge. The student of economic fluctuations who plans to introduce capital into his or her analysis would be well advised to examine both the perpetual inventory and census-style series.

Finally, experiments with the perpetual inventory series show that decisions concerning whether capital should be measured gross or net, as well as decisions regarding the appropriate service life and system of capital consumption, affect the level of the measured capital stock much more than they do the rates of change. This is particularly true with long-lived capital, but also holds, to a lesser degree, with respect to the relatively short-lived manufactured producers' durables. The greatest deviations occur across periods in which the flow of investment varied very widely—such as the decade of the 1860s, during the beginning of which capital formation was unusually low, and at the end of which it was unusually high. Under these circumstances, the differences in the weighting schemes between a declining-balance series and a B.L.S. concave series will give rise to fairly marked differences in computed rates of change between the two series. But such unusual circumstances aside, the rates of change traced by the various types of series are remarkably similar, a fortunate result since it means that more conclusive findings with respect to the evolution of the capital stock can be obtained than would otherwise be possible.

Appendix

The Gallman annual estimates of the flows of producers' durables and "other construction" into the U.S. economy (1834–1859, 1869–1909), in constant prices, were interpolated and extrapolated to the missing years 1860–1868 and 1791–1833 on Berry's annual estimates of gross private domestic investment, also in constant prices. All calendar-year estimates (Berry, throughout; Gallman, 1869–1909) were first converted to an approximation of census-year values by computing moving two-year averages. Each series was then decomposed into a time trend and a cyclical component. Through least-squares analysis the Gallman trends and cyclical movements were associated with the like characteristics of the Berry series. The predicted values of the cyclical part of the Gallman series were then combined with the predicted values of the trend relationship to produce estimates for the years missing in the original Gallman series. All these calculations were carried out in logarithms. The last step consisted of taking the antilogarithms.

The new series were then used to generate perpetual inventory capital stock estimates in the manner described in the text. The tables that follow exhibit the various annual cumulations produced. The net series are intended to be net of capital consumption, including all casualty losses, *except for war losses*. The only important losses of this type during the period considered were Civil War losses. Capital was also destroyed during the war by neglect—lack of maintenance—especially in the South. This type of loss was also left out of account in the assembly of the tables.

TABLE 7.8
Annual Perpetual Inventory Cumulations, Midyears of Years Specified
(Millions of 1860 dollars)

		Net		
Midyear of	Gross	Str. line	B.L.S. concave	Decl. bal.

PANEL A: MANUFACTURED PRODUCERS' DURABLES, SERVICE LIFE OF 13 YEARS, RETIREMENT AGE OF 13

Midyear	Gross	Str. line	B.L.S. concave	Decl. bal.
1840	305.8	193.3	225.8	138.2
1841	319.5	195.3	230.9	136.4
1842	338.2	201.9	240.2	140.5
1843	357.6	208.3	249.1	145.1
1844	376.6	215.0	257.8	150.4
1845	403.1	230.4	274.8	164.7
1846	433.4	250.4	296.2	182.7
1847	470.3	277.0	324.4	206.0
1848	513.8	316.3	365.6	239.8
1849	548.7	345.4	398.0	260.1
1850	583.5	369.7	426.7	274.6
1851	625.9	400.5	462.7	295.3
1852	678.9	439.2	507.8	323.3
1853	754.4	489.6	565.7	361.9
1854	841.2	543.9	628.5	403.0
1855	934.1	603.3	697.2	447.5
1856	1044.8	674.5	778.7	502.4
1857	1165.3	748.8	864.4	558.2
1858	1259.2	797.5	925.0	586.2
1859	1332.3	824.7	964.3	594.3
1860	1405.5	855.4	1006.1	609.2
1861	1462.0	879.2	1040.5	619.2
1862	1447.6	821.0	992.9	549.9
1863	1408.1	736.7	915.2	466.9
1864	1379.4	675.4	854.7	419.2
1865	1421.2	697.9	873.4	461.8
1866	1687.7	957.7	1127.7	735.5
1867	2075.4	1327.9	1500.6	1087.0
1868	2438.4	1655.3	1843.0	1357.0
1869	2703.7	1876.2	2090.3	1495.1
1870	2916.5	2035.7	2285.0	1565.0
1871	3153.2	2186.4	2478.0	1630.0
1872	3485.1	2399.8	2739.3	1764.6
1873	3902.5	2682.2	3075.8	1967.3
1874	4268.1	2879.5	3335.5	2077.7
1875	4621.3	2958.7	3481.3	2081.5
1876	4999.8	3008.7	3589.3	2084.2
1877	5386.8	3058.1	3680.8	2113.6
1878	5737.1	3122.7	3767.6	2177.6
1879	5918.1	3231.4	3880.1	2285.4
1880	6144.0	3502.2	4153.1	2540.3
1881	6617.5	3990.1	4657.3	2981.2
1882	7318.0	4590.0	5295.8	3490.7
1883	8087.5	5164.1	5928.2	3932.5
1884	8711.0	5540.5	6377.2	4150.3

TABLE 7.8 (*continued*)

Midyear of	Gross	Net Str. line	Net B.L.S. concave	Net Decl. bal.
1885	9122.0	5737.4	6651.8	4189.7
1886	9621.0	6085.2	7079.6	4398.2
1887	10528.0	6749.7	7838.0	4925.0
1888	11622.0	7441.3	8643.9	5452.9
1889	12726.0	8056.8	9384.9	5887.0
1890	13899.0	8684.9	10141.5	6332.9
1891	15148.0	9343.7	10927.9	6811.1
1892	16468.5	10049.0	11756.7	7335.5
1893	17654.0	10693.7	12519.3	7789.4
1894	18378.0	11020.2	12961.2	7914.2
1895	18928.0	11265.5	13319.9	7980.1
1896	19794.0	11812.5	13983.8	8375.6
1897	20755.5	12249.9	14551.7	8660.4
1898	21683.5	12448.3	14874.7	8733.5
1899	22698.5	12844.8	15362.9	9050.7
1900	23790.5	13595.3	16182.7	9716.6
1901	25033.5	14509.8	17176.8	10496.4
1902	26515.5	15575.6	18341.6	11372.8
1903	28316.5	16944.0	19829.2	12490.6
1904	29943.0	18120.3	21158.2	13332.0
1905	31432.5	19177.0	22398.7	14008.5
1906	33619.0	20857.1	24296.1	15289.9
1907	36619.0	22955.5	26668.1	16925.2
1908	39029.0	24207.7	28241.5	17631.9
1909	40579.0	24758.4	29116.5	17664.8

PANEL B: MANUFACTURED PRODUCERS' DURABLES, AVERAGE SERVICE LIFE OF 13 YEARS, DIVERSE RETIREMENT AGES (WINFREY DISTRIBUTION)

1840	300.2	190.2	222.0	136.2
1841	313.5	191.9	226.6	134.5
1842	330.9	198.2	235.3	138.7
1843	347.6	204.5	243.6	143.3
1844	364.2	211.3	252.2	148.7
1845	388.8	227.2	269.5	163.1
1846	417.8	247.8	291.8	181.1
1847	453.6	275.0	321.1	204.3
1848	503.3	315.1	363.9	238.3
1849	544.2	344.2	397.1	258.4
1850	581.9	368.0	425.7	272.5
1851	627.6	397.9	461.0	293.1
1852	683.6	435.4	504.3	320.8
1853	753.9	483.9	559.4	358.6
1854	832.2	536.5	619.6	398.6
1855	919.8	594.6	686.2	442.3
1856	1023.6	664.8	765.8	496.3
1857	1135.4	738.3	850.2	551.0
1858	1226.2	786.5	910.6	578.3
1859	1297.9	813.4	949.7	586.1
1860	1372.7	843.7	991.5	601.5
1861	1440.5	867.3	1025.6	612.6

TABLE 7.8 (*continued*)

Midyear of	Gross	Net Str. line	Net B.L.S. concave	Net Decl. bal.
1860	1372.7	843.7	991.5	601.5
1861	1440.5	867.3	1025.6	612.6
1862	1423.6	807.9	975.6	543.7
1863	1372.2	723.3	895.8	461.6
1864	1332.7	663.5	835.4	415.9
1865	1367.0	689.3	856.6	460.6
1866	1634.0	953.1	1115.9	735.7
1867	2024.1	1326.7	1494.4	1084.5
1868	2395.2	1655.6	1842.6	1349.0
1869	2684.7	1875.6	2094.9	1482.1
1870	2933.5	2030.9	2290.4	1548.9
1871	3189.9	2172.9	2476.7	1611.4
1872	3528.4	2374.5	2724.4	1743.9
1873	3955.0	2642.6	3040.4	1944.9
1874	4319.0	2823.0	3271.7	2052.7
1875	4577.7	2884.6	3382.8	2049.6
1876	4811.8	2925.0	3465.4	2045.9
1877	5046.2	2977.3	3551.5	2071.2
1878	5288.0	3058.0	3658.6	2135.3
1879	5562.5	3192.9	3814.2	2255.8
1880	5969.3	3484.1	4123.9	2525.8
1881	6577.0	3978.8	4643.1	2973.7
1882	7303.1	4574.3	5277.6	3477.7
1883	8034.8	5139.7	5900.8	3907.2
1884	8609.4	5507.4	6344.0	4112.2
1885	9039.0	5696.9	6617.4	4146.7
1886	9637.8	6034.1	7037.7	4359.6
1887	10575.3	6677.6	7766.6	4885.6
1888	11585.6	7342.3	8527.9	5402.4
1889	12569.7	7934.3	9226.3	5822.6
1890	13606.3	8546.4	9950.6	6256.5
1891	14711.7	9198.6	10720.4	6724.3
1892	15899.0	9907.0	11552.5	7240.1
1893	17056.3	10564.0	12339.3	7692.1
1894	17908.9	10903.9	12812.6	7825.4
1895	18658.2	11153.2	13186.8	7907.1
1896	19675.1	11690.5	13836.0	8316.2
1897	20568.0	12106.8	14361.2	8597.2
1898	21215.1	12290.2	14646.3	8655.1
1899	22043.9	12696.3	15138.1	8969.6
1900	23219.1	13473.0	15995.2	9649.5
1901	24557.9	14406.7	17021.0	10436.0
1902	26064.0	15483.1	18207.8	11307.6
1903	27903.6	16858.4	19716.6	12417.7
1904	29621.6	18033.6	21062.6	13249.0
1905	31286.7	19076.3	22304.3	13920.3
1906	33626.6	20724.0	24168.0	15197.2
1907	36489.3	22769.4	26466.4	16803.5
1908	38665.7	23967.4	27964.6	17471.2
1909	40243.9	24475.9	28780.2	17495.0

TABLE 7.8 (*continued*)

Midyear of	Gross	Net Str. line	Net B.L.S. concave	Net Decl. bal.
PANEL C: MANUFACTURED PRODUCERS' DURABLES, SERVICE LIFE OF 17 YEARS, RETIREMENT AGE OF 17				
1840	341.6	225.3	259.9	166.7
1841	359.4	230.7	268.9	167.2
1842	382.3	240.8	282.3	173.2
1843	405.4	250.7	295.4	179.5
1844	429.1	261.1	308.9	186.6
1845	461.7	280.2	331.1	202.8
1846	500.2	304.1	358.0	223.1
1847	547.1	334.5	391.7	249.3
1848	607.4	377.8	438.6	287.1
1849	658.1	410.7	475.6	312.3
1850	703.9	438.5	507.8	331.5
1851	756.6	472.8	546.7	356.9
1852	811.5	515.2	594.2	389.1
1853	880.4	570.1	655.2	432.2
1854	961.0	630.6	723.2	478.8
1855	1051.8	698.1	799.8	530.3
1856	1161.0	779.4	891.6	593.1
1857	1288.6	865.8	990.6	658.7
1858	1401.4	928.3	1067.1	698.4
1859	1494.3	969.9	1123.2	717.7
1860	1595.0	1015.1	1182.7	743.1
1861	1692.8	1053.3	1234.7	763.4
1862	1702.6	1007.9	1202.4	702.4
1863	1678.6	934.8	1139.2	622.7
1864	1665.7	883.1	1093.4	574.1
1865	1718.9	913.8	1127.0	613.3
1866	2019.4	1181.7	1397.6	887.5
1867	2452.9	1563.0	1787.5	1252.6
1868	2864.3	1905.7	2147.8	1550.3
1869	3185.8	2145.7	2412.9	1724.9
1870	3450.7	2325.8	2623.0	1831.8
1871	3713.4	2497.8	2828.5	1929.8
1872	4045.3	2735.4	3102.9	2093.6
1873	4452.7	3047.9	3457.4	2326.6
1874	4795.5	3283.5	3743.2	2471.5
1875	5064.7	3408.9	3925.3	2506.2
1876	5346.1	3516.5	4091.5	2534.7
1877	5647.0	3636.0	4269.5	2586.9
1878	5994.1	3782.8	4474.8	2676.5
1879	6489.8	3980.2	4730.0	2829.2
1880	7188.8	4324.5	5126.3	3137.2
1881	8102.3	4862.1	5710.2	3633.1
1882	9082.6	5494.5	6386.3	4199.0
1883	9850.6	6097.2	7034.0	4693.7
1884	10349.0	6516.3	7510.4	4968.3

TABLE 7.8 (*continued*)

Midyear of	Gross	Net Str. line	Net B.L.S. concave	Net Decl. bal.
1885	10729.0	6774.5	7840.3	5071.8
1886	11370.0	7192.9	8339.1	5347.9
1887	12407.0	7928.6	9163.1	5941.1
1888	13533.5	8700.3	10036.3	6543.6
1889	14587.0	9413.7	10863.8	7060.0
1890	15643.5	10162.6	11740.2	7591.6
1891	16874.0	10970.4	12694.2	8169.6
1892	18337.0	11848.3	13734.2	8811.0
1893	19843.0	12681.2	14738.2	9399.3
1894	21093.5	13198.4	15430.0	9672.1
1895	22273.5	13616.6	16013.7	9876.2
1896	23726.5	14309.4	16856.7	10389.6
1897	24960.5	14873.8	17562.0	10772.7
1898	25795.0	15200.5	18022.0	10918.7
1899	26750.5	15747.6	18696.6	11302.9
1900	28110.0	16670.6	19753.3	12060.6
1901	29856.0	17761.6	20995.1	12964.2
1902	31980.5	18996.8	22390.9	13990.7
1903	34339.0	20523.6	24077.4	15270.4
1904	36289.0	21858.2	25582.7	16284.6
1905	38147.5	23083.5	27000.1	17148.7
1906	40736.0	24937.6	29065.1	18623.1
1907	43813.5	27225.8	31597.3	20460.5
1908	46154.5	28717.6	33379.7	21404.2
1909	47837.0	29555.6	34534.2	21683.4

PANEL D: MANUFACTURED PRODUCERS' DURABLES, AVERAGE SERVICE LIFE OF 17 YEARS, DIVERSE RETIREMENT AGES (WINFREY DISTRIBUTION)

1840	335.7	221.6	255.4	163.9
1841	352.6	226.6	263.7	164.2
1842	374.3	236.2	276.5	170.2
1843	396.1	245.8	289.0	176.5
1844	418.6	255.9	301.9	183.6
1845	450.0	274.8	323.5	199.8
1846	486.4	298.5	350.0	220.0
1847	530.0	329.0	383.4	245.9
1848	587.6	372.4	430.3	283.3
1849	636.2	405.5	467.7	307.9
1850	680.9	433.5	500.6	326.7
1851	733.2	468.0	540.4	351.7
1852	795.0	510.6	588.8	384.1
1853	870.7	565.0	649.8	427.2
1854	955.0	624.5	716.9	473.3
1855	1049.1	690.6	791.8	524.0
1856	1160.0	769.8	880.9	586.0
1857	1281.0	853.9	976.4	650.1
1858	1383.6	914.0	1049.4	688.0
1859	1469.0	953.7	1102.6	706.0

TABLE 7.8 (*continued*)

Midyear of	Gross	Net Str. line	Net B.L.S. concave	Net Decl. bal.
1860	1560.6	997.2	1159.4	730.5
1861	1647.1	1034.1	1209.3	750.2
1862	1650.7	988.2	1175.8	688.9
1863	1623.1	915.1	1112.1	610.3
1864	1610.5	864.2	1066.5	563.8
1865	1672.9	896.3	1100.5	605.8
1866	1969.2	1165.3	1370.7	881.0
1867	2389.4	1547.1	1760.5	1243.4
1868	2788.9	1890.0	2121.7	1535.6
1869	3103.5	2129.6	2389.1	1704.3
1870	3371.9	2309.0	2602.2	1806.5
1871	3642.6	2479.3	2810.0	1901.2
1872	3991.2	2713.8	3084.9	2063.0
1873	4430.6	3020.9	3436.7	2294.8
1874	4808.4	3247.4	3713.6	2438.1
1875	5088.8	3360.3	3879.5	2469.8
1876	5362.3	3454.0	4025.1	2495.8
1877	5658.8	3560.0	4179.4	2547.3
1878	5990.4	3693.5	4357.2	2636.7
1879	6375.2	3878.5	4582.9	2783.0
1880	6901.2	4216.8	4959.4	3080.9
1881	7629.0	4758.3	5542.2	3565.6
1882	8485.6	5404.6	6239.9	4124.2
1883	9351.8	6027.5	6927.1	4623.2
1884	10055.4	6460.0	7435.6	4905.6
1885	10597.0	6718.9	7776.1	5012.3
1886	11277.7	7127.7	8266.6	5282.1
1887	12278.3	7850.0	9075.0	5866.0
1888	13359.5	8607.4	9932.3	6458.7
1889	14435.6	9306.1	10744.5	6969.8
1890	15586.4	10036.4	11598.4	7499.9
1891	16828.6	10817.3	12512.2	8069.2
1892	18177.3	11665.0	13502.0	8692.3
1893	19529.6	12471.5	14460.2	9260.2
1894	20612.4	12968.8	15116.0	9516.3
1895	21614.0	13376.0	15677.5	9709.2
1896	22897.9	14068.2	16516.9	10215.4
1897	24078.2	14641.4	17237.8	10598.9
1898	25037.0	14980.6	17721.0	10757.5
1899	26204.8	15534.7	18407.5	11160.6
1900	27716.2	16452.8	19451.9	11928.5
1901	29370.8	17529.4	20660.2	12819.2
1902	31191.5	18754.4	22030.1	13817.0
1903	33359.4	20287.0	23725.2	15076.7
1904	35401.8	21635.4	25261.7	16089.0
1905	37380.4	22866.7	26701.2	16950.8
1906	40010.2	24716.6	28776.0	18412.5
1907	43132.7	26992.4	31309.4	20229.0
1908	45580.9	28460.6	33079.7	21153.0
1909	47450.9	29263.1	34202.1	21424.0

TABLE 7.8 (*continued*)

Midyear of	Gross	Net Str. line	Net B.L.S. concave	Net Decl. bal.

PANEL E: IMPROVEMENTS (OTHER THAN CANALS, RAILROADS, FARM LAND CLEARING, AND IMPROVEMENTS CONSTRUCTED WITH FARM MATERIALS), SERVICE LIFE OF 50 YEARS, RETIREMENT AGE OF 50

Midyear	Gross	Str. line	B.L.S. concave	Decl. bal.
1840	1354.8[a]	1070.3[a]	1246.0[a]	878.71[a]
1841	1465.8[a]	1154.2[a]	1345.6[a]	945.75[a]
1842	1576.8	1235.9	1444.2	1009.46
1843	1679.4	1308.7	1534.9	1063.16
1844	1784.8	1382.2	1627.3	1116.95
1845	1919.3	1482.9	1747.9	1197.3
1846	2081.3	1608.5	1894.9	1301.3
1847	2265.9	1754.3	2063.7	1423.4
1848	2447.2	1893.5	2228.1	1536.5
1849	2621.1	2021.6	2383.3	1636.5
1850	2794.9	2146.5	2537.1	1731.7
1851	3002.9	2302.3	2723.5	1856.5
1852	3250.9	2494.3	2948.2	2015.3
1853	3538.6	2721.4	3210.7	2206.4
1854	3860.1	2977.1	3504.9	2422.0
1855	4201.2	3246.8	3816.8	2647.3
1856	4569.7	3538.1	4154.1	2889.8
1857	4975.6	3860.7	4526.7	3158.7
1858	5361.0	4155.7	4876.3	3394.5
1859	5698.1	4394.6	5174.1	3570.2
1860	6035.0	4626.6	5468.3	3736.8
1861	6360.5	4840.2	5747.3	3883.6
1862	6579.5	4941.7	5916.9	3917.3
1863	6722.7	4963.5	6007.9	3874.0
1864	6847.6	4965.4	6078.7	3815.7
1865	6977.5	4972.1	6153.6	3767.4
1866	7203.6	5075.0	6324.2	3820.2
1867	7550.9	5298.1	6615.6	3994.7
1868	8036.4	5654.9	7043.5	4301.1
1869	8636.1	6117.7	7581.9	4707.6
1870	9365.6	6698.0	8243.9	5223.4
1871	10040.8	7209.3	8844.8	5659.1
1872	10826.4	7818.7	9550.4	6184.4
1873	11736.7	8538.2	10374.1	6809.2
1874	12586.1	9180.2	11129.7	7343.3
1875	13402.9	9775.2	11846.6	7820.4
1876	14202.9	10339.3	12540.0	8258.8
1877	14964.3	10851.3	13188.3	8639.2
1878	15699.6	11326.0	13805.2	8978.3
1879	16433.0	11787.1	14413.7	9301.1
1880	17185.2	12254.9	15034.0	9629.1
1881	18066.6	12842.5	15778.7	10074.9
1882	19064.6	13533.8	16633.9	10619.6
1883	20093.2	14243.0	17514.7	11174.4
1884	21232.1	15050.7	18502.2	11819.7

[a] Assuming no investment in the year 1791 (July 1, 1790–June 30, 1791), and 1792 (July 1, 1791–June 30, 1792), years for which we have no data.

TABLE 7.8 (*continued*)

Midyear of	Gross	Net Str. line	Net B.L.S. concave	Net Decl. bal.
1885	22435.1	15895.8	19536.9	12493.1
1886	23726.8	16819.7	20660.0	13234.8
1887	25200.5	17893.9	21944.5	14115.8
1888	26680.2	18971.1	23244.5	14983.1
1889	28128.3	20003.7	24513.8	15790.8
1890	29946.0	21367.0	26127.1	16918.4
1891	32017.4	22950.4	27978.1	18246.1
1892	34299.6	24703.3	30018.9	19718.3
1893	36633.1	26455.1	32081.0	21161.4
1894	38684.2	27880.6	33838.1	22253.1
1895	40739.4	29298.5	35603.2	23321.4
1896	42638.1	30546.4	37214.8	24205.4
1897	44473.4	31716.4	38763.1	25003.3
1898	46368.4	32906.4	40345.5	25818.0
1899	48133.4	33921.1	41766.1	26455.0
1900	49967.6	34970.0	43230.0	27129.7
1901	52010.2	36224.9	44908.6	28011.1
1902	54255.2	37681.7	46801.4	29088.0
1903	56509.1	39142.8	48715.0	30156.8
1904	58639.9	40469.8	50512.7	31080.6
1905	60786.3	41790.3	52321.2	31992.7
1906	63057.8	43221.5	54258.5	33010.9
1907	65457.4	44773.9	56337.7	34141.7
1908	67826.6	46228.4	58343.5	35167.3
1909	70287.5	47678.9	60367.0	36188.9

PANEL F: IMPROVEMENTS (OTHER THAN CANALS, RAILROADS, FARMLAND CLEARING, AND IMPROVEMENTS CONSTRUCTED WITH FARM MATERIALS), AVERAGE SERVICE LIFE OF 50 YEARS, DIVERSE RETIREMENT AGES (WINFREY DISTRIBUTION)

Midyear of	Gross	Net Str. line	Net B.L.S. concave	Net Decl. bal.
1890	29219.0	20943.9	25534.1	16614.7
1891	31248.0	22505.0	27356.0	17923.8
1892	33480.4	24233.6	29366.4	19374.0
1893	35750.0	25958.8	31396.8	20791.4
1894	37729.1	27355.4	33122.7	21855.1
1895	39730.4	28743.4	34857.2	22897.5
1896	41589.0	29960.1	36437.1	23757.2
1897	43398.2	31097.8	37952.3	24533.6
1898	45254.1	32254.9	39499.3	25327.8
1899	46960.2	33235.8	40882.3	25944.3
1900	48720.2	34251.0	42307.7	26599.5
1901	50707.0	35473.3	43948.5	27464.3
1902	52925.1	36896.6	45802.6	28526.9
1903	55177.5	38322.8	47674.2	29582.6
1904	57327.8	39612.8	49424.2	30494.6
1905	59493.1	40893.7	51177.0	31395.2
1906	61792.1	42283.7	53049.4	32402.6
1907	64243.2	43792.2	55054.6	33524.6
1908	66623.7	45199.8	56971.4	34538.0
1909	69033.5	46601.3	58893.8	35542.8

SOURCES: See text.

Thus all the values in the tables are gross of important elements of capital consumption that took place during the Civil War. The values computed on the assumption that average service lives of seventeen years and fifty years characterized producers' durables and improvements, respectively, are also almost certainly too large (see the text); the other estimates (associated with average service lives of thirteen and forty years) may or may not be too low. (Only the thirteen-, seventeen-, and fifty-year estimates are shown here.)

I publish the annual data with some misgivings—in view of the weaknesses of the evidence on which they are based—and refuse to warrant them for any particular purpose. Future users are on their own and are asked not to blame me if the series do not perform up to expectations. On the other hand, I am willing to accept the credit if they do.

8

Sources of Savings in the Nineteenth Century United States

JOHN A. JAMES *and* JONATHAN S. SKINNER

I. Introduction

The dramatic rise in savings and capital accumulation uncovered by Robert Gallman almost two decades ago has remained one of the most intriguing and important topics in nineteenth century American economic history. The ratio of net investment to NNP doubled over the last two-thirds of the century, rising from 9.5 percent during 1838–1848 to 19.7 percent during the century's last decade.[1] Savings rates have never before, and never since, accounted for such a large portion of national income. During the sixty years following 1840, the ratio of capital to NNP also, as a result, more than doubled, rising from 1.7 to 3.8. Moreover, this increase in domestic savings and investment took place during a period of secular decline in the real interest rate. This shift toward greater capital-intensity represented a fundamental transformation in the structure and pattern of American economic growth over the mid-nineteenth century.

A wide variety of factors have figured in studies accounting for this unprecedented rise in the savings rate. First, the traditional measurement of saving excluded improvements of land, an activity that accounted for a significant proportion of the farmer's investment activities in the antebellum United States. The shift toward greater nonagricultural employment over time implied a corresponding shift from "unconventional" forms of savings, such as land improvement, toward

---

We are especially indebted to the insightful and genteel discussion of the paper by Robert Gallman, as well as to the very useful comments and suggestions from Stanley Lebergott, and Mark Thomas. However, all remaining errors are, of course, the responsibility of the authors.

[1] Robert E. Gallman, "Gross National Product in the United States, 1834–1909," in *Output, Employment, and Productivity in the United States After 1800,* Conference on Research in Income and Wealth, Studies in Income and Wealth, vol. 30 (New York: National Bureau of Economic Research, 1966), pp. 3–76; Lance Davis and Robert Gallman, "Capital Formation in the United States during the Nineteenth Century," in *Cambridge Economic History of Europe,* vol. 7, part 2 (New York: Cambridge University Press, 1978), p. 2.

more conventional forms.[2] Gallman's estimates of these activities suggest that the shift from unreported land improvement into financial savings accounted for approximately 3.5 percentage points of the total rise in savings.[3]

A second reason for the shift in savings was the change in the age composition of the population. Fertility rates, which were very high during the first part of the nineteenth century, declined markedly over the period between 1830 and 1900. Various investigators have attributed a portion of the savings increase to differences in family size, as parents shifted from investing in large families (and then depending on their large families to support them in retirement) into conventional sources of saving. David, in particular, perceived the fall in fertility rates as a rational response to changing relative values of investments. Parents, reacting to the increased relative rate of return on capital caused by labor-augmenting technical change, found children less attractive as investments and cut back on the number of offspring. In this sense, the shift in the savings rate and the decline in the fertility rate were both part of a massive portfolio reallocation between human and physical capital.[4]

The substantial decline in prices of investment goods also played a prominent role in explaining the increase in capital accumulation.[5] The 22-percent drop in the relative price of investment goods between 1840 and 1900 would have caused a rise in capital accumulation as a forgone dollar of current consumption bought a greater (physical) quantity of new capital.[6] As individuals held more capital units (and may or may not have saved more in terms of consumption), the economy would have moved down the stable investment demand curve, resulting in higher capital-income ratios and a lower marginal productivity of capital.

[2] Davis and Gallman, "Capital Formation"; Peter Temin, "Causal Factors in the 'Great Traverse,'" unpublished manuscript.

[3] Following Temin, ibid., we assume that the ratio of gross unconventional investment to total gross investment is equal to the ratio of net unconventional investment to total net investment.

[4] Davis and Gallman, "Capital Formation"; Frank Lewis, "Fertility and Savings in the United States: 1830–1900," *Journal of Political Economy* 91 (Oct. 1983), pp. 825–840; Paul A. David, "Invention and Accumulation in America's Economic Growth: A Nineteenth-Century Parable," in *International Organization, National Policies and Economic Developent,* Karl Brunner and Allan Meltzer, eds. (Amsterdam: North Holland, 1977), pp. 179–228.

[5] Davis and Gallman, "Capital Formation"; David, "Invention and Accumulation"; Temin, "Causal Factors"; Jeffrey G. Williamson, "Inequality, Accumulation, and Technological Imbalance: A Growth-Equity Conflict in American History?" *Economic Development and Cultural Change* 28 (Jan. 1979), pp. 231–253.

[6] David, "Invention and Accumulation," p. 209.

Sources of Savings

Williamson suggested that the rising level of inequality was another potential source of increased savings rates. If the wealthy had had a higher propensity to save, then increased inequality would have led to higher aggregate savings rates. However, his calculations did not ascribe a major role to inequality differences in explaining the savings increase. General changes in income levels also caused shifts in saving behavior, but only to the extent that saving was income elastic could the rising capital-income ratio be attributed to general gains in income.[7]

A final major factor that influenced the growth of savings, stressed by Davis and Gallman, was the growth of financial intermediation over the late nineteenth century. Organized financial markets reduced the costs of matching lenders and borrowers, reduced risk, and hence provided higher returns to savers and lower costs to borrowers. Intermediation, therefore, unambiguously caused an increase in savings rates. However, it is difficult to measure directly the effect of intermediation, or to separate it from other supply and demand shifts occurring contemporaneously.[8]

Whether shifts in the supply of savings or in the demand for investment were the principal cause of the increase in the savings rate, or what role their respective price elasticities played in what transpired, are not of central interest to this paper. Instead, we shall follow Lebergott's emphasis on wealth and wealth accumulation and focus on the more limited but related question: What were the sources of savings?[9] More than 95 percent of total investment during this period originated from domestic sources.[10] Thus we address the questions of who was doing the saving and how their savings behavior changed over the forty years between 1850 and 1890.

In this paper a technique is applied for determining the extent to which the rapid changes in wealth accumulation were responses to shifting patterns of childbearing, to occupational and wage differences, and to economy-wide shifts in the rate of return or in reinvestment behavior. Specifically, information collected by the census on total wealth, by state, for census years 1850, 1870, 1880, and 1890 allows us to measure empirically how capital accumulation depended on these different factors over a period that encompasses most of the upward shift in the savings rate. Given the parameters estimated in the combined cross-section time-series regression model, we can decom-

[7] Williamson, "Inequality, Accumulation, and Technological Imbalance."
[8] Davis and Gallman, "Capital Formation."
[9] For example, Stanley Lebergott, *Wealth and Want* (Princeton: Princeton University Press, 1975).
[10] Gallman, "Gross National Product," pp. 14–15.

pose the sources of the increasing level of wealth in the later portion of the nineteenth century. The results suggest that demographic effects had only minor influences on the total change in the capital stock. However, the trend toward occupations with greater propensities to save—such as trades, manufacturing, and personal and professional services—was a significant factor in the increased level of wealth accumulation.

The paper is organized in the following way. In section II, past studies of savings behavior in the nineteenth century are reviewed, and the process of life-cycle wealth accumulation is described. In section III, we show that the accounting identity relating statewide aggregate wealth to a variety of demographic attributes allows the estimation of important parameters of life-cycle savings. Results are presented from both linear (OLS) and nonlinear regressions and are used to measure the relative contribution of factors such as demographic, occupational, and income changes on the stock of real wealth. In section IV, we conclude with a discussion of the results. A discussion of the data follows in an appendix to the chapter.

II. Savings and Investment in the Nineteenth Century

Since we focus on asset accumulation in a life-cycle framework, we shall use a more comprehensive measure of assets than just reproducible capital, on which the literature on the shifts in the savings rate has centered. Our measure of real wealth—which includes land plus improvements and personal property, as well as reproducible capital—will be discussed in section III. The trends in the growth of real wealth and income over our period, however, are presented in Table 8.1. Note that the growth in the more broadly defined wealth-income ratio, 0.65 percent per annum, is not nearly as dramatic as the upward shift of 1.58 percent per year in the traditional savings rate. In part, the lower growth rate of wealth can be attributed to beginning in 1850 with a larger base of total wealth. Moreover, since our measure of total wealth already includes land plus improvements, any shift from "unconventional" to more conventional forms of savings would appear only as a change in form and not as increased saving. Nevertheless, the positive growth in the wealth-income ratio is consistent with Paul David's "Grand Traverse" of the nineteenth century toward a more capital-intensive economy.

Real interest rates declined substantially over the period 1840–1900, although the pattern of decline was not uniform. Davis and Gallman presented various indices of real rates based on different characteriza-

TABLE 8.1

Annual Growth Rates of Total Real Wealth and Income, by Decade, 1850–1890

(1860 prices in percent)

	$\dot{W}$	$\dot{W}$ (per capita)	$\dot{Y}$	$\dot{Y}$ (per capita)	$\dot{W} - \dot{Y}$
1850–60	4.3%	1.3%	4.5%	1.5%	−0.2%
1860–70	2.6	0.6	2.5	0.5	0.1
1870–80	7.0	4.4	4.7	2.1	2.3
1880–90	4.3	2.0	3.9	1.6	0.4

SOURCES: The construction of the measure of W is discussed in the appendix at the end of the chapter. The growth rates for net national product were calculated from Lance Davis, et al., *American Economic Growth* (New York: Harper and Row, 1972), p. 34; U.S. Bureau of the Census, *Historical Statistics of the United States* (Washington, 1975), p. 8.

tions of price expectations, and all showed a rise in the periods immediately before and after the Civil War.[11] Nevertheless, the trend in the postbellum period is unambiguous. Both real and nominal rates fell continuously, so that by 1900 real rates were at least 70 percent below their 1840 values.

In a comprehensive survey of the change in savings behavior, Davis and Gallman suggested a number of reasons why the savings increase occurred. On the demand side, they proposed three reasons for the increase in the use of capital. The first was that capital-intensive sectors of the economy grew at a faster rate than less intensive sectors, thus causing the aggregate capital-output ratio to grow. The second was that, owing perhaps to biased technical change, only a few sectors became capital-intensive, leaving other sectors unaffected. However, neither of these explanations gained much support from the data. Rather, the third hypothesis, that capital-intensity rose consistently across most sectors of the economy, received the strongest empirical confirmation.[12] This general rise in capital-output ratios, combined with the declining marginal product of capital, suggests an important role for shifts in the supply of savings. As a result, we now turn to a more detailed examination of the four factors emphasized by past studies as important to the increased level of savings: the falling relative price of capital goods, changes in demographic structure, changes in the degree of inequality, and improvements in financial intermediation.

One notable characteristic of this period was the sharp decline in the relative price of investment goods, owing primarily to a fall in the relative price of machines and tools. Although prices of other capital goods, such as nonbusiness housing and churches, rose during some decades in

[11] Davis and Gallman, "Capital Formation," p. 23.
[12] Ibid.

the later nineteenth century, the weighted index of relative investment-goods prices dropped 22 percent between 1840 and 1900.[13] Following the convention of past studies, we shall take this as a shift in the supply-of-savings curve, although we can equivalently model it as a shift in investment demand. It will be useful to follow the analysis of Williamson and Temin in outlining the effect of a drop in the price of investment goods.[14] We can write the equilibrium equation relating the interest rate, r, with the marginal product of capital, MPK, as follows

$$P_y(MPK) = P_k r \qquad (1)$$

where P_k is the price of a unit of new capital and P_y is the price of a unit of output. The left-hand side is the value of the marginal product of capital, while the right-hand side is the rental cost of capital. Without loss of generality, we define P_y as the numeraire and initially set the price of a new unit of capital, P_k, equal to 1. Capital is defined to be a homogeneous unit with geometric depreciation at the rate δ. Thus the output of an n-year-old unit of capital will be $MPK(1 - \delta)^n$. The initial equilibrium is pictured in Figure 8.1, where the flow of investment (per year) is measured on the horizontal axis, and the value of the marginal product is on the vertical axis. (Since $P_y = 1$, we simply write MPK.) For purposes of simplicity, the savings curve is initially drawn as perfectly inelastic with respect to the interest rate.

If the price of investment goods, P_k, were to drop from $1.00 to $0.80, the savings curve would shift to the right from S to S' by $(1/0.8 - 1)$ or by 25 percent, since the consumer can purchase 25 percent more investment goods for the same (dollar) savings. Recall that the savings schedule is measured in physical units of capital, so as to ensure comparability with investment demand. Although the savings schedule may shift to the right, savings in dollar terms may not change, owing to the cheaper price of investment goods. In fact, the consumer must finance 25 percent more investment goods to attain the same nest egg, since each unit of capital is also less valuable for resale purposes. However, from equation 1, it is clear that there will be upward pressure on the interest rate, since holding MPK constant, a unit of capital (which will produce MPK in the following year) cost only $0.80 rather than $1.00. Of course, MPK will also fall as more investment goods are added to the capital stock.

For the example in Figure 8.1 with vertical supply schedules for sav-

[13] David, "Invention and Accumulation."

[14] Williamson, "Inequality, Accumulation and Technological Imbalance"; Temin, "Causal Factors."

Sources of Savings

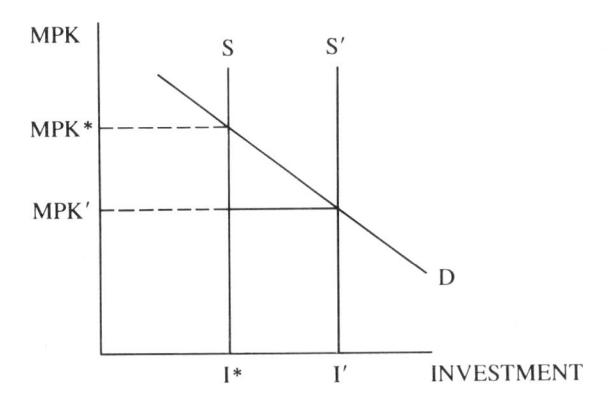

FIGURE 8.1 Savings and Investment: Perfectly Inelastic Savings Schedule

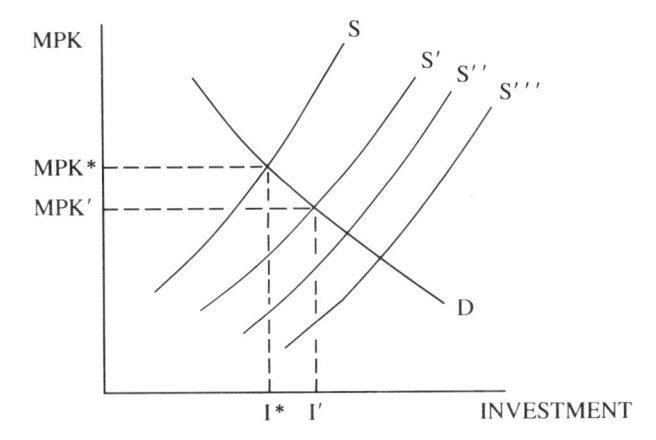

FIGURE 8.2 Savings and Investment: Positively Sloped Savings Schedule

ing, variations in interest rates will not affect savings. Therefore, the decline in *MPK* from *MPK** to *MPK'* depends only on the elasticity of the demand curve. If the elasticity were 1.0 (corresponding to a Cobb-Douglas production function), the *MPK* would fall by 25 percent, and the interest rate would not change. However, if the elasticity were less than 1.0, as David has argued, then the *MPK* would fall by more than the drop in investment prices.[15] Thus a fall in the price of investment goods can be consistent with falling interest rates (as well as a declining *MPK*) when the demand schedule is inelastic.

An example in which savings does respond to interest rates is presented in Figure 8.2. Initially, the 25-percent change in P_k affects the

---
[15] David, "Invention and Accumulation."

supply curve by shifting it to the right by 25 percent (from S to S'), which brings the *MPK* down from *MPK** to *MPK'*. If the percentage change between *MPK** and *MPK'* is less than the initial 25-percent change, then the interest rate must rise. Since the vertical axis measures *MPK* rather than r, the increase in r means that the supply curve will shift to the right even further (i.e., to S''). If, on the other hand, the percentage change in *MPK* exceeds 25 percent, a possibility if the supply curve or the demand curve is highly inelastic, then the supply curve will shift to the left in response to the drop in interest rates.

To summarize, we have shown that the secular decline in the relative price of investment goods is consistent with increased savings. Although *MPK* certainly falls, the effect on the interest rate is uncertain and depends on relative values of supply and demand elasticities. Note further that while the fall in the price of investment goods increases savings, it may not affect the value of wealth. For example, if the interest rate were held constant, the *MPK* would fall by the same proportion as the relative fall in the investment-goods price (from equation 1). Although the savings schedule shifts to the right by the same percentage, the increased physical capital holdings of individuals are offset by a proportional drop in capital's productivity, leading to an unchanged value of total capital holdings.

An additional effect of falling investment-goods prices is that inframarginal owners of capital will suffer a loss, either anticipated or unanticipated, in the value of their wealth. (For owners of home computers purchased for $2,000 and now selling for $500, this point should be clear.) To provide an example, assume that a unit of capital was bought for $1.00, and a 9-percent depreciation rate has, over two years, reduced its value to $0.83. Now the price of a new unit of capital falls to $0.80; holding r constant, the market value of this two-year-old unit drops to $0.66, and the owners suffer a capital loss. To the extent that consumers anticipated this drop in investment prices, the net interest to savers, reflecting the capital loss as well as the return on capital, would have been that much less. Calculations by Davis and Gallman, including the capital loss in their rate-of-return measure, indicated that the net return on savings was still positive.[16] Nevertheless, for price-inelastic savings schedules, consumers would respond to the anticipated decline in asset prices by saving more, since savings represent expenditures on future consumption which, relative to current savings, has become more costly. For an unanticipated fall in the investment

---

[16] Davis and Gallman, "Capital Formation."

price, the capital loss suffered by savers would have led to an unambiguous rise in savings.[17]

For illustrative purposes, we assume a perfectly inelastic savings function. Between 1840 and 1900, the composite investment-price index fell by about 0.4 percent each year.[18] If the components of investment (which included nonbusiness structures as well as equipment) were the same as the components of the existing capital stock, the market value of existing capital would have fallen by the same proportion. Assuming a capital-output ratio of 2.5, on average almost 1 percentage point of the net savings rate per year would have been necessary to restore the value of capital holdings back to its original figure. Using the assumption of inelastic savings, then, the supply curve shifts to the right, from S'' to S''', although for elastic supply schedules, the supply curve could shift to the left as well.

The effect of changes in fertility rates on saving was an important, although unmeasured, factor in explaining the shift in savings rates in the studies by David and Davis and Gallman. The argument is based on a life-cycle model of savings, which emphasizes different saving behavior for families of different ages and sizes. The stylized life-cycle individual dissaves while young, accumulates during his or her peak working years, and either spends down the accumulated nest egg during retirement or preserves assets for bequest purposes. The implication of this hypothesis is that the age composition, as well as standard factors such as income and interest rates, is an important explanatory variable in determining the aggregate supply of saving.[19]

Lewis measured the effects of changes in fertility in the context of a

---

[17] This point comes from Auerbach and Kotlikoff in the context of investment tax credits. Saving is likely to increase because in a life-cycle model, a 1-percent drop in asset value will cause a ψ-percent fall in lifetime wealth, where ψ is the ratio of assets to total resources (assets plus present value of earnings). As long as savings are not income inelastic, consumption will drop by at least ψ percent, but income will only fall by $r\psi'$, where ψ' is the ratio of assets to current income. Since ψ will almost certainly exceed $r\psi'$, savings, or income minus consumption, will rise. Alan Auerbach and Laurence Kotlikoff, "Investment Versus Savings Incentives: The Size of the Bang for the Buck and the Potential for Self-Financing Business Tax Cuts," in *The Economic Consequences of Government Deficits,* Lawrence H. Meyer, ed. (Boston: Martinus Nijhoff, 1983).

[18] David, "Invention and Accumulation."

[19] Davis and Gallman also suggested that rising life expectancy in the latter part of the nineteenth century would have led to higher levels of savings. However, in a model of consumption and bequests, risk-averse consumers will respond only marginally to a rising probability of attaining old age. Jonathan Skinner, "The Effect of Increased Longevity on Capital Accumulation," *American Economic Review,* 75 (Dec. 1985), pp. 1143–1150.

life-cycle model instituting children as an alternative form of investment. That is, children represented a method by which parents could save for retirement, since as Lewis reports, children during the nineteenth century often repaid 40–50 percent of their rearing costs to their elderly parents. Children therefore affected the pattern of capital accumulation by causing parents to cut back on saving, both while the children were being raised and later as parents depended on children for their retirement income. The movement during the nineteenth century toward smaller families provided an incentive to increase financial investments (and reduce human ones). Lewis used microeconomic data on rearing costs of children to estimate that the change in demographic composition between 1830 and 1900 accounted for somewhat less than 1.5 percentage points in the savings rate, or only a fourth of the 6-percentage-point change in rates measured in current dollars.[20]

Paul David extended the demographic argument further by suggesting that the decline in fertility rates was an endogenous response to lower returns from investing in children and higher returns on financial investments.[21] Rather than portraying saving as responding passively to changes in family size, as Lewis did, David viewed the decline in family size as a symptom of a portfolio shift from human to nonhuman wealth. Consider, for example, a family that is deciding whether to save by having a child or to invest the money in the bank. If the family could expect to receive, say, 40 percent of the costs of raising the child from repayments by the child once grown, then their investment (net of enjoying the extra child) is 40 percent of child costs. If they did not choose to have the child, they would invest that 40 percent, and spend the 60 percent on other consumption goods. Thus each child would be associated with a reduction in family holdings of 40 percent of rearing costs. Conversely, for a fixed level of total saving, a decline in family size by one child must be associated with a rise in financial savings equal to 40 percent of the costs of raising the child. That is, imposing the 1830 family sizes on the 1890 population (as Lewis did) should have depressed the counterfactual 1890 capital stock to a level comparable with the 1830 capital stock. Lewis's result that demographics accounted for a small proportion of the total change therefore casts some doubt on the simple portfolio-rearrangement explanation.

Williamson provided some evidence on the role of inequality in ex-

---

[20] Expressed as a percentage of the rise in net saving rates in constant dollars, demographic effects account for only 15 percent of the total change. Lewis, "Fertility and Savings," p. 836.

[21] David, "Invention and Accumulation."

plaining the rise in savings rates. Although the nineteenth century was characterized by increasing earnings differentials between skilled and unskilled workers, he concluded that shifts in the distribution of income could play only a minor role in the higher level of savings. With the extreme assumption that labor earners save nothing and owners of capital account for all national savings by reinvesting some of their proceeds, he suggested that the rising factor share of capital could have accounted for at most about a 3-percentage-point rise in savings rates between the 1830s and the 1890s.[22]

Finally, Davis and Gallman placed substantial emphasis on institutional change in financial markets and on the growth of financial intermediation in accounting for a "substantial portion" of the observed rise in the savings-income ratio.[23] By reducing search or negotiation costs between borrowers and lenders, intermediaries both increased returns to savers and reduced net borrowing costs to investors, thereby shifting both the demand and the supply for investment goods to the right. Commercial banks were the most important type of financial intermediary, and the growth of the banking system over this period was indeed dramatic. Between 1834 and 1900 total assets of the banking system increased twenty-seven-fold. Moreover, new institutions, such as organized securities markets, were also developing to promote capital mobilization and accumulation. Mutual savings banks, and later savings and loan associations and insurance companies, facilitated the accumulation and transfer of funds into businesses, while mortgage brokers and commercial-paper houses greatly aided the transfer of capital across geographical and industry barriers. In short, the rapid growth of intermediaries providing both higher returns on savings and diversified risk was very likely to have been quite important in stimulating domestic savings.

In the next section, we develop a model that focuses on demographic and occupational influences on capital accumulation. The effects of fertility on savings, as stressed by David and Lewis, can be evaluated in this empirical model, while the effect of occupational differences provides some evidence on changing patterns of income inequality. In addition, year-specific variables will be introduced that measure the combined effects, over time, of the generalized shifts in financial intermediation, interest rates, and investment-goods prices.

---

[22] Williamson, "Inequality, Accumulation, and Technological Imbalance."
[23] Davis and Gallman, "Capital Formation," p. 65.

III. The Empirical Model of Capital Accumulation

Many of the studies of the rising rate of capital accumulation have focused on changes in the savings rate. For purposes of measuring sources of saving, however, it is useful in this model to measure capital directly, and then, from the changes in the capital stock over time, to infer real net rates of savings.

A Model of Wealth Accumulation

In the life-cycle model of consumption and savings, the individual plans consumption (both for self and for children) subject to earnings, inheritances, planned bequests, and anticipated repayments from the children. We assume a model of perfect certainty for the sake of simplicity, although the introduction of uncertainty about length of life will not affect the basic results of the model.[24] Given these consumption plans, savings will simply be the difference between current income and current consumption. Wealth accumulation for an individual will be the sum of past savings or dissavings, augmented by accumulated interest. The pattern of capital assets over the life cycle will typically differ from the pattern of savings, which measures the change in the level of assets.

The aggregated stock of capital assets in a population can be separated into that wealth owned by people of age i and occupation j. That is, for state s, we can write aggregate wealth stock as

$$W^s \equiv \sum_{i=1}^{m} \sum_{j=1}^{n} P_{ij}^s K_{ij}^s \qquad (2)$$

where W^s is aggregate wealth owned by residents of state s, K_{ij}^s are average assets held by individuals of age i, $i = 1, \ldots, m$, and occupation j, $j = 1, \ldots, n$, and P_{ij}^s is the total number of (i,j) individuals in state s. If we assume that the premium paid to occupation j, b_j, is constant across all ages and states, and that K_{ij}^s, for the moment, is also constant across states, (i.e., $K_{ij}^s = K_i + b_j$ for all s), then we can write the aggregate stock of wealth in state s as

$$W^s = \sum_{i=1}^{m} \sum_{j=1}^{n} P_{ij}^s (K_i + b_j) \qquad (3)$$

---

[24] Jonathan Skinner, "The Effect of Variable Lifespan and the Interest Elasticity of Consumption," *Review of Economics and Statistics*, 67 (Nov. 1985), pp. 616–623.

Rearranging,

$$W^s = \sum_{i=1}^{m} \theta_i^s K_i + \sum_{j=1}^{n} V_j^s b_j \qquad (4)$$

where θ_i^s *is the number of individuals in state s* within category i, and V_j^s the state-specific number of individuals in occupation j. Asset accumulation differs by a constant amount over the life cycle for the m occupations, owing to differences in income, the propensity to spend out of income, and other factors.

All population groups are included in this equation, even though infants, for example, hold no capital of their own. However, the degree to which capital stock responds to the population under age ten indicates the extent to which the parents must vary their own asset (and savings) decisions in response to the children. Therefore, the expected family assets for a husband of occupation j, age i (equal to the wife's age), and with two children, age i' and i'', would be

$$W = K_i + b_j + K_{i'} + K_{i''} \qquad (4')$$

We expect that the introduction of children would lead to lower levels of wealth accumulation, other things being equal, both because of the additional expenses for children (estimated to be $80 per year for a child of seven in 1889–90)[25] and because the anticipated repayments by children will depress current saving. The only exception to this general rule could be when the children were young. If the children were anticipated, savings (and hence asset accumulation) for a family would be higher than for those families who did not anticipate children, since the expectant family would have to save for the extra expenses of the children to come. Once the children arrive, savings will drop, and asset accumulation will, over the child-raising years, fall to a level below that of the benchmark childless couple. As the children leave the household, assets will rise, until the point when the elderly begin to dissave. Thus in the first few years of retirement, total wealth may be at its maximum, even while the family is dissaving. If the family has a bequest motive, there may be further incentives to maintain high levels of assets (equivalently, contingent bequests) during the elderly or retirement years.

One reason why the coefficients for wealth accumulation may differ across states is that there are regional variations in income. We correct

[25] Lewis, "Fertility and Savings."

for this effect by including proportional variables, μ_ℓ, for the Northeast, the South, the Midwest, and the West. By including these proportional measures, we capture general (percentage) differences across regions, principally in level of income, or, because of self-selection, for example, the effects of immigrants to the West who carried little wealth with them. Furthermore, since the data measure the value of capital in a particular state (i.e., investment in that state) and our model explains the value of capital held by inhabitants of that state (or per capita savings), the reported data will bias upward the wealth holdings of westerners and bias downward the holdings of easterners who invest in the West. The regional variable therefore corrects for these differences, as well as measuring the effect of geographical variation in the interest rate.[26]

We would also expect that the level of aggregate capital accumulation would vary over the nineteenth century in response to other factors in addition to just demographic and occupational effects. In particular, the forty years between 1850 and 1890 were characterized by falling relative prices of investment goods, increasing financial intermediation, and variations in interest rates. We therefore introduce year-specific proportional dummy variables to account for these temporal effects. Note that the proportional specification implies that wealth holdings over time will vary in equal proportions, regardless of the individual's age, occupation, or region. Note also that the variable coefficients by themselves cannot distinguish among the factors of intermediation, interest rate, investment-goods price, and income, although outside evidence will shed some light on their relative contributions.

Because slaves were held as assets in 1850 (but not counted as a part of the population because, by definition, they could hold no wealth), we introduce an additional interactive term equal to the number of farmers in southern states in 1850, but equal to 0 in other states or in other years. This term will measure the incremental "wealth" that slaveholders held before the Civil War. Including an error term, e, to reflect mismeasurement and random error, we write the empirical model as

---

[26] Separating out the many influences that are captured in these regional dummy variables is not an easy task. For example, Easterlin's measure of regional income and the number of years since statehood (to allow for differences in past wealth accumulation across states) were both added as additional explanatory variables to the basic regression equation. Both estimated coefficients, however, proved not to be, statistically, significantly different from zero, while the dummy variables continued to be so. As a result, the variables were dropped from the final specification. Richard A. Easterlin, "Regional Income Trends, 1840–1950," in *American Economic History*, Seymour E. Harris, ed. (New York: McGraw-Hill, 1961), pp. 525–547.

$$W_t^s = \left(\sum_{i=1}^{m} \theta_i^s K_i + \sum_{j=1}^{n} V_j^s b_j \right) \left(\prod_{\ell=1}^{3} (1 + \mu_\ell D_\ell) \right) \left(\prod_{t=0}^{3} (1 + \alpha_t D_t') \right) + \varepsilon_t^s \tag{5}$$

where D_ℓ is a dummy variable for region, D_t' is the dummy variable for time (D_0' measures the southern farmer effect), α_t is the time-specific coefficient, and W_t^s is subscripted by year, $t = 1, \ldots, 4$. This equation is a hybrid of an additive function (the population and occupation variables) and a multiplicative function (the year and regional variables). Therefore we shall use a nonlinear technique to estimate the relevant coefficients.

The Data

The basic wealth estimates employed in this paper are the figures for total wealth, by state, compiled by the census for the period 1850–1890, a valuable source of data that has gone relatively unrecognized and underutilized. Wealth is measured as the total value of real property and personal property. This measure is particularly useful because it goes significantly beyond the reproducible capital stock in encompassing other important aspects of the accumulation of personal wealth. For one thing, it includes real estate and improvements, the form in which the majority of total wealth was held in the nineteenth century United States. Land improvements, such as land clearance, and building improvements represented an important form of saving in rural areas. This wealth measure therefore captures any change from "unconventional" to "conventional" forms of saving that occurred over the period.[27] The inclusion of personal property in the total means that consumer durables are counted in total wealth as well. The value of slaves is also counted in wealth for the years before the Civil War.

The unadjusted census wealth estimates are also roughly consistent with figures based on data compiled by Davis and Gallman, who report ratios of the capital stock, including land and inventories, to GNP in current prices by census years.[28] Applying GNP estimates in current prices to these ratios allows us to compute wealth estimates in current prices comparable to those reported in the census. Gallman's GNP estimates are in real terms and must first be adjusted by a price index. Comparisons among the census totals and the numbers implied by adjusting by Gallman's GNP deflator, the Warren-Pearson wholesale price index, and the David-Solar retail price index, respectively, are pre-

[27] Davis and Gallman, "Capital Formation," pp. 50–51.
[28] Lance Davis and Robert Gallman, "The Share of Savings and Investment in Gross National Product During the 19th Century in the U.S.A.," in *Fourth International Conference of Economic History*, F. C. Lane, ed., (Paris: Mouton, 1973), p. 457.

TABLE 8.2
U.S. Wealth Estimates Compiled by the Census, in Comparison with Alternative Measures, in Current Prices, 1850–1890
(In $ millions)

	Unadjusted census figures	Modified census estimates	Estimates based on Davis and Gallman wealth-income ratios		
			Gallman GNP deflator	Warren-Pearson wholesale price index	David-Solar retail price index
1850	$ 7,135	$ 7,510	$ 9,131	$ 7,854	$ 8,174
1860	16,160	16,916	16,310	15,990	15,990
1870	30,068	30,068	27,343	28,762	31,107
1880	43,642	43,642	37,916	39,201	44,842
1890	65,037	64,096	55,025	53,315	65,909

SOURCES:

Census figures: U.S. Eleventh Census, 1890, "Wealth, Debt, and Taxation," Part II, p. 14.

Capital-output ratios: Lance Davis and Robert Gallman, "The Share of Savings and Investment in Gross National Product during the 19th Century in the U.S.A.," in *Fourth International Conference of Economic History*, F. C. Lane, ed. (Paris: Mouton, 1973), p. 457.

GNP estimates: Unpublished; obtained through private correspondence from Robert E. Gallman.

Price indices: Lance Davis and Robert Gallman, "Capital Formation in the United States During the Nineteenth Century," in *Cambridge Economic History of Europe*, vol. 7, part 2 (New York: Cambridge University Press 1978), p. 27.

George F. Warren and Frank A. Pearson, *Prices* (New York: John Wiley and Sons, 1933), pp. 10–13.

Paul David and Peter Solar, "A Bicentenary Contribution to the History of the Cost of Living in America," in *Research in Economic History*, vol. 2, Paul Uselding, ed. (Greenwich, Conn., JAI Press, 1977), pp. 16–17.

sented in Table 8.2. The series fit fairly closely. The deviation from the implied Davis-Gallman figures is never more than about 20 percent and is really relatively large only in one year, 1890. Furthermore, when GNP is adjusted by the David-Solar price index, the difference averages only 4.5 percent and never exceeds 9 percent. The accuracy of the wealth data is discussed further in the appendix to this chapter, but it would appear that the census figures are reasonably consistent with other estimates of wealth for the period.[29] Moreover, since wealth appears on the left-hand side of the equation the possible bias on coefficients caused by errors in the wealth data is reduced.

One important nontangible form of wealth not counted was government debt. To include federal debt in personal wealth holding, the amount of debt held domestically was first calculated by subtracting Simon's estimates of public debt held abroad from the total outstanding.[30] Then, to eliminate double counting, federal bonds held by banks, insur-

---

[29] To be sure, our comparisons have just involved examining the aggregate value for the United States by census year. We have no way of assessing the accuracy of the totals by state. Here we must make the leap of faith that if the aggregate values are reasonably accurate, the wealth totals by state will be also.

[30] Matthew Simon, *Cyclical Fluctuations and the International Capital Movements of the United States, 1865–1897* (New York: Arno Press, 1978), p. 71.

ance companies, and trust companies were subtracted off. Personal bond holdings were assumed to have been distributed across states for each census year in the same proportions as reported for 1880 in a detailed study by the census. State bond holdings were treated similarly.[31]

A potentially serious problem with this data is that it is not possible to distinguish between investment in a state and savings by inhabitants of the state. The census figures reflect the value of all wealth by state, including that owned by foreigners. Since we attribute all the wealth in the state to the residents, the neglect of foreign holdings means the magnitudes held by domestic residents will be somewhat overstated. Similarly, interstate holdings of capital are also neglected. In our specification, therefore, migration is the only means of transferring capital across states. Although perhaps a bit extreme, such a characterization is not wildly unreasonable for most of the nineteenth century. With the possible exception of railroad investments, the opportunities for interregional investment were quite limited. The wide divergences in interest rates across regions were often noted at the time and have been taken as evidence that barriers to interregional capital mobility existed, even after taking risk into account.[32] A well-developed interregional mortgage market and stock market for industrial securities did not develop until approximately the 1890s. Our time period, which ends in 1890, therefore captures almost all of the rise in the savings rate, but stops before the period when structural change greatly facilitated interregional capital flows, which would have increased the distortions in our measure of wealth holding by state residents. Nevertheless, the census warned that a much larger percentage of property in western states was held by nonresidents, exaggerating the wealth of individuals in those states.[33] As mentioned above, we allow for systematic regional differences in wealth holding across states by including regional dummy variables in the regression equation.[34]

---

[31] *U.S. Tenth Census*, 1880, "Report on Valuation, Taxation, and Public Indebtedness," pp. 478–510; B. U. Ratchford, *American State Debts* (Durham, N.C.: Duke University Press, 1941).

[32] Lance Davis, "The Investment Market, 1870–1914: The Evolution of a National Market," *Journal of Economic History* 25 (Sept. 1965), pp. 355–399; John A. James, "The Development of the National Money Market, 1893–1911," *Journal of Economic History* 36 (Dec. 1976), pp. 878–897.

[33] *U.S. Eleventh Census*, 1890, "Report on Wealth, Debt, and Taxation," Part II, p. 10.

[34] The 1880 census, which reported estimates of the "true valuation" of property owned by state residents, might give some idea of the magnitude of the differences between wealth by location and by ownership across regions. The data show property owned by easterners to have had a value 11.4 percent greater than the value of wealth located in the East. On the other hand, the South, Midwest, and West were all net wealth importers, with wealth owned by residents 6.5, 6.6, and 13.9 percent, respec-

Data on the age and occupation structure of the free male population, by census year, are taken from the respective censuses of population. The occupational categories basically follow those compiled by the census, with some modifications. Personal and professional services (PERS) was purged of less skilled occupations. Domestic servants, soldiers, and launderers were subtracted off and grouped in an urban unskilled category with nonagricultural laborers (NAGLAB). The essentially intermediate skill category (TRADE) included those employed in Trade, Transportation, Mining, and Manufacturing (plus fishermen from Agriculture and hotel keepers and barbers from Services). Among agricultural occupations, farm owners (FARM) are separated from agricultural laborers (AGLAB). We omit 1860 from the pooled regression because even though individual occupations were reported by the census in that year, they were not grouped into occupational categories. The population classifications separate males into those aged under 10 (POP1); those between 10–19 (POP2), 20–39 (POP3), and 40–59 (POP4); and those aged 60 and above (POP5).

Not all of the equivalent information for females was reported by the census, so we limit ourselves to age and occupation distributions of males and take males as the heads of households.[35] Thus the interpretation of the coefficients for workers over, say, age 20 would relate to total family wealth holdings. It is in the interpretation of the coefficients for children that one must make adjustments, since the number of male children under age 10 or ages 10–19 will be almost exactly matched by an equal number of female children. We assume that female children cost the same to raise as male children. Since accounting for female children would double POP1 and POP2, the "true" effect of a child on asset accumulation will be measured as half the measured coefficients. Whether those over age 60 should be considered a family unit, or as widowers, is less clear.

Empirical Results

Results from the cross-section time-series data are presented in Table 8.3. States and territories for which complete data existed were in-

---

tively, lower than the value of wealth located there. However, these figures should be viewed with a great deal of caution. The census itself warns that the "results can be accepted as true only in a very general way and as subject to large qualifications." Moreover, since the reported total value of property owned by state residents equals the total value of wealth, the implied value of wealth owned in the U.S. by foreigners must be zero, which is clearly contrary to fact. U.S. Tenth Census, "Report on Valuation," p. 13.

[35] In particular, occupational classifications for working women were not provided in the 1850 census.

TABLE 8.3
Nonlinear and Linear Wealth Regressions, 1850 and 1870 to 1890[a]

Variable	(1) Nonlinear (all years)	(2) Nonlinear (1850 excluded)	(3) OLS (all years)
POP1	4357 (2.47)	3458 (1.90)	3528 (1.91)
POP2	−1131 (0.50)	−755 (0.33)	−6100 (2.57)
POP3	2806 (1.81)	1166 (0.70)	573 (0.41)
POP4	4469 (1.42)	1103 (0.35)	8028 (3.17)
POP5	7111 (2.22)	7420 (2.41)	4976 (1.47)
TRADE	−1434 (1.01)	493 (0.32)	750 (0.66)
AGLAB	−5501 (4.96)	−4100 (3.45)	−2170 (2.25)
NAGLAB	−5824 (3.32)	−3283 (1.63)	−4064 (2.44)
FARM	−4027 (3.32)	−1926 (1.09)	−1066 (0.86)
PERS	33291 (5.91)	34900 (5.89)	35415 (5.79)
STHFA50	−0.004 (1.83)		−1678 (1.06)
SOUTH	−0.181 (2.76)	−0.153 (2.06)	0.016[b] (1.24)
MIDW	−0.01 (0.27)	0.01 (0.23)	0.409[b] (2.84)
WEST	0.246 (2.42)	0.293 (2.70)	0.018[b] (1.60)
YR50	−0.348 (4.81)		−0.014[b] (1.39)
YR70	−0.192 (3.26)	−0.152 (2.43)	−0.039[b] (3.62)
YR80	0.078 (1.65)	0.091 (1.84)	−0.010[b] (1.09)
CONSTANT			−20301 (0.46)
R^2	0.95		0.93
N of Observations	161	126	161

[a] Absolute values of t-statistics in parentheses.
[b] Expressed as a percentage of mean state wealth ($3,535,000).

cluded in the set of observations, so that as new ones were formed, more data became available in later years. The results of equation 5, estimated with 161 observations and weighted to correct for heteroscedasticity in the error term, are presented in column 1 of Table 8.3.[36] Recall that these population data include males only, so that the effect of children (POP1, POP2) would be approximately half of the reported coefficient, since the male population of children understates the total population by a factor of one-half. The calculated R^2 is 0.95, although some of the low t-statistics suggest potential multicollinearity.

The pattern of population coefficients is broadly consistent with the theory of life-cycle consumption. If we assume that the economic household begins at age 20, then a farmer between the ages of 20 and 40 would have negative assets of $-\$1,221$, using 1890 as a benchmark, or $-\$795$ for a farmer in 1850 (in constant dollars, but deflated by the 1850 dummy). Assets rise to $660 between ages 40 and 60, and finally attain $3,084 ($2,009 at 1850 levels) at ages over 60. The substantial assets accumulated by the oldest group lend support either to a strong bequest motive or to high levels of precautionary saving in the event of a long life. Children have a marked impact on life-cycle patterns of wealth accumulation. The introduction of a child under age 10 (male or female) is associated with a rise of $2,178 ($1,419 in 1850), while children between ages 10 and 19 cause assets to fall by $565. While the sign of the POP1 coefficient is consistent with the life-cycle hypothesis that families anticipating children will have saved for the event, the magnitude of the effect seems high.[37] The predicted wealth,

---

[36] The residuals exhibited heteroscedasticity with respect to total state population. Therefore, a first-stage regression was estimated, which provided a vector of residuals. These residuals were then squared, and regressed on total population, POP: $(\varepsilon_i^s)^2 = a + d(\text{POP})$. Each observation was then weighted by $(\hat{\varepsilon}_i^s)^{-1/2}$ and estimated in the second stage. The coefficients did not differ significantly between the two stages.

[37] The predicted effect of a newborn child on assets, $2,178, appears puzzling. However, regardless of the specification of the model or the time period, the result that the POP1 coefficient was positive, and the POP2 coefficient negative, persisted. Regressing the ratio POP1/POP2 on occupation and year dummy variables suggested that farm families were more strongly associated with children under 10 than children over 10. One potential explanation for these results is that population is measured more accurately than occupation, so that small children are a partial proxy for the true number of farmers. This would explain why the "representative" farmer's wealth is predicted accurately, but the marginal effects of children are not. In the decomposition exercise, we attempt to correct for this bias by setting the coefficient on POP1 equal to the coefficient on POP2, resulting in extreme dissaving for families with children. In his study of factors influencing the distribution of landholding (the principal form of wealth) in colonial North Carolina, Gallman finds the estimated coefficient of the variable, children under 8 years old, to be unexpectedly positive as well, similar to the result here. He attributes the positive relationship possibly to gifts or bequests from

however, of a representative farming family whose head is aged 30, with two children under age 10 and one over age 10, living in the Northeast in 1870, is $2,075. This estimate is very close to the estimate of an 1870 farmer's family wealth (approximately $2,000 at age 30) in Soltow's careful study of wealth holdings during the period 1850–1870.[38]

The occupation coefficients indicate moderate differences in savings among trade and manufacturing workers, farmers, and laborers. For a 40–60-year-old head of household, wealth ranged from −$822 for a nonagricultural worker and $660 for a farmer, to $3,253 for an individual in trade or manufacturing. A surprising result is the large and strongly significant effect of those in personal and professional services (PERS) on aggregate savings. The difference exceeds what one might expect on the basis of differences in income. The PERS coefficient, in combination with the age effect for an individual aged 40–59, is $37,977, or a value of $24,738 in 1850.[39] This coefficient suggests that professional and personal service occupations accounted for a large portion of aggregate savings.

The findings in Soltow's study, while more detailed, suggest that this coefficient is not implausible. The occupations that he found listed for eighty-seven men who, in 1850, had estates greater in value than $60,000, included a "farmer, planter, tobacconist, gentleman, landlord, hotel keeper, engineer, doctor, lawyer, clergyman, corn merchant, lumber merchant, grocer, . . ." and other merchants. Substantial wealth could have been held by the bourgeoisie (or at least the upper bourgeoisie) as well as the capitalist elite. In addition, the proportion of PERS in 1870, for example, is roughly consistent with the wealth distribution of the same year. Again using Soltow's wealth-distribution figures, the average wealth of the top 2.1 percent of the

---

grandparents. Robert E. Gallman, "Influences on the Distribution of Landholdings in Early Colonial North Carolina," *Journal of Economic History* 42 (Sept. 1982), pp. 568–569.

[38] Lee Soltow, *Men and Wealth in the United States, 1850–1870* (New Haven: Yale University Press, 1975). There are, in fact, two important sources of savings or accumulation, life-cycle accumulation and inheritances. Since the timing of changes in wealth depends in part on the timing of inheritance, the pattern of accumulation inferred from the data does not represent solely a response to the life-cycle factors. However, the data do suggest that significant wealth accumulation was associated with middle age, a pattern certainly consistent with the life-cycle model.

[39] An alternative specification was to allow occupational variables to enter the equation proportionately, so that wealth holdings of families in, say, trade differed from assets of families in farming by a constant proportion. This specification resulted in an unstable likelihood function. It also assumes that dissaving caused by children differs by the same constant proportion across occupational groups.

population was $37,571, while for the top 5.4 percent, it was $20,426. The estimated value of wealth for 1870 from the regression (deflated by the 1870 coefficient) was $30,670 for the 3.3 percent who were under the category of PERS. If indeed the PERS group were the wealthiest group, then the coefficient is broadly consistent with the distributional evidence.

One shortcoming with the regression technique is that one cannot establish causality; were those in PERS attracted by wealth, or did they themselves accumulate the wealth? For example, accountants, lawyers, engineers, and doctors may all have been attracted to large wealthy cities, which would lead to a spurious correlation between savings and professional and personal services. While this possibility cannot be ruled out, the magnitude of PERS across states does not support the theory that these higher-level occupations congregated only in urban, high-wealth areas. In 1850, for example, the proportion of PERS in New Hampshire (0.035) was higher than that in New York (0.034). By 1890, that ordering had changed; the proportion of PERS in New York had risen to 0.077, while in New Hampshire, it was 0.045. However, the proportion had risen as well in other states (Florida, to 0.054, Ohio, to 0.063, and Missouri, to 0.057), although southern states lagged behind the rest of the country (Mississippi was only 0.028). Thus the personal service and professional occupations were not confined exclusively to urban areas.

The regional variables measure the proportional differences across geographic areas. Relative to the Northeast, wealth holdings were 18 percent lower in the South, 25 percent higher in the West, and 1 percent lower in the Midwest. The high relative wealth in the West is consistent with the view that easterners held property in the West. The interactive term for southern farmers in 1850, designed to capture the effect of the antebellum slave economy, is negative and insignificant. This result is surprising in light of Soltow's findings, which indicated that southern farmers were wealthier than their northern counterparts before the Civil War, but poorer after the war.[40]

The year-specific variables indicate a general, proportional shift in saving that cannot be attributed either to demographic or to occupational effects. Holding these latter two factors constant, the regression indicates that wealth grew by 24 percent from 1850 to 1870 [(0.348 − 0.192)/(1 − 0.348)] and by 33 percent from 1870 to 1880, but that it fell by 7 percent from 1880 to 1890. The overall rise of these nonspecific factors accounted for a growth in net wealth formation of

[40] Soltow, *Men and Wealth*.

1.07 percent per year, which translates (for a total capital-income ratio of 2.5) to a net rise of 2.6 percentage points in the savings rate.

Table 8.3 also presents results from alternative specifications of the estimating equation. Column 2 provides coefficients from the nonlinear model excluding 1850, perhaps the most questionable wealth data (see the appendix at the end of the chapter). The POP3 and POP4 coefficients are less than those estimated in the full sample, but the occupation coefficients are somewhat higher, so the combined effect of, say, a farmer aged 20–39 is not substantially different between the two equations. Both the positive coefficient for POP1 and the strong effect of PERS are maintained in this regression. Limiting the data set to the postbellum period therefore does not seem to affect significantly the pattern of accumulation implied by the original regression. Column 3 of Table 8.3 provides an OLS regression (with all years included) for comparison with column 1. The region- and year-specific variables were entered in the regression as simple dummy variables (so that, for example, the WEST coefficient was $65,074); to provide a basis for comparison with the other two equations, we deflated these coefficients by average state wealth. The OLS equation indicates similar demographic effects (once again, POP1 is positive), although POP5 falls relative to POP4. Nonagricultural laborers are estimated to be poorer than agricultural workers, while the coefficient on professional and personal services is strongly positive. Note that the OLS equation does not reflect the accounting identity (equation 2) but instead measures wealth that cannot be attributed to any specific group in the state population. (For example, a midwestern state with no inhabitants is predicted to be worth $124,250 in 1890.) We now return to the original equation to examine how demographic and occupational changes affected capital accumulation.

Decomposing Changes in the Capital Stock

The total change in real wealth can be decomposed into its components in a series of steps in which all variables but one are held constant, so that the contribution of that one factor can be evaluated. It should be cautioned that owing to the nonlinearities of the estimated equation, not all factors are independent of each other, and thus the order in which each variable is changed does make a difference to the result.[41] The total decomposition is shown in Table 8.4.

[41] For example, the effect on the capital stock of shifting the population from the 1850 distribution to that in 1890 will be greater if the year-specific effects have already been changed, since the counterfactual 1850 population will have had its initial capital stock already upgraded to 1890 levels.

TABLE 8.4
Component Factors of the Total Change in Real Wealth per Head,
1850–1890

Type of effect	Percent
Effects of change in the age distribution	13%
Effects of change in the occupational distribution	34
Effects of increasing real income per capita	44
Residual effect—increases in financial intermediation	9
TOTAL CHANGE	100%

First of all, we can measure the effect of demographic changes on wealth accumulation by comparing the predicted wealth value from the fitted regression using the 1850 population distribution with that from a similar specification but based on the 1890 population distribution. We adopt the Northeast as the base-line region, and hold the occupation distribution constant by imposing the occupational mix in 1870 on both the 1850 and the 1890 labor forces.[42] Changing the population causes the aggregate capital stock to rise, primarily because the population more than doubled over the forty years. However, the wealth per male over age 20 actually declines, from $3,482 to $2,905.[43] The wealth attributed to children under 10 (who comprised 29 percent of the population in 1850 but only 24 percent in 1890), was a factor in this decline. To test the sensitivity of our results to the coefficient on POP1, we imposed the POP2 coefficient (−$1,131) on the POP1 variable, and adjusted the POP3–POP5 coefficients upward to ensure that the predicted aggregate capital stock in 1850 did not change. The implied coefficients were therefore −1,131 for each child under age 20 (which accords well with the child-rearing costs reported by Lewis), $5,320 for POP3, $8,870 for POP4, and $13,450 corresponding to POP5. Clearly this configuration of coefficients represents an extreme pattern of age-specific life-cycle asset accumulation. However, the implied change in the capital stock, while positive, is not large even when these coefficients are used to predict how assets would vary if only population had changed between 1850 and 1890. The figure for assets per male over age 20 was predicted to rise from $3,482 in 1850 to $3,802 in 1890, only 13 percent of the total change in real wealth per

[42] We use the 1870 occupation mix as a representative midpoint distribution. A shortcoming of the 1850 occupation mix is that, because slaves are not included in agricultural workers, the proportion of AGLAB to total workers is less than that in 1870.

[43] The 1850 per capita figure, $3,482, is not adjusted by the 1850 dummy variable capturing income and other effects; the comparison holds income, occupation, and other factors constant.

household head. Changes in fertility rates and family size are likely to have shifted the savings-supply curve to the right, but alone, these factors accounted for only about one-eighth of the increase in real wealth.

The effect of the changing distribution of occupation had more impact on the rise in total capital accumulation.[44] The proportion of those working in personal and professional services rose from 3.9 percent of the population in 1850 to 5.4 percent in 1890. At the same time, the proportion in trade rose from 31 to 41 percent, while the low-saving nonagricultural laborers dropped from 21 to 12 percent.[45] Changing the distribution of the total 1890 work force in the fitted regressions from the counterfactual 1850 mix to the actual 1890 mix (e.g., shifting the number of PERS workers from 3.9 to 5.4 percent of the 1890 work force), while in effect holding relative wealth levels across occupations constant, involves a shift toward occupations that held more assets. This shift in the occupational distribution caused the total capital stock to increase by 34 percent of the total change over the forty-year period.[46] One could possibly interpret this occupational shift toward jobs such as trade, manufacturing, and professional services as part of the movement toward increased inequality along the lines discussed by Williamson,[47] although the effect of these occupational changes on general levels of inequality is not immediately evident. It is clear, however, that these shifts contributed significantly to the growth in real wealth between 1850 and 1890.

Finally, the year-specific variables accounted for 53 percent of the rise in capital stock per head.[48] The year-specific effects will be appor-

---

[44] The effect of occupation on wealth depends crucially on whether one evaluates the change at 1850 levels (with the 1850 dummy imposed) or at levels corresponding to 1890. The number reported here, a rough average of the two extreme measures, was actually calculated as the residual of the total change, net of income and demographic effects.

[45] The ratio of agricultural laborers (not counting slaves) to total workers in 1850 was 0.11; following the Civil War, it rose to 0.24 in 1870, and then fell to 0.15 in 1890. Farmers dropped from 0.34 in 1850 to 0.28 in 1890.

[46] A portion of the rise in wealth could also have been caused by an increase in aggregate income as workers simply moved into higher-paying jobs. However, the effects are not large. For example, if those in TRADE and FARM were paid three times as much as laborers, and those in PERS three times as much as TRADE and FARM workers, the effect of occupational shifts would have increased average income by 7.5 percent, or only 13 percent of the total increase in per capita income between 1850 and 1890.

[47] Williamson, "Inequality, Accumulation, and Technological Imbalance."

[48] There are additional effects caused by changing geographical distributions. That is, westward migration would imply that the aggregate stock was increasing, since the WEST coefficient is positive. We hold the geographic distribution constant in this comparison, since it is not clear whether the geographic coefficients measure differences in wealth or differences in out-of-state investments.

tioned among three principal factors: changes in real income per capita, the effects of increased financial intermediation, and the decline in capital goods prices. During the forty years following 1850, real income per household head (males over age 20) rose at an annual rate of 1.2 percent, compared with the 1.8 percent growth rate in capital stock per head of household. Davis and Gallman suggest that 0.3 percent of the income rise was caused by the increased capital accumulation. To separate out the endogenous effect on income of the rising capital stock, we subtract the "endogenous" 0.3 percent increase, and assume that "exogenous" income was rising at 0.9 percent per year. If we further assume that 13 percent of the rise in income was caused by the shift in the occupation mix toward higher-paying jobs (see footnote 46), then the rise in "exogenous" per capita real income is 0.78 percent per year. Therefore, if a 1 percent increase in income is associated with a 1 percent increase in wealth (i.e., savings has a unitary income elasticity in the long run),[49] then growth in real income per head alone accounts for 0.78 percent growth per year in real wealth, or 44 percent of the 1.8 percent growth rate of the capital stock.

These calculations suggest that of the 1.8 percent annual rise in capital stock per household, 53 percent can be attributed to general year-specific shifts in asset holdings; of that 53 percent, 44 percent was caused by rising levels of income. The residual 9 percent of the total increase in wealth per head was caused by the declining price of investment goods, increased financial intermediation, and a decline in interest rates. As argued in section II, the fall in capital goods prices, while increasing savings rates, could have had relatively little or no effect on the value of the capital stock.[50] The increased quantity of real capital units held by the saver was likely to have been offset by the decreased *MPK* (holding interest rates constant). The decline in real interest rates

---

[49] The precise decomposition of the year-specific variables depends on this not particularly controversial assumption. Friedman's permanent income hypothesis, for example, takes the long-run income elasticity of savings to be unity. Davis and Gallman, however, are a bit more guarded in only ruling out the possibility that savings were income elastic. While in principle it is possible actually to compute the implied cross-sectional wealth elasticity from the regional wealth and income data, it has not been possible in practice because too many other factors are not being held equal. Davis and Gallman, "Capital Formation," p. 50.

[50] Interest rates fell during the period 1850–1890, suggesting that the *MPK* dropped by more than the fall in the relative price of investment goods. If the proportional fall in the *MPK* were more than the proportional increase in units of physical capital, then the yearly factor income (*MPK · K*, where *K* is units of physical capital) would decline. However, the value of capital is the present value of the future income stream; since r has fallen, the discount rates applied to future income also falls. The direction of change in the value of capital is uncertain.

over the 1850–1890 period would have reduced rather than increased savings and hence the total capital stock. If we neglect the negative effects of declining real interest rates and the positive effects of inframarginal capital losses on saving, both of which are assumed to be small, then we can attribute 9 percent of the increase in total wealth per head (or 17 percent of the growth in the total capital-income ratio) to financial intermediation. This measure, however, is a lower bound for the role of financial intermediation because the higher savings of professional and service occupations may also reflect better access to banks and other financial institutions. We caution that this estimate should be revised upward or downward depending on whether the magnitude of the interest rate effect exceeds or falls short of the investment price effect.

IV. Summary and Conclusions

This paper uses an unexploited data source, the U.S. Census figures for real and personal property by state, to examine empirically the sources of the dramatic increase in savings during the mid-nineteenth century. The focus in this paper is somewhat different from that of most others that have attempted to explain the "Grand Traverse." Rather than examining changes in savings flows over time, we concentrate instead on changes in the stock of total wealth holdings over the period 1850–1890. In a life-cycle framework, the emphasis is on the pattern of accumulation of assets by individuals. Because we are interested in the broader problem of total wealth accumulation, the census data, which are very comprehensive, are especially valuable. They include not only reproducible capital, but also other important categories of wealth holding such as land plus improvements and personal property. The transition from "unconventional" to conventional forms of savings associated with the shift from rural to urban employment, for example, would appear as a rise in the conventional savings rate based on the accumulation of reproducible capital. Here, it is captured in our more inclusive measure of wealth holding. As a result, the accumulation of total wealth proceeded at a rate slower, relative to income growth, than that of reproducible capital. Between 1850 and 1890 the capital-income ratio for the total capital stock, including land and inventories, rose from 3.3 to 4.6 in current prices, while the reproducible capital stock–output ratio increased from 1.3 to 2.2.[51]

We develop an estimating equation based on an accounting identity

---

[51] Davis and Gallman, "The Share of Savings and Investment," p. 457.

attributing total wealth in a state to residents of the state (here males, who are taken to have represented heads of households), based on age and occupational characteristics. The equation was estimated as a cross-section time series by state over the years 1850, 1870, 1880, and 1890, a period that encompasses most of the rise in the savings rate. This equation allows us then to decompose the total increase in wealth holding into particular component effects. Assuming a unitary income elasticity, increases in real income accounted for 44 percent of the increase in real wealth per head over the period. The remaining 56 percent of the total rise therefore represents an increase in the comprehensively defined capital-income ratio.

This 56-percent component of the rise in real wealth holding may be apportioned among three principal influences: demographic changes, occupational changes, and as a residual, primarily increases in financial intermediation. First of all, the regression results indicate that changes in the age distribution over time accounted for at most only about 13 percent of the increase in wealth holdings per household head. The changing demographic pattern over the nineteenth century due to the declining fertility rate does not appear to have been the principal factor behind increased wealth accumulation. This conclusion from our estimated equation is consistent with that of Lewis, who simulated the effects of declining fertility in a life-cycle model with exogenously specified rearing and pay-back costs.[52]

Once we are past the strong conclusion that demographic influences were not central in the Grand Traverse, attributing the remaining effects becomes a bit more uncertain. The changing occupational distribution appears to have been by far the single most important factor, accounting for around 34 percent of the total increase in real wealth (34/56 = 61 percent of the increase in the wealth-income ratio). The residual, here taken to represent primarily the effects of increased financial intermediation, accounts for only 9 percent of the increase in real wealth. These shifts in the occupational distribution are a phenomenon that has not been previously discussed in the literature and their interpretation and implications are not necessarily straightforward. For one thing, if the change in the occupational mix was indeed one aspect of increasing inequality over the period, then a significant influence of inequality on increased accumulation cannot yet be ruled out. In addition, to the extent that high-saving occupations had better access to financial markets and institutions, at least part of the influence of the changing occupational mix on asset accumulation may be attributed to improved financial intermediation.

[52] Lewis, "Fertility and Savings."

Data Appendix

As noted in section III, the wealth data used in this paper are the measures of the total value of real property and personal property, by state, reported by the census. In this appendix, the potential accuracy of the census figures will be briefly discussed and some minor adjustments to them will be described.

Estimates of the "true valuation" of real property and personal property were obtained by adjusting the assessed values upward. The "true value" was supposed to represent the cash or market value of the property. In 1850, 1860, and 1870, estimates of "true" wealth were left to U.S. marshals in the different districts who were instructed to add to the local assessments "such a percentage as they thought the assessors themselves would have added for estimates of real wealth."[53] In later years, the Census Office itself did the revaluations. Because of the vagueness of the instructions and apparent arbitrariness of the adjustments, the wealth estimates, especially for 1850, 1860, and 1870, were held in rather low regard by the late nineteenth century. The 1870 census itself admitted that certain duplications may have occurred by including the value of both mortgages held and the property represented by them.[54] The 1850 estimates were the most harshly criticized, called "old, crude, and utterly incorrect," primarily because the increase in wealth between 1850 and 1860 (see Table 8.2) was thought to have been too large to have been plausible. As a result, the 1850 figures must have been too low. David A. Wells put the true rate of increase between 1850 and 1860 at 65 percent, or at most 80 percent, in contrast to the 126-percent rise in the unadjusted census numbers.[55]

Opinion about the accuracy of these earlier estimates had changed, however, by 1900. In a detailed reassessment of the wealth data, the 1900 census concluded, "Hence it is deemed probable that the Census estimates for 1850 represent fairly the market values of the tangible property of the nation at that time."[56] The report found that the distribution of total wealth across states corresponded fairly well with the reported distribution of agricultural and manufacturing wealth. Moreover, even the unadjusted census figure falls within a reasonable margin of error of the figure implied by Goldsmith's estimate, built up from separate estimates for the components of reproducible

---

[53] *U.S. Eleventh Census*, "Report on Wealth, Debt, and Taxation," p. 9.

[54] The possibility of double counting extends further. Stocks and bonds, for example, might appear as holdings of individuals, while the physical assets they represent might have been counted as property of corporations. If the degree of indirect finance had been constant across states and over time, the overstatement of wealth would simply be proportional and not affect the reported results. The use of indirect finance was growing over the period, however, so it is possible that the influence of the year-specific dummy variables on changes in wealth may be overstated.

[55] *U.S. Eleventh Census*, "Report on Wealth, Debt, and Taxation," p. 9; Raymond Goldsmith, "The Growth of Reproducible Wealth of the United States of America from 1805 to 1950," in *Income and Wealth of the United States—Trends and Structure*, Simon Kuznets, ed. (Baltimore: Johns Hopkins University Press, 1952), p. 257.

[56] *U.S. Twelfth Census*, 1900, "Report on Wealth, Debt, and Taxation," p. 29.

tangible wealth for 1850. Similarly, the adjusted Kuznets estimate for 1880 wealth, built up also from its components, of $45.1 billion compares quite closely with the census figure of $43.6 billion, which was based on adjustments to assessed values.[57] The discussion in section III indicates that the unadjusted census-of-wealth estimates are also roughly consistent with figures implied by data compiled by Davis and Gallman.

However, some adjustments to the reported figures are necessary to assure consistency across years. The year 1890 was the first to include the value of vacant federal land and property on Indian reservations. Since Indians are not included in our population totals and to ensure comparability among census-year observations, those values are subtracted from the 1890 wealth totals. The estimates for 1850, 1860, and 1870 were limited to taxable real property and personal property, excluding property held by charitable, religious, and educational institutions, all of which were included in subsequent years. To adjust for the omission of nontaxable property, the ratio of nontaxable to taxable property reported for 1890 was assumed to have prevailed for 1850 and 1870 as well, and the wealth estimates were raised proportionally by state. The totals for earlier years may still be somewhat understated in view of the "increasing tendency toward exempting personal property" over the century, as noted by Gallman and Howle.[58] The aggregate wealth figures with these adjustments appear in the second column of Table 8.2.

In 1860, the census takers reported both their estimates of "true valuation" of real property and a total reflecting owner valuations. Gallman and Howle believed the owner-based estimates to have been more accurate than those done by the marshals. As a result, they argue, the 1850 and 1860 census figures for real property valuation were understated.[59] Following their adjustment, we also mark up the 1850 real property valuation by the 1860 ratio of owners' valuation of real property to the census "true" valuation. It should be noted, however, that since all the state wealth totals are marked up proportionally in 1850, the change will affect only the coefficient of the 1850-year dummy in the regression and none of the other estimated coefficients.

Finally, the wealth figures in current prices, by census year, must be deflated to a real wealth series for the pooled cross-section time-series regressions. The choice of a suitable deflator is not an obvious one. Gallman's capital stock price index does not seem appropriate because it falls during the period, primarily because of a large decline in the price of equipment. Machinery and inventories constitute only 4.7 percent of our total wealth estimate for 1890, while real estate and improvements account for 61 percent. While many land improvements are represented in the Gallman deflator, the value of land itself is not. Therefore, the decline in the price of capital may be too large for our measure of total wealth, since it seems unlikely that the relative price of im-

---

[57] Goldsmith, "The Growth of Reproducible Wealth," pp. 258, 306, 310.
[58] Robert E. Gallman and Edward S. Howle, "Fixed Reproducible Capital in the United States," unpublished manuscript, p. 64.
[59] Ibid., p. 55.

proved land declined as substantially as that of physical capital. Instead, we opt for a more generalized measure of purchasing power with which to adjust the nominal wealth figures, Gallman's GNP deflator.[60] To the extent, however, that we have understated the decline in the "true" price index, we shall also have understated the increase in the real value of wealth over time. Our figures imply a 4.54-percent-per-annum growth rate of real wealth between 1850 and 1890, which therefore may be a lower bound to the true value.[61] Although we implicitly assume in our specification that there were barriers to the free mobility of capital across regions over the period, we nevertheless do not adjust for possible regional differences in the prices of capital and improved land. Our guess is that this regional price variation may not be a severe problem, but in any case, there does not exist a satisfactory regional deflator, for our purposes, with which to take the effect explicitly into account. The Coelho and Shepherd regional price index based on a survey of local retail commodity prices from the Weeks Report, for example, covers neither the whole time period (ending in 1880) nor all the regions in the country.[62]

[60] Davis and Gallman, "Capital Formation," p. 27.

[61] Interestingly, and perhaps somewhat suggestively, the calculated growth rate of our measure of real wealth fits quite closely with those implied by other, although rather different, measures of real capital. Abramovitz and David's Divisia index of reproducible and nonreproducible capital grows at an average rate of 4.6 percent per year between 1853–1857 and 1889–1892, although that measure excludes consumer durables and slaves. The Gallman measure of real capital, which includes land improvements but excludes land, grows at an average annual rate of 4.5 percent between 1850 and 1890 as well. Moses Abramovitz and Paul David, "Reinterpreting American Economic Growth: Parables and Realities," *American Economic Review, Papers and Proceedings* 63 (May 1973), 431; Robert E. Gallman, "The U.S. Capital Stock in the Nineteenth Century," National Bureau of Economic Research Working Paper no. 1541, p. 47.

[62] Philip Coelho and James Shepherd, "Differences in Regional Prices: The United States: 1851–1880," *Journal of Economic History* 34 (Sept. 1974), pp. 551–591.

9

Economies of Scale and Efficiency Gains in the Rise of the Factory in America, 1820–1900

JEREMY ATACK

Sometime between 1820 and 1870 production processes were mechanized in many American industries. This change had a significant impact upon the country's industrial structure. Mechanized production emerged first in the cotton textile industry at the start of the nineteenth century, and it spread only slowly to other industries before mid-century.[1] Thus, in 1820 most manufacturing was carried on either in households or small artisan shops. By 1870, however, home manufactures had all but disappeared and manufacturing was increasingly concentrated in large factories.

This paper investigates the supply forces that brought about these changes. It explains the transition to large-scale mechanized production techniques in terms of productivity gains and scale economies. Despite the cost reductions afforded by these techniques, they were not adopted by all firms in each industry and the size distributions of firms became increasingly bimodal. A few large firms produced a substantial and growing proportion of each industry's total output, while seemingly inefficient and uneconomically small firms proliferated. It is argued that this structure was a result of improving transportation in the developing and expanding economy of nineteenth century America.

The study was sparked by the apparent contradiction between the meager total factor productivity gains for the transition from handicraft to factory production, reported by Kenneth Sokoloff for the first half of

---

I wish to thank Fred Bateman, Stanley Engerman, John James, Larry Neal, and William Parker for their helpful suggestions and comments on an earlier draft of this paper. They are not implicated in any remaining errors. Research assistance was provided by Susan Hotopp through a grant from the University of Illinois Research Board and part of the research was funded through an IBE grant from the College of Commerce of the University of Illinois. All statistical analyses were made on an IBM PC using DBASE III, LOTUS 123, MICROSTAT, and user-written programs.

[1] Victor S. Clark, *History of Manufactures in the United States* (New York: McGraw-Hill for the Carnegie Institution, 1929); vol. 1, pp. 448–455.

the nineteenth century and Carroll D. Wright's late nineteenth century estimates of multifold labor productivity gains from mechanization.[2] Factories, by definition, employed mechanized production techniques, but while Sokoloff's results suggest little competitive advantage from the adoption of factory production, Wright's research implied strong incentives to mechanize.

Despite the fairly small productivity gains from its adoption, the factory spread quite rapidly during the period. It is in search of an economic explanation for this that the paper investigates the possibility of scale economies favoring larger producers. Whereas total factory productivity gains may be thought of as a downward shift in the firm's entire average cost curve as the structure and organization of production is changed, scale economies result in falling unit costs along a given curve as the level of plant output is expanded. The former case assumes scale elasticity to be constant and equal to unity, but production methods change; the latter case rules out shifts in the production function, and hence in the average cost curve, but any change in unit input requirements that are scale related are captured in the elasticity coefficient. Economies of scale proved to be quite substantial. It is argued that they helped to determine the level of production while the productivity differences between factories and other forms of large-scale production explain the choice of techniques.

Most historians have used the term *factory* rather loosely, probably because they have not sought to quantify the extent of the phenomenon. For this essay, however, a precise definition is needed, despite a certain amount of oversimplification that such a definition introduces to the analysis. Factory production depended upon steam or water power to drive machinery. Artisan shops, sweatshops, and manufactories, on the other hand, relied upon hand tools. Human muscle was sufficient for their power needs. Artisan shops, sweatshops, and manufactories are therefore differentiated from factories by the absence of an inanimate source of power, and from one another by the size of their labor forces.[3]

---

[2] See Kenneth L. Sokoloff, "Was the Transition from the Artisanal Shop to the Small Factory Associated with Gains in Efficiency?: Evidence from the U.S. Manufacturing Censuses of 1820 and 1850," *Explorations in Economic History* 21, (Oct. 1984), pp. 351–382; U.S. Department of Labor, *Thirteenth Annual Report of the Commissioner of Labor. 1898. Hand and Machine Labor*, 2 vols. (Washington D.C.: U.S. Government Printing Office, 1899). Wright's study was brought to my attention by Stanley Lebergott's recent text, *The Americans: An Economic Record* (New York: Norton, 1984), p. 135.

[3] The categories are Artisan shops: No steam or water power, 1–6 employees; Sweatshops: No steam or water power, 7–25 employees; Manufactories: No steam or water power, over 25 employees. This classification is similar to that developed by

However, reliance upon inanimate power alone is not sufficient to identify a factory since plants using inanimate power were called mills from the very beginning. Consequently, this paper uses a second distinguishing characteristic of a factory, namely, that labor was specialized. Despite Victor Clark's observation that "it is impossible to define precisely at what point the . . . mill became a factory,"[4] it is assumed for the purposes of this essay that specialization could not be practiced extensively unless a mill had more than twenty-five workers. A factory is therefore defined as an inanimately powered plant employing over twenty-five workers.[5] The division between mills and factories is arbitrary but was selected so that it was more likely that some factories would be misclassified as mills than vice versa.

The transition from shop to factory was of great importance for later development. Alfred D. Chandler, Jr. regards the factory as the genesis of the modern business enterprise. In his analysis, the factory emerges as a response to new power sources and widening market opportunities.[6] To Marxists, the factory represents a further decisive step in the divorce of the worker from control over the means of production. Craftsmen in the artisan's shop often owned their own tools, and they worked at their own pace; in factory production the rhythm was determined by machines that the laborers operated but did not own.

This aspect of the subject has received widespread attention in British economic history,[7] but the emergence of the factory in America has been little explored, especially by cliometricians.[8] Only three recent

---

Bruce Laurie and Mark Schmitz, except that they call all mechanized establishments factories. See Bruce Laurie and Mark D. Schmitz, "Manufacture and Productivity: The Making of an Industrial Base, Philadelphia, 1850–1880," in *Philadelphia: Work, Space, Family, and Group Experience in the Nineteenth Century,* Theodore Hershberg, ed. (New York: Oxford University Press, 1981), pp. 43–92. Their category of "factories" has been divided into mills and factories on the basis of employment. See text and note 5 below.

[4] Clark, *History of Manufactures,* p. 447.

[5] Mills: Steam and/or water power, 1–25 employees; Factories: steam and/or water power, over 25 employees.

[6] Alfred D. Chandler, Jr.: *The Visible Hand: The Managerial Revolution in American Business* (Cambridge, Mass.: The Belknap Press of Harvard University Press, 1977).

[7] See, for example, George Unwin, "Transition to the Factory System," *English History Review* 36 (1922), pp. 206–218 and 383–397; Paul Mantoux, *The Industrial Revolution in the Eighteenth Century,* rev. ed. (London: Jonathan Cape, 1961), especially pp. 220–270; Jennifer Tann, *The Development of the Factory* (London: Cornmarket Press, 1970); S. D. Chapman, *The Cotton Industry in the Industrial Revolution* (London: Macmillan for the Economic History Society, 1972).

[8] Even those economic historians who rely more upon qualitative data and historical narrative, such as Stuart Bruchey or Thomas Cochran, have not given the subject much attention. See, for example, Stuart Bruchey, *Growth of the Modern American Econ-*

quantitative studies have looked at aspects of this issue. Bruce Laurie and Mark Schmitz have studied the transformation of manufacturing from nonmechanized to mechanized production in Philadelphia between 1850 and 1880. Their results show significant diseconomies of scale and contradict those presented here and elsewhere.[9] Kenneth Sokoloff has published estimates of the efficiency gains realized in the transition from artisan shops to small factories between 1820 and 1850.[10] He found the gains to be small. John James examined the shifts in the range of scale economies between 1850 and 1890, and found evidence both that they can help explain the rise of big business and that nonneutral technological change may play a role in this explanation.[11] To the extent that the estimates by Sokoloff and James overlap with those presented here, they are mutually consistent, despite the differences in our approaches and techniques.

This study, like Sokoloff's and James's, focuses upon the supply. This is the conventional wisdom. The role of technological change in both factor saving and shifting industry cost curves is stressed. Declining market prices for commercially manufactured goods coincided with an expanding industrial sector and this is consistent with a fixed downward-sloping demand curve and rightward shifts in the supply curve. We do not, however, need to assume that demand is invariant but only that rightward shifts in the industry supply curve are greater than any increases in market demand.

In the case of cotton textiles, however, Robert Zevin has argued that more than half of the industry's expansion may be attributed to demand changes alone, particularly after 1833.[12] Between 1815 and 1833, out-

---

omy (New York: Dodd, Mead, 1975), or Thomas C. Cochran, *The Frontiers of Change: Early Industrialism in America* (New York: Oxford University Press, 1981). An exception is Alfred D. Chandler, Jr., "Anthracite Coal and the Beginnings of the Industrial Revolution in the United States," *The Business History Review* 46 (Summer 1972), pp. 141–181, and Chandler, *The Visible Hand*.

[9] See Laurie and Schmitz, "Manufacture and Productivity," especially pp. 74–75 and 82–83, and compare with Jeremy Atack, "Estimation of Economies of Scale in Nineteenth Century United States Manufacturing and the Form of the Production Function," (Ph.D. diss., Indiana University, 1976), or Jeremy Atack, "Returns to Scale in Antebellum United States Manufacturing," *Explorations in Economic History* 14 (Oct. 1977), pp. 337–359, and Sokoloff, "Transition from the Artisanal Shop." Laurie and Schmitz make explicit mention of the conflict between their results and those in Atack, "Returns to Scale." See Laurie and Schmitz, pp. 86–87.

[10] Sokoloff, "Transition from the Artisanal Shop."

[11] John A. James, "Structural Change in U.S. Manufacturing, 1850–1890," *Journal of Economic History* 43 (June 1983), pp. 443–460.

[12] Robert B. Zevin, *The Growth of Manufacturing in Early Nineteenth Century New England* (New York: Arno Press, 1975), p. 10-6. Note that the pagination in this book is inconsistent. In the sections of the book cited herein, page numbers are hyphenated. The specific reference here is to page 6.

put of cotton goods expanded at an average annual rate of 16.3 percent; growth from 1833 to 1860 was slower, averaging 5.2 percent. Population growth averaged about 3 percent, but urban and western populations rose by about 5.5 percent up to the 1830s and this then accounts for about one-third of the expansion in the industry before 1833. Further, if we assume an income elasticity in the neighborhood of unity, then per capita income growth from the 1820s to 1860, as estimated by Paul David, would increase demand by another 1-2 percent.[13] Zevin also attributes a 1-percent annual increase in demand to the effects of transportation improvement.[14] He argues that most of the change occurred before 1824, but the early growth of the railroad, beginning in the 1830s in the East and continuing later into the Midwest, as well as the increasing use of the steamboat on the Mississippi and Ohio rivers make it certain that the effects were not all concentrated before 1833.[15] Zevin argues that in the post-1833 period, each of these factors grew more slowly and consequently demand expanded by only 3-4 percent a year.[16]

The primary sources of data for this study are the manuscripts of the federal censuses of manufactures for 1820, 1850, 1860, and 1870. These sources are supplemented where necessary by data drawn from published census volumes for the period after 1870, but these contain only aggregated data and have many arithmetic mistakes and extensive printer's errors.[17]

[13] Paul A. David "The Growth of Real Product in the United States before 1840: New Evidence, Controlled Conjectures," *Journal of Economic History* 27 (June 1967), pp. 151-197, especially p. 155.

[14] Transportation costs may be represented by shifts in either the supply or demand curves. Zevin's argument is that transportation improvements reduce the price to consumers and lead to a greater quantity demanded at the mill, that is, a movement along a given demand curve. See Zevin, *Growth of Manufacturing*, 10-14. Transportation improvements, however, shift the c.i.f. supply curve and hence it is usually treated as a supply-side factor.

[15] See, for example, Albert Fishlow, *Railroads and the Transformation of the Antebellum Economy* (Cambridge, Mass.: Harvard University Press, 1965), on railroads; and James Mak and Gary M. Walton, "Steamboats and the Great Productivity Surge in River Transportation," *Journal of Economic History* 32 (Sept. 1972), pp. 619-640, on steamboats.

[16] Zevin, *Growth of Manufacturing*, 10-28 and 10-32. Zevin also argues that demand became less responsive to price as time passed, with the short-run price elasticity of demand falling from 2 or 3 to about 1.5 because one-time luxuries came to be regarded as necessities. This further retarded the rate of industrial expansion by reducing the increase in quantity demanded in response to increased supply, which reduced prices.

[17] For an appraisal of the errors in the 1850-1870 censuses, see Jeremy Atack, *Estimation of Economies of Scale in Nineteenth Century United States Manufacturing* (New York: Garland Publishing, 1985), pp. 49-56; Fred Bateman and Thomas Weiss,

The sample from the 1820 census was drawn by Kenneth Sokoloff. It covers New England and the Middle Atlantic states.[18] Southern industry was also enumerated at the 1820 census, but it was not sampled by Sokoloff. The limited extent of southern manufacturing at the time and allegations of serious errors in the enumeration suggest that this omission is of minor consequence.[19]

Many scholars have regarded the 1820 census data with great skepticism. Much of this doubt derives from the jaundiced views of contemporary observers. For example, the editors of *Niles Register* were arguing that "the returns will be so imperfect that it will have been better if the subject of the inquiry had been altogether omitted," even before the 1820 census was taken.[20] Afterward, many congressmen complained that the published summaries omitted the returns for their districts.[21] A supplement was subsequently published, but even so, many errors remain.[22] Most of the objections, however, concern the published "Digest."[23] Anyone who has worked with the original documents cannot but be impressed by the quality, detail, and consistency of the information reported.

The samples for 1850, 1860, and 1870 were drawn by Fred Bateman and Thomas Weiss. They cover every state in the Union for which the data were available.[24] These censuses, particularly those for 1850 and 1860, are generally thought to be reliable, with consistent data from state to state in a given census year.[25] Among the data collected were information on capital, employment, wages, inputs, and outputs.

A Deplorable Scarcity: The Failure of Industrialization in the Slave Economy (Chapel Hill: University of North Carolina Press, 1981), pp. 169–171.

[18] See Kenneth L. Sokoloff, "Industrialization and the Growth of the Manufacturing Sector in the Northeast, 1820–1850," (Ph.D. diss., Harvard University, 1982).

[19] For a general discussion of the nineteenth century manufacturing censuses, including that for 1820, see Meyer H. Fishbein, "The Censuses of Manufactures, 1810–1890," *Reference Information Paper No. 50*, (National Archives and Record Service, 1973.)

[20] *Niles Weekly Register*, 18 (26 Aug. 1820), p. 450.

[21] Congressional complaints over inaccuracies and omissions appear in *Annals of Congress*, 17th Cong., 2d sess., 1823, pp. 887–901. For a summary, see Fishbein, "Censuses of Manufactures."

[22] *American State Papers*, Finance, vol. 4, "Digest of Manufactures (Supplementary Returns)," Doc. 675, 17th Cong., 2d sess., 1823, pp. 291–299. See especially pp. 298–299 for list of counties from which no returns were received.

[23] *American State Papers*, Finance, vol. 4, "Digest of Manufactures," Doc. 662, 17th Cong., 2d sess., 1823, pp. 28–223.

[24] See Bateman and Weiss, *Deplorable Scarcity*, especially pp. 165–184. Since their study was concluded, new data have been added, including Illinois (1850), Kansas (1870), and Michigan (1850, 1860, and 1870). Data for the Pacific states (California and Oregon) has been excluded for this study.

[25] Beginning with the 1840 census, for example, Robert Gallman in his Presidential

There are three issues that might be raised regarding these data for the purposes of this study. First, the census officials repeatedly expressed doubts about the comparability of data on capital from census year to census year.[26] Second, the census did not report until 1880 how many months per year each firm worked. These data show considerable variations from firm to firm, industry to industry, and region to region, even at the end of the century, and it is likely that the length of the work-year was even more variable at midcentury.[27] We have not tried to adjust for this. Moreover, the greater the ratio of fixed to variable cost, the greater the incentive for year-round production, so that we would expect factory production to be more stable and regular than artisan production, and this should be reflected in higher productivity.[28]

---

address to the Southern Economic Association declared, "I want to notify both [economists and historians], in the strongest possible terms, that the 1840 income estimates [based upon the 1840 census] are composed of data worthy of respect assembled within a relevant theoretical structure." See Robert E. Gallman, "Slavery and Southern Economic Growth," *Southern Economic Journal* 45 (April 1979), pp. 1007–1022. The quote is from n. 8, p. 1009.

[26] See, for example, remarks by Francis A. Walker, a propos the inquiry for 1870, in *Statistics of Wealth and Industry of the United States at the Ninth Census* (Washington, D.C.: U.S. Government Printing Office, 1872); p. 381. See also similar remarks by successive Superintendents of the Census at the Twelfth Census (*U.S. Twelfth Census, 1900,* "Manufactures"), and at the Fourteenth Census ("Manufactures, 1919.") See also S. N. D. North, "Manufactures in the Federal Census," in *The Federal Census: Critical Essays* (New York: Macmillan for the American Economic Association, 1899), pp. 257–302, especially pp. 283–298.

Most researchers, following Creamer, have assumed that the census capital figures represent gross book value at original cost. See David Creamer, *Capital in Manufacturing and Mining* (Princeton: Princeton University Press for the NBER, 1960). However, in his essay in ch. 7 of the present volume, "Investment Flows and Capital Stocks: U.S. Experience in the Nineteenth Century," Robert Gallman argues that the capital stocks were valued at reproduction cost; that is, these data are current net replacement costs.

[27] There were questions at both the Tenth (1880) and Eleventh (1890) Censuses regarding the number of months of operation at different capacity levels. See Carroll D. Wright, *History and Growth of the United States Census* (Washington, D.C.: U.S. Government Printing Office, 1900), especially p. 315 and p. 363.

[28] Data collected by the 1820 census on the related issue of recent and current business conditions are ambiguous with respect to the hypothesis that factory employment was more stable than that in artisan shops.

A little over half of the largest establishments in 1820 described business conditions variously as "in decay," "limited," or "not worth pursuing," while a marginally smaller faction (44 %) of the smallest establishments made similar comments. Larger establishments would thus seem to have been marginally more pessimistic than smaller businesses. On the other hand, almost 30% of the factories described business as being good or better, whereas only a little more than a fifth of the smallest plants thought so. Indeed, most of the factories that described conditions as "good" thought that they were "very good."

Employment statistics from Philadelphia between 1816 and 1819, however, present a very sanguine picture of employment stability in the factory. In the depression of 1818–1819, manufacturing employment fell almost 75%, from 9,672 employees to

Third, the samples were collected by state and are unweighted. The aggregation of these separate samples to regional and national samples, therefore, gives too much weight to plants in newer areas of settlement and in the less industrialized states. The possible effects of these problems will be noted where appropriate.

The number of separately identified industries expanded at each successive census. This study, therefore, focuses upon sixteen industries that were among the more important in the nineteenth century. Some of these industries, such as meat packing, tobacco, and clothing, do not appear in the sample from the 1820 census. For the most part, however, there were an adequate number of observations in each year on which to base empirical estimates.[29] All regressions had at least thirty degrees of freedom, and where other sample statistics are based on fewer observations, this is noted.

At the 1850, 1860, and 1870 censuses, manufacturers were asked about the firm's source of motive power. This information is crucial to our study for differentiating mills and factories from other establishment types, and it seems to have been reported quite reliably. Regrettably, however, data on motive power were rarely recorded in the 1820 census. It is doubtful that the lack of mention reflects no use of inanimate motive power, for while steam engines were almost certainly always recorded, waterwheels were probably often overlooked.

Only about 10 percent of all establishments in the 1820 sample mentioned the use of water or steam or gave information from which such use could be inferred. The sample shows wide variations from industry to industry. These may reflect differences in use and in the accuracy of reporting. For example, virtually all flour mills and cotton mills probably used inanimate power, but only about three-quarters of the flour mills reported a waterwheel, while less than 10 percent of the cotton mills reported a source of power. It is debatable whether or not the percentage of plants using power to grind grain is too low, but there can be no doubt that the estimate of the percentage of steam- or water-powered cotton mills is too low.

I

Data on power use by proportion of plants in each industry (not to be confused with the share of value-added associated with differing power

---

2,137. The most precipitous decline was in the cotton textiles industry, which was wholly mill organized and dominated by factories. There, employment fell from 2,325 to only 149. See *Niles Weekly Register* 17 (23 Oct. 1819), p. 117.

[29] Sample sizes, by industry and year, are given in Appendix 8-A at the end of the chapter.

TABLE 9.1
The Changing Use of Steam and Water Power by American Industry,
1850–1870
(Ranked in order of decreasing power use in 1850)

	1850		1860		1870	
Industry	% using inanimate power	% using water power	% using inanimate power	% using water power	% using inanimate power	% using water power
Cotton goods	95%	78%	79%	71%	93%	79%
Woolen goods	95	86	94	80	90	80
Flour milling	94	86	95	71	95	70
Lumber milling	93	74	96	47	99	47
Iron	90	59	96	50	95	24
Liquor	58	17	56	21	64	5
Leather	15	11	31	21	26	13
Furniture	13	8	34	20	25	7
Brewing	10	0	31	0	38	5
Meat packing	9	3	10	0	22	0
Wagon making	5	4	10	4	7	2
Sheet metals	4	4	5	2	5	1
Clothing	1	1	4	4	2	2
Boots & shoes	0.2	0.2	3	2	3	1
Saddlery	0	0	2	2	2	2
Tobacco	0	0	2	2	3	1

SOURCE: Calculated from the Bateman-Weiss samples.

sources) between 1850 and 1870 give some idea of the spread of mill and factory production (Table 9.1).[30] These represent lower-bound estimates of the extent of inanimate power's adoption, because the industrially less progressive parts of the country are overrepresented in these data and because where the source of power was not clearly specified, we assumed that the plants used an animate power source. In 1850, inanimate-power use by industry fell into three quite distinct divisions. There were those industries, such as textiles, iron, flour, and lumber, in which the use of steam or water power was almost universal.[31] In another group of industries, such as cigar making, saddlery, and boots

---

[30] The census collected the manufacturing data we use by establishment, not by firm. Where a sampled establishment was owned by a firm that owned more than one plant in the same location, data on each establishment were collected. All analytical results reported herein are for plant data.

[31] A small percentage of lumber and flour mills reported wind or animal power (2–4%), and one or two establishments in each of the industries still claimed to depend on hand power. The balance of power use was "mixed," which could mean steam and water, or steam and hand, hand and animal, etc. Short of going back to the original manuscripts, there is now no means of distinguishing these combinations from one another in the machine-readable sample. As a result, we treated establishments that reported "mixed" sources of power as using "animate" power. This biases our estimates of the use of inanimate power downward.

and shoes, power was hardly, if ever, used. Lastly, there were a number of industries, such as tanning, wagon making, furniture, brewing, and meat packing, where inanimate power had made limited inroads. The liquor industry is an outlier in this scheme. Its use of power was far from universal, but also hardly limited. A small majority of liquor firms were using power in 1850. The industry is unusual in another way: it alone was dominated by steam rather than water power.[32] Indeed, except for the liquor industry and, to a much lesser extent, iron, steam had made little progress; the rapid expansion of steam-power use was not to occur until later in the century.[33]

Inanimate power use increased quite slowly between 1850 and 1870, and the essentials of the tripartite division of industries by power use remained.[34] Except for the more extensive adoption of steam and water power by firms in tanning, furniture, brewing, and meat packing, the percentage of power-using firms shows no dramatic changes. Steam power, however, displaced water in most industries, although textiles and flour milling clung to their water rights.[35]

---

[32] The use of heat in the distillation process made the use of steam power by the liquor industry the natural choice whenever mechanical power was needed. Indeed, this is one industry singled out by Chandler as exemplifying the newer energy-intensive, larger-scale continuous production processes: ". . . the distilling and refining industries lent themselves more readily to mass production . . . enlarged stills, superheated steam, and cracking techniques all brought high volume, large-batch, or continuous-process production of products . . . in the distilling of alcohol and spirits." See Chandler, *Visible Hand*, p. 243.

[33] See Jeremy Atack, Fred Bateman, and Thomas Weiss, "The Regional Diffusion and Adoption of the Steam Engine in American Manufacturing," *Journal of Economic History* 40 (June 1980), pp. 281–308.

[34] Probably no importance should be attached to the small declines in power use, as in the textile industry, from 1850 to 1860 and 1860 to 1870. First, the percentage of "mixed"-power users tended to increase. Most of them were combined steam- and water-power users, but they are treated as non-power-using plants so as to maintain a consistent downward bias in the estimates. Second, the percentage of nonreporting plants tended to increase in 1870. Third, given the small sample sizes (under 100 observations) in some of these industries, minor sample variations and errors induce large swings in this statistic.

Aggregate statistics on power use were not reported until the Ninth Census (1870). See U.S. Census Office, *Statistics of Wealth*. The marked increase in nonreporting of motive power biases early estimates of the use of power downward. This has led some authors to drastically overstate the increase in power use between 1870 and 1880. See, for example, Allen H. Fenichel, "Growth and Diffusion of Power in Manufacturing, 1838–1919," in *Output, Employment, and Productivity in the United States After 1800*, Conference on Research in Income and Wealth, Studies in Income and Wealth, vol. 30, (New York: National Bureau of Economic Research, 1966), pp. 443–478.

[35] See Atack, Bateman, and Weiss, "Regional Diffusion." The persistence of water power in the textile industry may be attributed to the monopoly rents accruing to long-term water-right contract holders especially in the face of rising efficiency in water use with turbines. For this reason, many textile mills installed steam engines when they needed more power while maintaining their use of existing water power until dis-

TABLE 9.2
The Percentage of Industry Value-Added Originating from Different Methods of Production in Selected Industries, 1850 and 1870

Industry	Artisanal Shops[a] 1850	Artisanal Shops[a] 1870	Sweatshops[b] 1850	Sweatshops[b] 1870	Manufactories[c] 1850	Manufactories[c] 1870	Mills[d] 1850	Mills[d] 1870	Factories[e] 1850	Factories[e] 1870
Boots & shoes	39%	33%	23%	20%	38%	25%	0%	4%	0%	19%
Brewing	41	21	24	0	0	0	35	49	0	30
Clothing	13	16	39	24	48	42	0	0	0	18
Cotton goods	0	0	0	0	4	3	16	1	79	96
Flour milling	7	5	0	0	0	0	91	95	2	0
Furniture	50	18	13	6	7	8	10	26	19	41
Iron	0	0	1	1	32	0	22	10	44	89
Leather	54	20	16	7	0	0	26	29	4	43
Liquor	9	4	18	4	0	0	73	82	0	10
Lumber milling	3	1	1	2	0	0	88	63	8	34
Meat packing	34	31	10	0	19	8	11	69	25	0
Saddlery	62	71	19	20	19	27	0	1	0	0
Sheet metals	89	41	6	6	0	35	5	2	0	24
Tobacco	24	30	33	33	43	17	0	2	0	0
Wagon making	32	33	47	30	15	12	3	3	3	18
Woolen goods	0	4	1	0	0	0	39	7	60	77

SOURCE: Computed from the Bateman-Weiss samples.
[a] Artisan shops: no power, 1–6 employees.
[b] Sweatshops: no power, 7–25 employees.
[c] Manufactories: no power, more than 25 employees.
[d] Mills: water or steam power, 1–25 employees.
[e] Factories: water or steam power, more than 25 employees.

The Rise of the Factory

Despite the minor changes in the proportions of establishments using steam and water power between 1850 and 1870, the percentages of industry production accounted for by factories and mills expanded rapidly over the period.[36] In 1850, the factory system was dominant in only three industries, cotton, iron, and woolens, while mills dominated flour and lumber milling and liquor distilling. Nonmechanized plants dominated in the other ten industries. However, as a result of dramatic gains between 1850 and 1870, mills and factories had become the dominant production technique in most industries by 1870, displacing nonmechanized establishments (Table 9.2).[37]

In five industries (cotton, furniture, iron, leather, and woolens), factories were the leading source of output in 1870. The inclusion in this list of industries not usually associated with large-scale mechanized production in the nineteenth century, such as furniture and leather, is particularly noteworthy. In five other industries in 1870 (brewing, flour milling, liquor distilling, lumber milling and meat packing), mills were the most important source of industry supply.

Some industries experienced sharp increases in the fraction of factory value-added between 1850 and 1870. For example, no boots and shoes or men's clothing came from factories in the samples in 1850, but by 1870, 19 and 18 percent respectively of their industry value-added was accounted for by factory production[38] and a further 4 percent of boot and shoe value-added came from mills.

The structure of the boot and shoe industry in 1850 was much the same as it had been in 1820. There were many small shops, fewer mid-size firms, and a few large ones employing more than twenty-five.[39] Output was more or less evenly divided between the three groups, and

---

possessed of water by cities claiming eminent domain over this resource for domestic water supply and sanitation.

[36] I attach no great importance to the decline in mill and factory production of woolens in 1870 and the disappearance of meat-packing factories. The case of woolens simply reflects an increase in the fraction of plants for which no power source was given in 1870. For meat packers, it was the result of chance that none of the large meat packers were included in the random sample.

[37] Jeremy Atack, "Industrial Structure and the Emergence of the Modern Industrial Corporation," *Explorations in Economic History*, 22 (Jan. 1985), pp. 29–52, especially Tables 1 and 2 (pp. 34–35 and 40–41).

[38] These results are not directly comparable with those of Laurie and Schmitz, who describe the spread of the factory system by citing the increasing percentage of the industry's labor force employed in factories. The fraction of total industry value-added originating in factories has been used here. Furthermore, while they give results for 1850 and 1880, the data here are for 1850, 1860, and 1870.

[39] See Blanche E. Hazard, *Organization of the Boot and Shoe Industry in Massachusetts Before 1875* (Cambridge, Mass.: Harvard University Press, 1921), for a discussion of the early industry. Data on the size distribution of plants are from Sokoloff's sample.

none of the boot and shoe shops in the samples reported a source of power before 1860.[40] The industry began to mechanize in the 1850s, with the adaptation of the sewing machine to leather work. This is reflected in the statistics on motive power. Whereas in 1850, 39 percent of industry output had come from large nonmechanized shops, in 1860 their share of industry output fell to 24 percent and the output of factories (power-using establishments with over twenty-five employees) increased from zero to 17 percent, rising to 19 percent in 1870. The share of sweatshops changed relatively little, but manufactories declined even more than artisan shops. The boot and shoe factory thus replaced some larger nonmechanized shops and some very small artisan shops, but it did not drive all of them out of business. Indeed, as we will show below, very small plants continued to increase in number, if not in importance as a source of output, throughout the period.

In a few industries, there was little change in the proportion of mill- or factory-produced goods. These fell into two groups: those that were already extensively mechanized and those untouched by mechanization. In 1850, for example, flour and lumber were almost exclusively produced in steam- or water-powered mills and cotton and woolen textiles were factory products. These industries were the pioneers of mechanized production. No tobacco or saddlery and harness plants, however, were using inanimate power in 1850, but even in these industries things began to change.

Tobacco, the handicraft industry par excellence before the introduction of Bonsack's cigarette machine in the 1880s, had begun to mechanize by 1870. Of industry value-added in the sample that year, 2 percent came from mills, and these made up only 2.5 percent of the tobacco plants in the sample.[41] Moreover, in the Bateman-Weiss special samples of large establishments, four large tobacco plants employing 1,410 and producing over $4.5 million of output,[42] were using steam power in 1870 in New York.[43] Furthermore, the U.S. Census for 1870

[40] Bateman-Weiss sample data. A few of the larger boot and shoe shops reported the use of "mixed" power sources.

[41] This percentage certainly would have been higher if some of the larger plants in the industry had appeared in the 1870 sample.

[42] In 1870 dollars. Adjusting to 1860 prices, using the Warren-Pearson price index for farm products, to eliminate the residual effects of Civil War inflation would reduce output value by about 25%, to $3.375 million. See U.S. Department of Commerce, Census Bureau, *Historical Statistics of the United States from Colonial Times to 1970* (Washington D.C.: U.S. Government Printing Office, 1975), Series E53.

[43] The special "large firm" samples which Bateman and Weiss collected consist of at least the twenty largest establishments in each state, and sample sizes were often larger than this.

reported 178 steam engines and 26 waterwheels in use by the 5,204 establishments in the industry.[44] However, mill and factory production of tobacco products and saddlery and harness remained the exception in 1870. These industries were the only ones in which mill and factory organization had failed to make much headway.

The rise of the factory usually paralleled a decline in artisan shops. The percentage of industry value-added produced in artisan shops fell in nine of the sixteen industries in Table 9.2. It rose only in tobacco manufacture, woolen goods, clothing, saddlery, and wagons. Each of these latter industries also experienced an increase in mill and factory production, so that sweatshops and manufactories made little or no gains in their share of total industry value-added. Given sample sizes, the changes in their shares appear to be well within the margin of error.[45]

The mechanized mill and factory were thus growing in importance while handicraft industry stagnated or declined. This is hardly unexpected, but the question is, why? One important factor favoring the adoption of mechanization was the rise in labor productivity that it was expected to generate. Comparisons of labor productivity between hand and machine production by Carroll Wright at the end of the nineteenth century show tremendous gains in labor productivity from mechanization.[46] Thus, for example, in 1813 three workers took 236 man-hours with hand tools to produce 29,000 4^d cut nails. Labor costs were $20.24. In 1897, machines produced the same quantity of 4^d nails in less than 2 hours. Eighty-three workers were then involved in the production process; labor costs were only 29 cents.[47] For the particular industries in this study, Wright's estimates of the labor-productivity gains for specific tasks ranged from about a twofold increase in leather tanning to an almost sixtyfold increase in the case of lumber (Table 9.3).

---

[44] *U.S. Ninth Census*, 1870, "The Statistics of Wealth and Industry of the United States," p. 479. The percentage of power-using plants in the population was therefore greater than that in the sample, provided each plant had only one engine or wheel.

[45] The apparent gains by manufactories in the production of woolen goods and sheet metals are almost certainly a reflection of reporting errors in the power use. A few other industries, not included here (agricultural implements, iron foundries, and steam engine manufacture), also recorded increases in manufactory value-added. It is likely that these were assembly rather than manufacturing operations. See Atack, "Industrial Structure."

[46] Wright's estimates, unfortunately, are usually not for different techniques at the same point in time, although Wright tried to collect as much contemporary data as possible. Rather, they are for hand production of about the time of the Civil War versus machine production toward the end of the century. Nevertheless, Wright felt that they were indicative of the potential scale of labor productivity gains. See U.S. Department of Labor, *Thirteenth Annual Report*.

[47] Ibid., vol. 1, pp. 24–79.

300   Jeremy Atack

TABLE 9.3
Labor Productivity and the Switch from Hand to Machine Production, 1835–1897

Product	Technique and date	Number of operations	Number of workers	Man-hours worked	Labor Productivity[a]
Chewing tobacco	Hand, 1855	7	32	304.5	
(per 1,000 lbs. to 16-oz. plugs)	Machine, 1895	55	668	141.0	216
Cotton gingham	Hand, 1835	16	3	5,130.2	
(500 yds. by 27 in.)	Machine, 1895	40	166	119.2	4,304
Pine lumber boards	Hand, 1854	9	2	16,000.0	
(per 100,000 ft.)	Machine, 1896	38	78	272.7	5,867
Chairs	Hand, 1845	15	6	51.0	
(12 maple chairs)	Machine, 1897	33	26	7.9	646
Kid leather	Hand, 1860	20	226	1,506.9	
(1,200 skins)	Machine, 1896	30	116	773.3	195
Boots (cheap)	Hand, 1859	83	2	1,436.7	
(100 pairs)	Machine, 1895	122	113	154.1	932
Forged anvils	Hand, 1855	9	7	62.0	
(per unit)	Machine, 1896	6	10	15.0	413
Men's shirts	Hand, 1853	25	1	1,439.0	
(per gross)	Machine, 1895	39	230	188.2	765
Wash basins	Hand, 1840	31	1	96.0	
(per gross)	Machine, 1895	22	23	7.2	1,333
Farm wagon	Hand, 1848	37	5	242.0	
(per unit)	Machine, 1895	63	75	48.3	501

SOURCE: U.S. Department of Labor, *Thirteenth Annual Report*, vol. 1, pp. 24–79.
[a] (Hand man-hours ÷ machine man-hours) × 100.

Labor-productivity gains, however, are not automatically and immediately converted into total factor productivity gains. Mechanization required a new plant and more equipment. The capital component therefore tended to rise.[48] Moreover, mechanization tended to lower the skill level of the work force by substituting machine precision for human skills and experience.[49] As a result, the capital-labor ratio tended to

[48] Although crude capital-labor ratios for mechanized plants were generally higher than those for the nonmechanized, there were occasional exceptions. For example, in 1900 the capital-labor ratio in boot and shoe factories was only $875, compared with almost $1,000 in the handicraft sector of the industry. See *U.S. Twelfth Census, 1900*, "Manufactures," vol. 7, pt. 1, p. 4.

[49] Goldin and Sokoloff, for example, show that the larger, mechanized plants consistently employed a higher percentage of female and child labor than smaller artisan shops. See C. Goldin and K. L. Sokoloff, "Women, Children and Industrialization in the Early Republic: Evidence from the Manufacturing Censuses," *Journal of Economic History* 42 (Dec. 1982), pp. 741–774. See also A. Field, "Skill Intensity in

The Rise of the Factory

rise, and it is possible for virtually the entire labor-productivity gain to be absorbed by higher capital-labor ratios.

Suppose that the production process can be modeled by a Cobb-Douglas production function of the form

$$Q = AL^a K^b$$

where Q, L, and K are output (measured as value-added in constant dollars), labor, and capital respectively, and a and b are the output elasticities with respect to labor and capital. If $(a + b) = 1$, then total factor productivity, A, is output per unit of input, where in this multi-input case, a unit of input is measured by the geometric weighted mean of the inputs

$$A = Q/(L^a K^b)$$

Total factor productivity may also be expressed as the weighted average of the partial productivity indexes for labor and capital:

$$A = \{Q/L\}^a \cdot \{Q/K\}^b$$

This may be written as[50]

$$\{Q/L\} = A[\{K/L\}^b]$$

Thus, a rise in labor productivity may be offset by a rise in the capital-labor ratio.

Nevertheless, one would expect that dramatic labor-productivity gains would increase total factor productivity quite sharply. This would be consistent with the rapid adoption of mechanized production after 1850 in almost every branch of industry. However, rather than focusing upon the narrower issue of mechanization's effects upon labor productivity, this study examines its effects upon total-factor productivity because of the ambiguous role played by changes in the capital-labor ratio. Indeed, we will show below that it is impossible to reconcile our estimates of the total-factor productivity gains with the labor-productivity gains quoted by Wright, even after allowing for large changes in the capital-labor ratio.

Early Industrialization: The Case of Massachusetts," *Journal of Human Resources* 15 (1980), pp. 149–175.

This may, however, have been partially offset by an increase in labor-intensity as a result of machines setting the pace for human operatives, but there is no means of measuring this with the available data.

[50] By dividing through by $\{Q/L\}^b$ and using the result that since $a + b = 1$, $L^a \cdot L^b = L$ and $Q^a \cdot Q^b = Q$.

II

The relative total-factor productivity gains to be realized from changing the scale of operations and the organization of the plant are estimated by using the mean values of Q, L, and K for medium and large establishments and factories in the relationship

$$A = Q/(L^a K^b)$$

with productivity in artisan shops serving as numeraire.

For 1820, we estimate that larger plants were generally about 40 percent more efficient than the small artisan shops employing fewer than seven workers (Figure 9.1). This is similar to the result reported by Sokoloff.[51] No calculations of relative efficiency were made for tobacco in any year because of the small sample sizes.

Boot and shoe manufacturers, however, seemed to do much better than average. The largest boot and shoe shops were twice as efficient as small artisan shops in 1820 and about 12 percent more efficient than the midsize plants. One possible explanation for this would be that labor and capital might be understated in the larger boot and shoe manufactories if these relied heavily upon a "putting-out" system. Unfortunately, there is no means of ascertaining this from the census data.

Larger iron producers and sheet metal manufacturers, on the other hand, were less efficient than smaller ones. Their inferiority increased with size. Indeed, the very largest iron plants produced only about one-third as much value-added per unit of input (defined as the geometrically weighted average of labor and capital) as small plants with fewer than seven employees. Given the rural location of most iron works, the high transport costs for both raw materials and finished products, and the absence of technological economies of scale in furnaces, which might have offset these disadvantages at the time, rising unit costs for larger plants seem quite plausible. However, such plants could only continue to survive in the face of more efficient, smaller establishments if, for example, they produced a good that smaller

[51] Our assumptions differ slightly, notably in our treatment of child and female labor and the degree of aggregation in our results. The choices made here parallel those made in Atack, *Estimation,* especially Ch. 3. Whereas Sokoloff weighted women and children at 0.4 of an adult male equivalent, a weight of 0.5 for women and 0.33 for children was used here (see Sokoloff, "Transition from the Artisanal Shop," footnotes to Table 5, pp. 364–365). Second, whereas Sokoloff estimated a single production function across all industries except textiles and iron, separate production functions for each industry are made here (see Sokoloff, Table 6, p. 368). No effort has been made to replicate Sokoloff's results, but both our aggregate results and our estimates of labor's share in 1820 are similar. The differences, therefore, seem to be of little importance.

The Rise of the Factory

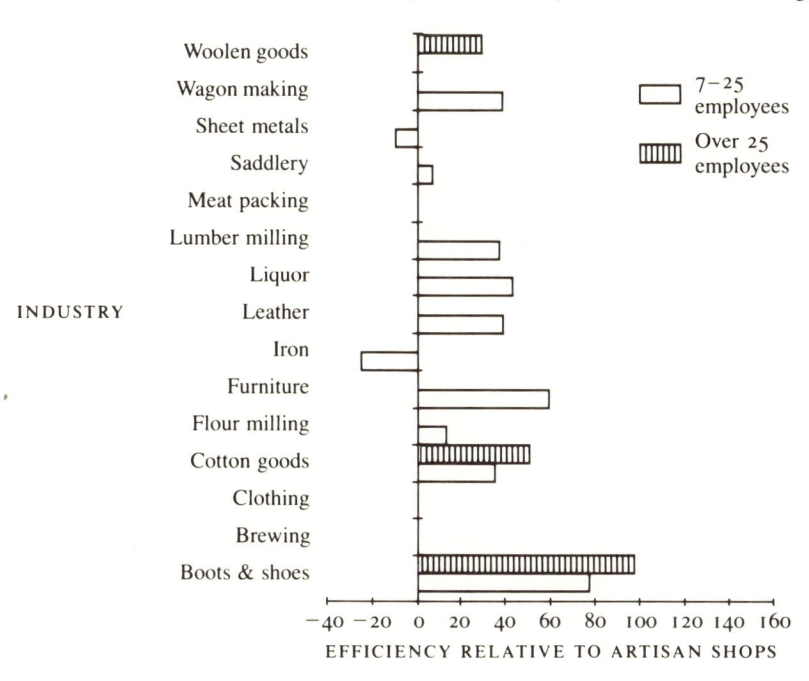

FIGURE 9.1 Relative Efficiency by Industry, 1820

SOURCE: Calculated from Sokoloff's sample from the 1820 census.

plants would have been incapable of supplying. Unfortunately, the product descriptions are usually quite generic, and we have no information about cross-elasticities of substitution.

After 1820, no relative efficiency figures are calculated for cotton goods, since no establishments met the criterion of six or fewer employees and no power. All plants in this industry were either mills or factories. No calculations for woolen goods were made after 1850 for the same reason. Furthermore, although relative efficiency estimates for flour and lumber milling are reported, the representativeness of the artisan-shop data in these industries is suspect. Both industries made early and extensive use of power, and it is probable that those plants identified as artisan shops simply neglected to report their power use to the census enumerators.

The efficiency gains realizable by larger plants and through mechanization seem to have been smaller in 1850 than in 1820 (Figure 9.2). Where mechanized (i.e., powered) plants were more efficient than artisan shops, the efficiency gains seem to have been less than 20 percent. Notice, however, that factories (i.e., those mechanized plants employing more than twenty-five) were usually less efficient than mills, and only woolen factories managed to be more efficient than the artisan

304   Jeremy Atack

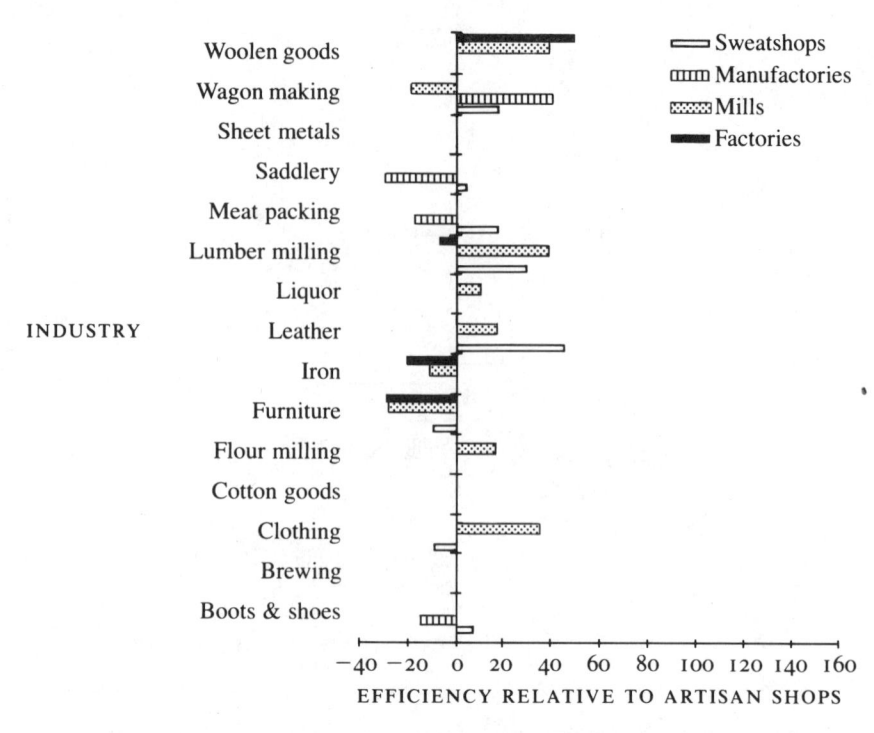

[Figure: horizontal bar chart showing Relative Efficiency by Industry, 1850, with categories Sweatshops, Manufactories, Mills, Factories. Industries listed: Woolen goods, Wagon making, Sheet metals, Saddlery, Meat packing, Lumber milling, Liquor, Leather, Iron, Furniture, Flour milling, Cotton goods, Clothing, Brewing, Boots & shoes. X-axis: −40 to 160, EFFICIENCY RELATIVE TO ARTISAN SHOPS]

SOURCE: Calculated from the Bateman-Weiss samples from the 1850 census.

FIGURE 9.2 Relative Efficiency by Industry, 1850

shops in the industry. This finding seems to contradict Sokoloff's conclusion, but the difference is almost certainly attributable to the breakdown of establishments between powered and nonpowered here. It may also reflect differences in our size categories.

Except for clothing and wagon and carriage manufacture, medium-sized nonmechanized plants seem to have been more efficient than large ones, and they were also generally more efficient than artisan shops. By 1870, however, sweatshops, even in these two industries, were more efficient than the manufactories in those same industries, and consequently the importance of manufactories should have been declining in each industry. With a few notable exceptions, this was the case (see Table 9.2 above).[52]

[52] There are two glaring exceptions—sheet metals and woolens. The former largely did remain a handicraft trade until the introduction of stamping machinery for sheet steel. See David A. Hounshell, *From the American System to Mass Production 1800–1932* (Baltimore: Johns Hopkins University Press, 1984), especially pp. 208–215 and n. 71, pp. 369–370. The results for woolens may reflect a coding or key-punch error whose seriousness is reinforced by the small sample size for this industry.

The Rise of the Factory

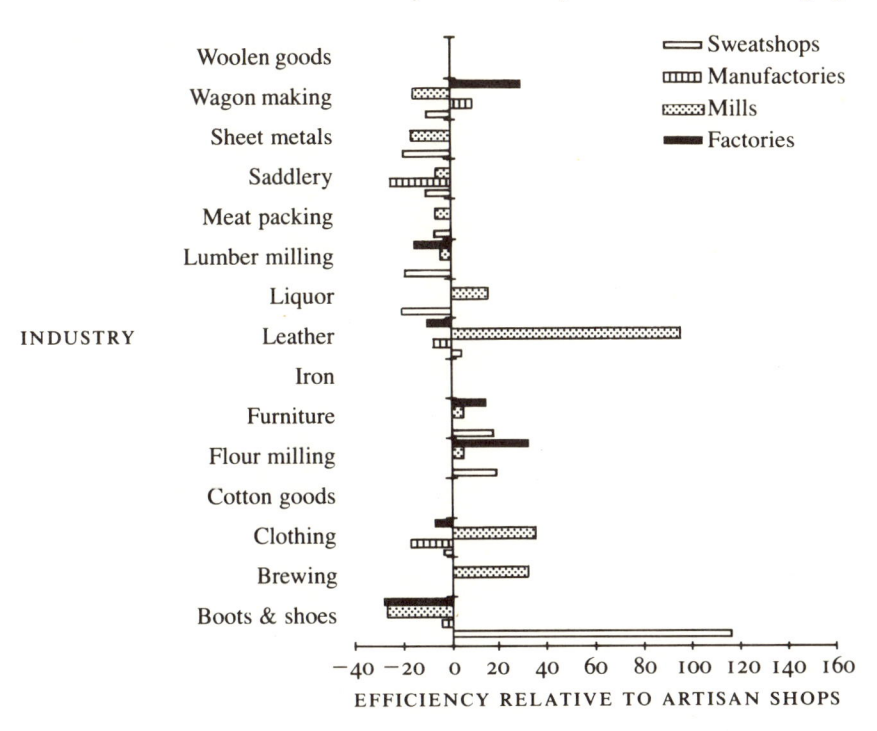

FIGURE 9.3 Relative Efficiency by Industry, 1860

SOURCE: Calculated from the Bateman-Weiss samples from the 1860 census.

In 1860, the potential efficiency gains, relative to artisan shops, to be realized from larger size and mechanization remained small, averaging less than 20 percent (Figure 9.3). Efficiency losses were also quite small, typically on the order of 5–10 percent. Plants in three industries, however, could have realized very substantial efficiency gains by changing organization and adopting mechanized production processes. Small mechanized tanneries, for example, were about 90 percent more efficient than very small nonmechanized leather tanneries; and very large, power-using meat-packing plants were some 220 percent more efficient than very small, nonpowered establishments. Whereas most sweatshops compared unfavorably with artisan shops in industry, in boot and shoe making they were the most efficient plants in the industry, producing about twice as much value-added per unit of input as artisan boot and shoe makers.

These estimates for 1860 present a much different picture than those for other years. About a quarter of the relative efficiency estimates in 1860 are in the opposite direction from results for 1850, and fully half of the results are counter to the 1870 results. On the other hand, there

FIGURE 9.4 Relative Efficiency by Industry, 1870

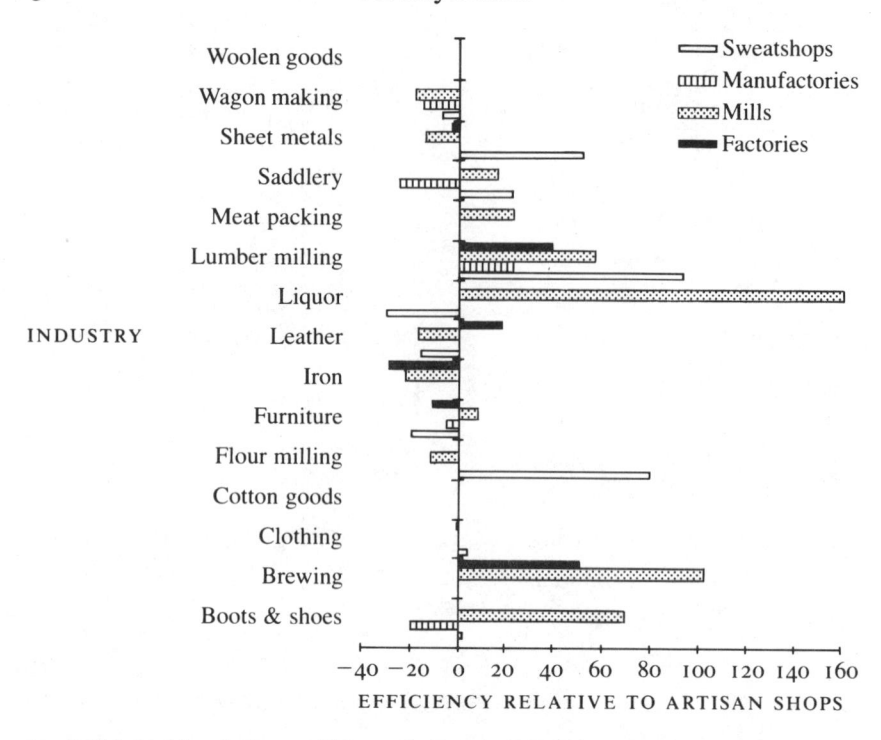

SOURCE: Calculated from the Bateman-Weiss samples from the 1870 census.

is a much higher degree of concordance between the 1850 and 1870 results, but the limited time-series observations with these data make such generalizations difficult.

By 1870, the efficiency gains to be made from a change in organization and structure were much more pronounced than in either 1850 or 1860 and were on a par with those for 1820 (Figure 9.4). Whereas for the 1850 and 1860 estimates it was hard to generalize about efficiency relative to that of artisan shops across industries, in 1870, as in 1820, the artisan shops were typically the least efficient producers. Furthermore, mechanized plants in 1860 and 1870 typically produced more output per unit of inputs of labor and capital than the nonmechanized establishment in the same industry.

If these data are reorganized within each industry by establishment type (between mechanized and nonmechanized and by the size of labor force) and by year, a different set of inferences can be drawn from them. First, the "manufactory" (i.e., a non-power-using establishment employing more than twenty-five) was clearly an inferior organization in terms of efficiency and should have been in decline. In most industries, the manufactory was less efficient than the artisan shop and this relative inefficiency grew over time. Labor supervision and control over management and entrepreneurial functions were perennial problems. In the artisan shop, the owner often labored alongside his employees and the intimacy of their physical location provided all the necessary control and supervision. Managerial and supervisory functions in larger businesses were usually delegated, with a resultant loss of some degree of control, but whereas in the mill and factory mechanization instilled discipline and interdependence and controlled the flow and pace of work, this element was lacking in the larger handicraft industries.

Second, relative efficiency declined in most sweatshops (nonmechanized establishments employing seven to twenty-five), although in many industries sweatshops were to remain as efficient, or somewhat more so, than artisan shops throughout the period. Sweatshop efficiency did, however, rise steadily in the one industry that was to become synonymous with this mode of production, namely, the needle trades, and by 1870 sweatshops in this industry were marginally more efficient (4 percent) than artisan shops.

Third, the relative efficiency of mechanized plants in virtually every industry grew over time, so that in many cases such plants had become more efficient than artisan shops by 1870. Where mills were more efficient than artisan shops, their advantage averaged about 40–50 percent. Where factories were more efficient, their advantage was considerably smaller, averaging perhaps 20 percent. These gains are similar to those reported by Sokoloff for more aggregated industry groups in 1820 and 1850.[53] There were thus some advantages to mechanization, although they seem to have been at least partially offset by other factors, perhaps managerial shortcomings, in larger plants. This pattern is most marked for furniture and lumber.

One particularly interesting pattern is apparent in these data: mechanized iron furnaces were uniformly less efficient than the very small nonmechanized producers, and their relative inefficiency grew over time. This result may well help explain the persistence in the United States of the small charcoal iron producer in competition with larger coke iron smelters. It is also consistent with Temin's argument that the cost differential between the two production techniques was not great

[53] Sokoloff, "Transition from the Artisanal Shop," pp. 368 and 377, shows productivity gains from larger plants of 30–35% in 1820 and 27–29% in 1850. The overall figures here are on the same order of magnitude, although, of course, there are some efficiency losses too, whereas Sokoloff found none with his more aggregated data.

enough to compensate for the lower market price commanded by coke-smelted pig iron, since the larger producers generated less value-added per unit of input.[54]

These estimates of total factor productivity change cannot be reconciled with Wright's estimates of the labor-productivity gains, even allowing for changes in the capital-labor ratio.[55] Consider, for example, the case of the boot and shoe industry. Mechanized producers in this industry had 25 percent higher capital-labor ratios in 1870 than artisan shops ($750/male equivalent versus $600/male equivalent).[56] We estimate that the output elasticity with respect to capital in the boot and shoe industry was 0.29 and that the switch from shop to factory led to a 5-percent gain in efficiency.[57] These data, then, imply a 40-percent gain in labor productivity from mechanization.[58] This is very small when compared with the ninefold increase shown in Table 9.3 above. Results for the other industries were similar. A partial explanation may be that capital quality, measured by its productivity per unit cost, improved with mechanization and/or that its quantity increased in real terms due to a decline in capital goods prices, and these changes were not captured in the capital estimates from the census.[59]

Small as the gains seem to have been, they still potentially overstate

[54] See Peter Temin, *Iron and Steel in Nineteenth Century America: An Economic Inquiry* (Cambridge, Mass.: MIT Press, 1964), Ch. 3.

[55] Inability to reconcile total and partial factor productivity estimates seems to be fairly common. See, for example, the Parker-Klein estimates of labor-productivity growth in American agriculture and compare with Robert Gallman's estimates of agricultural total factor productivity. William N. Parker and Judith L. V. Klein, "Productivity Growth in Grain Production in the United States, 1840–1860 and 1900–1910," in *Output, Employment, and Productivity in the United States After 1800*, Studies in Income and Wealth, vol. 30, pp. 523–582; Robert E. Gallman, "Changes in Total U.S. Agricultural Factor Productivity in the Nineteenth Century," *Agricultural History* 46 (Jan. 1972), pp. 191–210.

[56] Bateman-Weiss sample data.

[57] The output elasticity is estimated from a restricted Cobb-Douglas production function where the output elasticities were constrained to sum to unity: $\ln(Q/L) = \ln A + b\ln(K/L)$.

[58] Using the relationship $\{Q-L\}^* = A^* + b\{K-L\}^*$, where the variables are logged and * indicates rates of change. On the other hand, if one were to accept Wright's estimate of the labor-productivity gain and my estimate of the overall productivity gain and output elasticity with respect to capital, then the capital-labor ratio would have had to increase by a factor of 27, or my estimate of overall productivity gains is grossly in error, being less than one-eighth its true value.

[59] Some evidence consistent with this hypothesis is to be found in the movement of price indexes which generally fell during the period up to 1900, except for the Civil War interlude, but if Gallman's interpretation of the census capital estimates is right, then the hypothesis can be rejected. See *Historical Statistics*, Series E52–E134 for price data. See also Robert E. Gallman, "Investment Flows and Capital Stocks: U.S. Experience in the Nineteenth Century," Ch. 7 in this volume.

the benefits because of our treatment of entrepreneurial labor, which is biased against smaller plants.[60] One adult male equivalent has been added to each firm's labor force to represent the input of managerial and entrepreneurial labor. The figure is arbitrary but, as entrepreneurial and managerial labor contributes to output, it should be included, but the censuses omitted it.[61] The same adjustment was made in each year. This addition represents a proportionately greater increase in the labor input for very small establishments, many of which also probably only worked part-time or were only part-time interests of the owner, than for larger establishments. Consequently, small-firm value-added per unit of "inputs" is reduced relative to that of the larger producer, thereby lowering the small firm's relative efficiency.

The efficiency estimates, particularly those for factories, may also be biased downward by regional price variations. Value-added has not been deflated by local or regional product prices, and there may have been significant differences in these prices between regions. Certainly, the available price indexes show marked variations.[62] Average firm size was also correlated with regions, with smaller plants in the South and Midwest. As a result, western value-added may well be overstated, raising the relative efficiency of all small artisan shops, since these shops predominated in that region. Similarly, regional variations in competition would also bias estimates of value-added in favor of establishments in the less competitive markets. This would again favor plants in the West. The use of state dummy variables to account for such factors and any other unique differences was inconclusive.

The failure to realize the huge gains in labor productivity that contemporaries expected may have discouraged universal adoption of mechanized production. Certainly, in the boot and shoe industry, the changeover to factory production was leisurely, so that it is unlikely that the gains, whatever their absolute magnitude, were large. Nevertheless, although the gains may have been less than entrepreneurs anticipated, they should have been sufficiently attractive to ensure that

---

[60] See Atack, *Estimation*, especially pp. 77–78 and 133–135. Also Sokoloff, "Transition from the Artisanal Shop," pp. 367–370.

[61] Sokoloff's 1820 sample offers a superior method of dealing with the omission because he coded whether or not the firm was a partnership, sole proprietorship, or corporation and, if a partnership, how many partners, but comparable data were not coded for the other censuses.

[62] See, for example, Anne Bezanson, *Wholesale Prices in Philadelphia, 1852–1896* (Philadelphia: University of Pennsylvania Press, 1954), and compare with Thomas Berry, *Western Prices before 1861: A Study of the Cincinnati Market* (Cambridge, Mass., Harvard University Press, 1943). See also Philip R. P. Coelho and James F. Shepherd, "Differences in Regional Prices: The United States, 1851–1880," *Journal of Economic History* 34 (Sept. 1974), pp. 551–591.

mills and factories would have become relatively more dominant in industry structure through time.

III

The measurement of total factor productivity involved a strong assumption that the sum of output elasticities with respect to factor inputs sum to 1. This assumption is necessary so that factor payments exhaust the product. (Sometimes, authors have disguised it by asserting that market conditions are assumed to be perfectly competitive or that factors are paid the value of their marginal products.[63]) Making this assumption, we then compare the value-added per geometrically weighted average unit of factor inputs across different methods of organizing production. Although it is in the background and implicit in our separate identification of artisan shops, sweatshops, and manufactories and of mills and factories, scale as such plays no role in total factor productivity differences. Indeed, we go even further and assert, if only by default, that there are no scale economies, since they would be measured by the sum of the output elasticities with respect to inputs.

Estimates of the unconstrained production function indicate, however, that there were returns to scale to be realized in a growing number of industries.[64] The arbitrary restriction to constant returns to scale may therefore bias the output elasticities of the inputs, unless the monopoly elements were approximately equal in each market.

Total factor productivity and scale-economy arguments are thus theoretically incompatible with one another. On the one hand, constant returns to scale are assumed in order to exhaust the product; on the other, there is strong evidence that there were returns to scale for larger plants at least up to some output level. Indeed, the argument will be made that in a number of industries, nothing but increasing returns to scale are observed, and that in the absence of external constraints or

---

[63] The Cobb-Douglas production function may thus be rewritten as

$$Q = AL^a K^{(1-a)}$$

This is linear in logarithmic transformation, and the restriction upon the output elasticity with respect to capital was forced by regressing $\ln(L/K)$ on $\ln(Q/K)$ to obtain an estimate of a. The output elasticity with respect to capital was then estimated as $(1 - a)$.

[64] In 1820, cotton goods, flour, leather, and lumber exhibited increasing returns to scale. As the century wore on, increasing returns came to characterize more and more industries and, by 1870, the following industries had significant increasing returns to scale: lumber, beer, cotton, flour, liquor, meat, and woolens.

internal nontechnological constraints, such as managerial control, the optimum firm in these industries would have been a natural monopoly.[65]

Economists typically assume that well-behaved average costs are U-shaped, with returns to scale for small plants, then a range of constant returns, which eventually gives way to decreasing returns in very large plants. However, empirical studies of cost functions have not been particularly successful in validating this assumption. Most of the studies in a survey of cost functions by A. A. Walters suggest that the cost curve is perhaps better described as L-shaped than U-shaped; that is, average costs decline at first and then become constant over a wide range of alternative plants and ultimately begin to rise slowly.[66] Walters therefore concluded that "for 'competitive' industries, the U-shaped hypothesis does not inspire great confidence . . . [but] at least there is no large body of data which convincingly contradicts the hypothesis . . . and the fruitful results which depend upon it."[67]

The standard critique of the hypothesis that average costs are L-shaped is that a firm's output is a random variable. Consequently, the firm has no means of controlling its variation about the average value and the firm responds by searching for the best means to produce this distribution of outputs.[68] This has the effect of blurring the distinction between fixed and variable factors and, if there are no variable costs, then "a cross section study would show sharply declining average costs. When establishments are classified by actual output, essentially this kind of bias arises. The plants with the largest output are unlikely to be producing at an unusually low level; on average, they are clearly likely to be producing at an unusually high level, and conversely for those which have the lowest output."[69]

This argument is potentially devastating to the results presented here; for given the uncertainties of the time and the enumeration process, reported output probably did contain large random elements, and my data are cross-sectional. Further, we found precisely what Friedman

[65] The results reported by James from aggregate translog production-function estimates corroborate this conclusion. See James, "Structural Change," especially pp. 437–449.

[66] This argument is forcefully made by J. Johnston, *Statistical Cost Functions* (New York: Wiley, 1960).

[67] A. A. Walters, "Production and Cost Functions: An Econometric Survey," *Econometrica* 31 (Jan.–Apr. 1963), pp. 1–66, especially p. 52.

[68] Ibid., p. 48.

[69] Milton Friedman, "Comment," in *Conference on Business Concentration and Price Policy,* quoted by Walters. Note that the thrust of Friedman's argument is that unit costs for the "large" establishments are biased downward because of their unusually large output, while those of "small" establishments are, conversely, biased upward.

predicted would be found as a result of the regression fallacy, namely, average costs that were sharply declining, more closely approximating an L-shape than a U-shape. However, our interpretation of this result differs.

IV

The census questionnaires do not provide sufficient information from which to estimate a cost function directly. Data on capital costs, rents, incidental expenses, and so forth are missing. Instead, we estimate two nonhomogeneous production-function forms, as discussed in Appendix B at the end of this chapter. These have variable returns to scale and are consistent with unit costs dependent upon the size of the firm.

The full set of economically meaningful and statistically significant production-function estimates appears in Appendix C. Attention here is focused, for convencience, upon a subset. This subset was chosen so as to illustrate the salient points in all of the estimates.

Consider first the results from estimating translog production functions. The econometric results were mixed. They were not always statistically significant or economically meaningful. About half of the translog production-function estimates were rejected because none of the variables were significantly different from zero at the 10-percent level in a two-tailed test.[70] A number of other translog production-function estimates exhibited perverse behavior in the returns-to-scale parameter, which increased with firm size. Scale elasticity behaved similarly in most of the variable scale-elasticity estimators (VSE) estimates (see Appendix C at the end of the chapter), but the ability to interpret those results in terms of the "normal" behavior of unit costs via duality mitigated my concern with the phenomenon.

In all cases for which the translog results were statistically significant and economically sensible, scale elasticity declined along a constant capital-labor ratio expansion path. Returns to scale were quite large for small plants and decreased rapidly as firm size increased. Scale economies eventually disappeared and decreasing returns set in for larger plants. Establishments that were two or three times larger than the average were often producing under constant returns to scale. In a number of industries, however, scale economies were such that

[70] Changing the critical t-value to ± 1.28 (20% confidence interval) would result in 14 more regressions having one or more variables that were significantly different from zero. For an extended note on the unreliability of standard statistical tests (i.e., the t-test) on the squared terms in the equation, see Z. Griliches and V. Ringstad, *Economies of Scale and the Form of the Production Function* (Amsterdam: North-Holland, 1971), pp. 77–79.

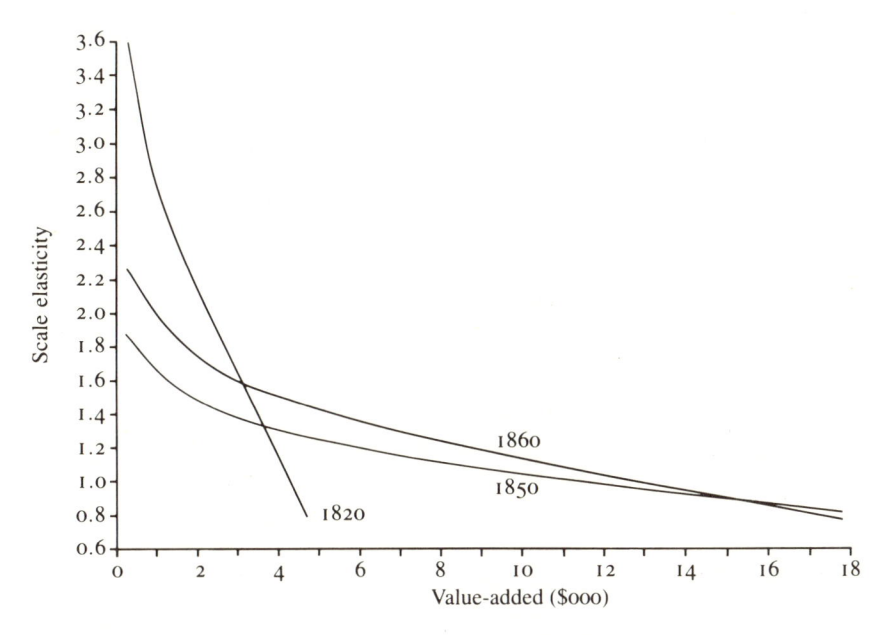

FIGURE 9.5 Translog Production Functions: Flour

constant returns were realizable only by the very largest of plants. In 1870, for example, the estimates predict that lumber mills producing less than $3,250,000 value-added would have been operating in the range of increasing returns to scale. No lumber mills that large existed then.

Where the scale elasticity estimates from the translog function for an industry could be examined over time, the scale elasticity curves for successive years lay above and to the right of those for earlier years (Figure 9.5). In 1820, for example, flour mills producing less than about $4,400 value-added were producing under increasing returns to scale. By 1850, flour mills had to be almost three times as large to produce under constant returns, and in 1860 those producing $12,600 or less were producing under increasing returns to scale. Returns to scale in 1860 were diminishing more rapidly than they had in 1850. Consequently, whereas small to medium-sized mills producing less than $16,000 value-added in 1860 had returns to scale greater than comparable plants in 1850, those producing over this amount had smaller returns to scale. In both cases, however, mills that large were producing in the range of decreasing returns to scale.[71]

---

[71] The estimate for 1870 was discarded, as scale elasticity was increasing with firm size.

In his study, John James found that the optimum size for flour mills before 1870 was very small, indeed close to zero; but in the estimates given here, flour mills had to be three to four times larger than the average to produce at about constant returns.[72] After 1870, James argues that there was a dramatic upward shift in the returns to scale. Our estimates show a similar upward movement, but from a much earlier date.

It is possible to make too much of these differences. Instead, the focus of attention should be on the similarity in the implications of our results. If the problem of increasing scale elasticity with plant size in the 1870 flour-milling estimate is overlooked, all of our scale-elasticity-curve estimates follow the same pattern that James describes, lying above and to the right of those for earlier years. The scale-elasticity estimate for flour mills in 1870 suggests that the most efficient plant would be one of infinite size, while in 1860 a mill producing only $12,600 value-added would have been most efficient. The scale differences that James noted between 1860 and 1880 in flour milling are on the same order of magnitude, given the size of the market.

Confirmation of James's finding of upward shifts in the returns to scale with the passage of time is significant. James's original results were not conclusive because he used aggregate state data in a pooled time-series–cross-section model in which time acted as a shift parameter. Hence, his results were in part forced by the particular form of the estimating equation. The estimates presented here, on the other hand, are independent of one another. Nevertheless, the same general pattern emerges from them, although without the symmetry that characterizes James's estimates. Plants grew larger over time partly because they had to if they wished to realize potential scale economies.

Many of the translog production functions were rejected because they contained no statistically significant regression coefficients.[73] No variable scale-elasticity estimates were rejected for this reason. Indeed, in the VSE equations, the regression coefficients and the conditional parameter, c_o, were usually significantly different from zero.[74] Just over half of the estimates were rejected because they implied unit costs that rose first and then fell, or that rose at a decreasing rate. These results arise because the equation predicts decreasing returns to

---

[72] James, "Structural Change," pp. 445–446.

[73] See Griliches and Ringstad, *Economies of Scale*, pp. 77–79.

[74] The distribution ($\ln L - \ln L_{max}$), where $\ln L$ is the estimated value of the log likelihood function, follows the a chi-square distribution such that ($\ln L - \ln L_{max}$) $< 0.5\chi^2$. The critical value of χ^2 for the 95% confidence interval is -1.94. See Ringstad, "Some Empirical Evidence," p. 93.

scale for even the very smallest plants.[75] Rejection of them is a denial that the average-cost-curve behavior that they imply is economically reasonable and plausible, rather than a rejection of the idea that there may be decreasing returns for very small plants. There is nothing to prevent production of a good under conditions of decreasing returns to scale, but we would not expect to observe plants growing under such circumstances, beyond the absolute minimum scale for production, unless and until conditions changed.

The conditional parameter, c_0, in the equation

$$\ln Q + c_0 (\ln Q)^2 = b_0 + b_1 \cdot \ln K + b_2 \cdot \ln L$$

allows scale elasticity to vary with plant size. It was almost always significantly negative, regardless of the year, industry, or magnitude of the regression coefficients for labor and capital. As a result, scale elasticity was not well behaved and increased with plant size over the observed range.

Consequently, we have interpreted these estimates in terms of the average-cost curve that is derived from the dual of the production function. Where the sum of the capital and labor coefficients was greater than unity and c_0 was negative, unit costs were declining across the entire observed range of plant sizes. The most efficient plant in the industry for a particular year was consequently larger than the largest plant in the data set, but there is no way of knowing how much larger.

In just two cases was c_0 positive.[76] Both were in 1870. Since the sum of the capital and labor coefficients was greater than one, these estimates imply an average-cost curve that was U-shaped. However, given the small (although significantly positive) values for c_0 and the magnitude of the sum of the labor and capital coefficients, the minimum point on the average-cost curves occurred at output levels in the billions of dollars. The estimates thus still imply that the optimum plant in these industries was one that supplied the entire market.

The level of each average-cost curve relative to the others gives no information about the relative cost between them from year to year. Absolute costs are determined by the magnitude of the constant, k, in each equation. It changed from year to year. Thus, although the curve

---

[75] When $m < 1$ and $c_0 < 0$, then the implied average-cost curve has an inverted U-shape, with rising unit costs for very small establishments, eventually decreasing for large establishments. For $m < 1$ and $c_0 > 0$, unit costs are increasing throughout, but at a decreasing rate for larger plants.

[76] Cotton goods and meat packing. See Appendix C at the end of the chapter.

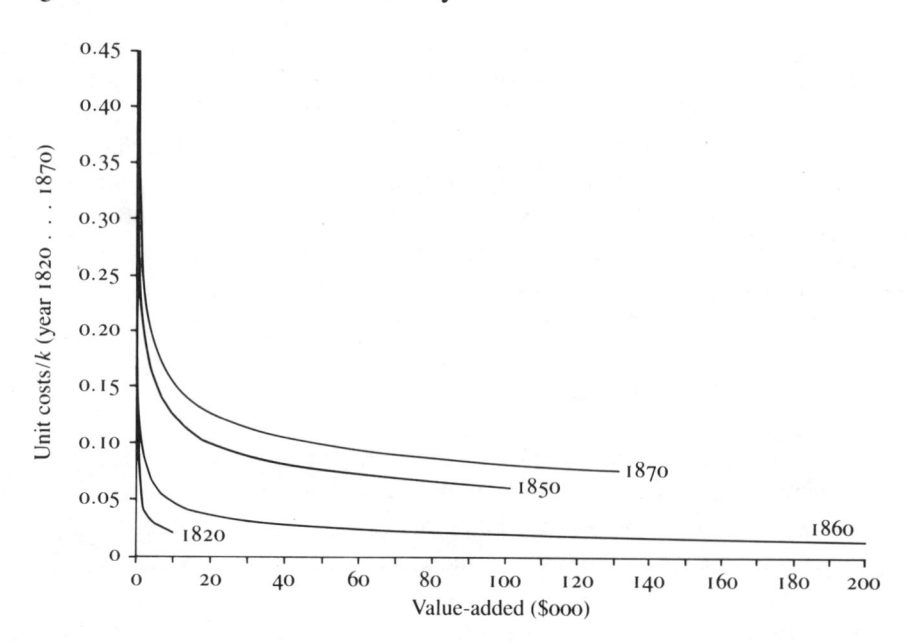

FIGURE 9.6 Flour-Milling Unit Costs, 1820–1870

for 1870 lies above the others, this does not imply that costs in 1870 were the greatest or that 1820 costs were lowest. Furthermore, since output is measured by value-added, not physical units, these results must be interpreted cautiously and the reader should recognize that product prices will affect the relative "unit" costs.

The estimates for flour milling in 1820, 1850, 1860, and 1870 (Figure 9.6) show the indeterminacy of the cost-minimizing scale of plant quite clearly. The unit-cost curves are declining throughout, and plant size was therefore theoretically bounded only by the size of the market. In practice, however, this was probably not the case. Plant size determinacy is restored by positive transport costs. This is discussed at length below. Moreover, although unit costs declined very steeply at first, the curve eventually flattened out. As a result, the average-cost curve had a very pronounced L-shape. The sharpest decreases in unit costs per unit increase in output were realized by the smallest plants. Further declines in unit costs once a certain scale had been attained were much smaller. Small plants beyond this critical point may therefore not have been placed at a serious cost disadvantage relative to those that were somewhat, or even much, larger.

The largest flour mill in the 1820 sample was quite small, producing less than $10,000 value-added, compared with the largest mill in the

1870 sample, which produced about $130,000 value-added.[77] In 1820, a flour mill half as large as the largest would have had unit costs some 40 percent higher. A mill of the same relative size in 1870 would have experienced costs only about 22 percent higher. A mill one-tenth the size of the largest would have unit costs almost three times greater than those for the largest in 1820. In 1870, the margin would have been about 80 percent.

These results are broadly consistent with those from the translog production function. Whereas the translog estimates predict that plant size, for constant returns to scale, was increasing over time, the cost functions derived from the VSE estimates imply falling average costs that did not begin to flatten as quickly over time, exhausting scale economies, as plant size increased.

Both the translog and VSE estimates lead to estimates of the optimum plant size (defined as that plant size with constant returns to scale, or minimum average cost) that are considerably larger than the average plant size by industry. In every industry for which sensible variable scale elasticity estimates were obtained, the optimum-sized plant would have enjoyed a natural monopoly. The translog estimates were somewhat more conservative. Optimum plant size was less than infinite, but with the exception of sheet metals in 1850 and the two estimates for 1820 (flour milling and wagon and carriage making), plants that were large enough to be producing at constant returns were more than twice the size of an average plant in the industry.

V

Another check upon these results was made using heuristic estimates, from the survivor technique of the optimal plant size.[78] The survivor technique uses the economic theory of long-run equilibrium adjustment to identify which plant sizes not only survived the rigors of market competition over time but also succeeded in increasing their share of total industry value-added. Competition guarantees that only the

---

[77] The average-cost curves in Figure 9.6 cover the range of observed plant sizes, from smallest to largest in the sample for each year. Note, however, that the largest firm in these samples was not the largest flour mill in the country in that year. In 1870, for example, the largest flour mill was the firm of Blecker Brothers, which produced $569,500 value-added.

[78] *Optimality* in this context is defined in terms of adjustment to market conditions rather than simple cost minimizing. Although the fundamental notion embodied in the survivor technique may be traced back to John Stuart Mill and Willard Thorp, it owes its modern revival to George Stigler. See George Stigler, "Monopoly and Oligopoly by Merger," *American Economic Review* 40 (May 1950), pp. 23–34, and "The Economies of Scale," *Journal of Law and Economics* 1 (Oct. 1958), pp. 54–71.

efficient plants survive in the long-run. Survival, through evolution if necessary, is the ultimate "market test" of efficiency.

A number of assumptions are implicit in the technique.[79] Shepherd, for example, argues that "survivor estimates for firm sizes are likely to be more valid for atomistic industries . . . than for highly concentrated ones," because the assumptions of atomistic competition ensure that, in the long run, market pressures force all surviving plants to operate at minimum long- and short-run average cost.[80] In perfect competition in a constant-cost industry, demand shifts affect only the number of establishments in the industry, but such changes under other market structures permanently alter the market solution, including the optimum plant size. The available evidence on market structure in the nineteenth century suggests that American industry was not perfectly competitive, but consisted of spatially separated local monopolies that were protected from competition with each other by high transportation costs.[81] As a result, the survivor results reflect both supply and demand changes, but to the extent that there were sharp discontinuities in the pattern of plant survivorship these will be attributed to supply-side changes, including changes in transport costs.

Profit-maximizing behavior ensures the survival of lower-cost plants regardless of market structure. Survivorship under imperfect competition, monopolistic competition, or oligopoly, however, no longer carries the implication of cost minimization. To avoid confusion between survivorship and the economic concept of optimum plant size, Leonard Weiss coined the term "minimum efficient scale" (MES) to characterize the smallest range of surviving plants.[82] This terminology is appropriate regardless of market structure, and it is used here.

Estimates of the range of surviving plants are consistent with the L-shape of the average-cost curve implied by the VSE production-function estimates and suggest that a considerable portion of the long-run average-cost curve may have been flat, or almost so. The ranges of surviving plants also generally overlapped the translog estimate of

[79] William A. Shepherd, "What Does the Survivor Technique Show About Economies of Scale," *Southern Economic Journal* 34 (July 1967), pp. 113–122. See also Thomas R. Saving, "Estimation of Optimum Plant Size by the Survivor Technique," *Quarterly Journal of Economics* 75 (Nov. 1961), pp. 569–607.

[80] Shepherd, "Survivor Technique," p. 115.

[81] See Fred Bateman and Thomas J. Weiss, "Comparative Regional Development in Antebellum Manufacturing," *Journal of Economic History* 35 (March 1975), pp. 182–208, and "Market Structure Before the Age of Big Business: Concentration and Profit in Early Southern Manufacturing," *The Business History Review* 49 (Autumn 1975), pp. 312–336.

[82] See Leonard W. Weiss, "The Survival Technique and the Extent of Sub-Optimal Capacity," *The Journal of Political Economy* 72 (June 1964), pp. 246–261.

plant size for unit scale elasticity.[83] A wide range of plants survived in almost every industry. In most, the MES plants were larger than average, and the upper range was usually many times greater than the mean establishment size. In only six industries (flour and lumber milling, saddlery, sheet metals, tobacco, and woolens) did the range of surviving plants include plants that were of average size. Plants in four of these industries, flour and lumber milling, saddlery, and sheet metal, were ubiquitous. These were local establishments that produced locally consumed goods. Surviving plants in the other ten industries were much larger than average.

There is, therefore, abundant evidence of increasing returns to scale and declining unit costs. The translog estimates suggest that to realize constant returns to scale, establishments had to be at least twice as large as average and sometimes much larger. Estimates of the unit-cost curve implied by the variable scale-elasticity results indicate continuous cost advantages for ever-larger plants. Both techniques, however, show the greatest gains would be realized by small establishments growing somewhat larger. Nevertheless, large plants could enjoy quite substantial cost advantages over smaller competitors. Similarly, the survivor results show that large plants, particularly factories, contributed ever-greater proportions of total industry production as time passed. They not only survived but prospered and grew even larger.

VI

Our empirical work, regardless of any inconsistencies between efficiency estimates and those for economies of scale, indicates persistent supply-side incentives for increasing plant size during America's industrialization. Sometimes the changes in the average scale of operation caused by the rise of the factory were quite dramatic. For example, whereas iron furnaces in 1820 had generally employed fewer than twenty workers and produced less than $6,000 value-added, by 1870 such establishments typically employed sixty or more workers and produced $45,000 value-added. Even cotton mills, which had been the first big business in American industry, underwent extraordinary growth. From establishments employing perhaps twenty adult equivalents and producing about $7,000 value-added on average in 1820, by 1870 the average mill employed more than a hundred and produced perhaps $100,000 value-added.

The dominant picture in the literature is one of increasing average

---

[83] See Appendix B at the end of the chapter.

plant size, with particular emphasis being given to the emergence of industrial behemoths. This began quite early, but the pace accelerated later in the century, and there was a sharp increase in the number of establishments producing a million dollars of output or more between 1820 and 1870. There were no plants this large in Sokoloff's 1820 sample. In 1850, there were 16 plants that produced more than a million dollars worth of output. By 1860, 56 plants produced this much, and by 1870, at least 138 plants were producing goods valued at more than a million dollars.[84] These data suggest an accelerating trend toward bigness in American industry.

Furthermore, big establishments were to be found in an increasingly wide range of industries. Whereas in 1850, million-dollar plants were confined to just four industries—cotton mills (11), sugar refineries (3), woolen mills (1), and steam engine factories (1)—by 1870, they were to be found in thirty-eight industries. Cotton mills and sugar refineries still dominated, but industries such as sewing machines, agricultural implements, iron, nails, clothing, boots and shoes, liquor, brewing, leather tanning, lumber milling, and flour milling, to name but a few, had representatives among the million-dollar-plus plants by 1870.

Given our results that show larger, mechanized plants to be economically superior to smaller, nonmechanized plants, the rise of very large plants is not surprising.[85] Nevertheless, average plant size remained small despite the establishment of more and bigger establishments in every industry. In some industries, scale differences were barely perceptible over the interval between 1820 and 1870. The ubiquitous lumber and flour mills, for example, changed hardly at all, except perhaps in switching from water to steam power. In 1820 the average flour mill employed 3.25 adult equivalents and generated about $2,000 value-added.[86] In 1870, the average mill in the sample employed 2.8 people to produce about $2,300 value-added.[87] The average boot and shoe establishment in 1870 was in fact only about half the size (in terms of employment and output) of its predecessor of 1820.[88]

[84] These figures are from the Bateman-Weiss large-firm samples. I believe that they are comprehensive, since each state sample that contained million-dollar plants also contained plants producing less than that. After adjusting output values by the Warren-Pearson price indexes, there would be 19 million-dollar plants in 1850, 57 in 1860, and 78 in 1870.

[85] Mechanized plants were larger in terms of value-added or output for comparable labor forces than nonmechanized plants in the same industry.

[86] Sokoloff sample data.

[87] Bateman-Weiss sample data. Average employment in flour mills in the published census is a little lower, 2.57 employees, and value-added a little higher, $3,400.

[88] Regardless of whether sample or published census statistics are used.

The Rise of the Factory

TABLE 9.4

Firms Producing Less than $1,000 Value-Added, by Industry, in 1850 and 1870, as a Percentage of All Firms in the Industry from the Samples and the Implied Number of Such Firms in the Population
(Rounded to nearest 10)

	1850		1870	
Industry	Percent of firms in sample	Number of firms in population[a]	Percent of firms in sample	Number of firms in population[a]
Boots & shoes	35%	3,960	60%	14,060
Brewing	14	60	65	1,280
Clothing	25	1,070	48	3,760
Cotton goods	2	20	0	0
Flour milling	43	5,110	68	15,350
Furniture	44	1,870	44	2,290
Iron	3	10	27	90
Leather	36	2,350	56	4,100
Liquor	47	460	51	370
Lumber milling	45	8,050	51	13,170
Meat packing	23	40	31	80
Saddlery	31	1,090	54	4,110
Sheet metals	23	520	34	2,260
Tobacco	42	600	31	1,600
Wagon making	41	1,730	52	6,160
Woolen goods	24	440	13	250

SOURCE: Percentages are computed from the Bateman-Weiss samples.

[a] Numbers are computed by applying these percentages to the published census estimates for the number of firms by industry.

Moreover, the median establishment in almost every industry (except textiles) was very small. To at least 1870, small-firm populations decreased in only three industries (cotton and woolen textiles and tobacco) that are considered here (Table 9.4).[89] These industries were overwhelmingly concentrated in cities and were confined to fairly narrow geographic areas. New England, for example, contained more than half of all the nation's cotton mills and accounted for about two-thirds of the production.[90]

In the other industries the fraction of very small plants seems to have been increasing. In particular, the percentage of plants producing less than $1,000 value-added increased and, in a majority of cases, the median plant size fell within this range. These populations, however, were not only increasing in relative frequency, they were increasing in absolute number. Between 1850 and 1870, for example, over 10,000 small

---

[89] Among the industries I study here. I have not yet examined the results for all industries.

[90] Francis A. Walker, *Statistics of Wealth & Industry*, p. 430.

boot and shoe establishments, 10,000 small flour mills, and 5,000 small lumber mills were established, not to mention perhaps 2,500 needle-trade establishments, 1,200 small breweries, and 1,000 cigar and tobacco establishments.[91] In aggregate, in the sixteen industries that are considered here, perhaps as many as 41,000 new small establishments, each producing less than $1,000 value-added, were established. As the increase in the entire population of business establishments between the two dates was only 130,000, between a quarter and a third of all new entrants were small plants in just these few industries.

Despite their overwhelming numbers in virtually every industry, small plants contributed relatively little in terms of total industry output, and although they were an increasing majority, they generally failed to maintain their share of industry production.[92] By simple arithmetic, one plant producing a million dollars of output contributes as much to total production as one thousand plants each producing a thousand dollars of output. The 122 additional large establishments with outputs exceeding a million dollars in 1870, over the number in 1850, thus probably generated a greater output value than the 41,000 new small plants in the sixteen industries studied here.

VII

This situation raises a potential economic paradox: If small plants were inefficient and uneconomic, how could they survive within a competitive environment? The answer, it is argued, lies in the role that transportation played in the development of the structure and location of industry in America.

First, consider the case of the very large-scale enterprise. According to our production-function estimates, plants faced monotonically decreasing average costs. As a result, optimum (that is, cost-minimizing) plant size was indeterminate. However, positive transport costs can make plant size determinate even under these conditions. Market boundaries for competing establishments are defined by equality in price to the consumer. That price covers not only production but also distribution costs. Internal production costs are therefore only part of the story, and it can be shown that the sum of a monotonically decreas-

[91] Estimated by applying the percentages in Table 9.4 to the number of plants reported by the census in those industries in 1850 and 1870.

[92] This might be inferred from the results shown in Table 9.2 above, which shows mills and factories gaining at the expense of other organizational forms.

ing average-cost curve and a monotonically rising transport-cost function may be U-shaped.

Consider now the situation that might have faced a small plant, say a typical flour mill. In 1820, such a mill, producing about $2,000 value-added, would have experienced unit costs that were about double those of the largest mills. In 1870, an average-sized mill (generating $2,300 value-added) would have been at an even worse competitive disadvantage. Its costs would have been about three times higher than those of the largest mills around then. With cost differentials of these magnitudes, mills of radically different size were unlikely to be located in close proximity to one another.

Positive transport costs, however, created protected markets for high-cost producers by providing a measure of protection to less efficient plants. A high-cost producer—provided it located at sufficient distance from a low-cost producer, such that the cost of transport exceeded the production-cost differential between them—could always survive.[93] This situation generates some interesting dynamics in the distribution of plant sizes, especially in a developing country.

Consider a large business located on a featureless plain in a developing economy. The situation is shown in Figure 9.7. The large plant is located at point C. Consumers are distributed across the plain, but population density decreases as we move away from C. Transportation costs are positive and equal in all directions. Let us examine the dynamics of competition along a radius, CD, from C. This is shown in the xy graph in the lower half of Figure 9.7. Distance from C is measured along the x-axis. The y-axis represents production and distribution costs per unit. The large plant at C produces a homogeneous good at a price of CA per unit. Adding to this the cost of transportation (a function of distance) represented by the line AB determines the selling price of the good at any location along the radius CD. Thus, at D, the good is available at price DB.[94]

Suppose now that small plants have higher unit-production costs

[93] Whether or not the measure of protection afforded by transport and information costs extended to cover costs that were two or three times greater is a different matter. One should not be too dogmatic about the specific cost ratio estimates presented here, provided that the existence of a "substantial" differential is recognized. How significant such cost differences were in affecting the competitiveness of establishments of varying size and location would depend upon transport costs.

[94] For similar ideas in the industrial organization literature, see Steven C. Salop, "Monopolistic Competition with Outside Goods," *The Bell Journal of Economics* 10 (Spring 1979), pp. 141–156; William J. Baumol, John C. Panzer, and Robert D. Willig, *Contestable Markets and the Theory of Industry Structure* (New York: Harcourt, Brace Jovanovich, 1982).

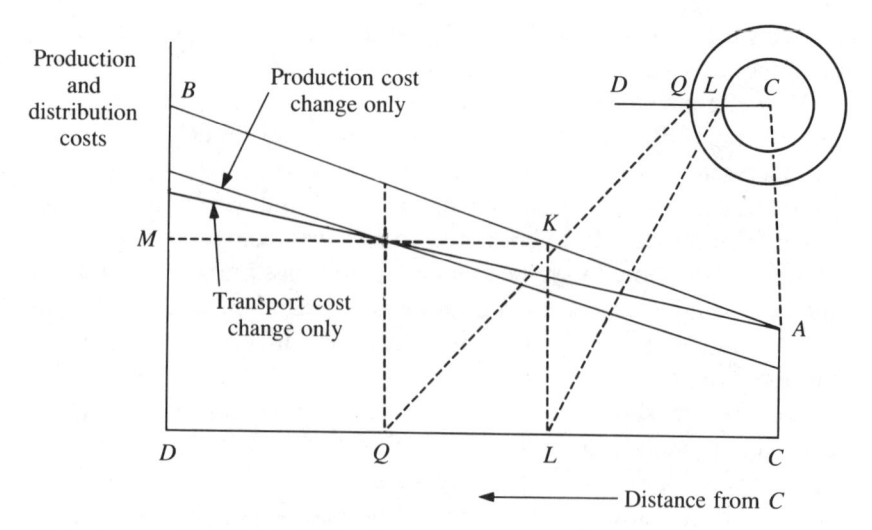

FIGURE 9.7 The Spatial Distribution of Competition

than large plants, say *LK*. Their cost levels are higher because of their inability to realize scale economies and because of the relative inefficiency of their handicraft organization. These costs, then, become the upper limit of the price that the larger enterprise can charge for its product. Assuming homogeneity, consumers will buy the cheapest product. Thus at distance *CL* from the large plant, consumers will be indifferent between the good sold by the large plant and distributed from *C* and that produced by a small enterprise located at *L*. To the extent that not many sales are required to realize unit costs of *LK*, the price of this product to any consumer will be given by the line *AKM*.

The circumference, *L*, around *C* represents the market boundary for the product made by the large-scale enterprise located at *C*. All consumers within this boundary purchase the good from *C*; all those without, purchase the good from a smaller plant located on or beyond the circumference. Sales for the larger-scale enterprise are determined by the number of consumers within the market boundary, which is a function of the area of the circle radius, *CL*, given the population density. Because population density decreases with distance from *C*, markets for the smaller plants are quite limited geographically, and market density is greatest closest to the circumference, *L*, of the large plant's market boundary. Most smaller establishments will therefore try to locate on or close to this circumference. Their number, given the volume of sales they needed to realize unit costs of *LK*, will be a function of the circumference of this market boundary centered on *C*. Other small establishments will locate beyond this boundary.

Over time, a number of changes took place that affected this situation. First, market density increased through population growth, migration, and rising incomes. Second, transport costs fell as transportation media improved. These two forces affect the transport costs for any given level of sales and hence alter the effective slope of *AB* as shown in Figure 9.7. Third, new production technologies were developed that were better suited to large-scale than small-scale production. Moreover, there is evidence of a quality differential between factory and handicraft goods that favored the former over the latter, thereby reducing the relative price of factory goods, adjusted for quality. Machine-made cloth, for example, was finer than the hand-loomed kind, and machine-made boots and shoes were more durable.

As a result, the market boundary for the large plant at *C* expanded to, say *Q*. Its sales expand as a function of the market area, that is at the rate of the square of the change in the market radius. This would further accelerate if population densities and income levels were to rise within the market boundary. Small establishments on the old market boundary, *L*, would be driven out of business by low-cost competition from the large producer. But there would now be a new, larger market boundary represented by the circumference *Q* around *C*. There might therefore be *more* small establishments than before, even after the exit of those establishments in the area between circumferences *L* and *Q*.

It is unlikely, however, that lowered transport costs would immediately lead to the displacement by factory products of rural handicraft production in the market between *L* and *Q*. Handicraft workers were often willing to squeeze their income levels to stay in business in competition with more efficient producers. To the extent that they were successful, the artisan shop represented an impediment to the spread of the factory by limiting the scope of market and the desirability of further investments. Alexander Hamilton in his "Report on Manufactures" clearly thought that this would be the case. In his opinion,

> the spontaneous transition to new pursuits, in a community long habituated to different ones, may be expected to be attended with proportionally greater difficulty. When former occupations ceased to yield a profit adequate to the subsistence of their followers; or when there was an absolute deficiency of employment in them . . . changes would ensue; but these changes would be likely to be more tardy than might consist with the interest either of individuals or of the society. In many cases they would not happen, while a bare support could be insured by an adherence to ancient courses, though resort to a more profitable employment might be practicable.[95]

---

[95] Alexander Hamilton, "Report on Manufactures, Communicated to the House of Representatives, December 5, 1791," in *American State Papers,* Finance, I, as re-

The lumpiness of factory investment compounded these problems. Each new factory or machine represented a significant, finite increase in industry supply. This increased pressure upon prices and was felt keenest in times of depression when factories were kept operating because of their relatively high ratio of fixed to variable costs. On the other hand, artisans either switched to part-time work or temporarily closed up shop.

Nevertheless, the long-term effects of such behavior seem to have been minimal as large establishments grew very rapidly. This growth seems to have been much faster than warranted by demand-side factors such as price and income elasticity and the growing number of potential consumers. Furthermore, to the extent that small establishments located in more remote rural markets,[96] which grew increasingly less remote, it is unlikely that the growth in their numbers might be attributed to complementarity between most handicraft and factory production.[97] Markets were segmented by heterogeneous access to transportation and by nonuniform transport costs, and as a result, there were areas where small establishments not only survived but prospered. Competition between producers with disparate costs took place on the boundaries of these markets.

---

printed in John C. Hamilton, ed., *The Works of Alexander Hamilton* (New York: Charles S. Francis, 1851), vol. 3, pp. 192–284. The quotation is from p. 217.

[96] Sokoloff has done some work mapping firm location in 1832 and 1850 and found that more large plants are located in urban areas and that plants tended to shun locations in close proximity to competitors. See Sokoloff, "Transition from the Artisanal Shop," pp. 359–362. More work on this topic, however, needs to be done.

[97] In this situation the handicraft sector would serve as a buffer for the factory, supplying extra goods beyond the factory's capacity in boom times and absorbing unemployment to maintain factory employment in times of slump. The scenario is akin to the microeconomic model of price leadership by a dominant firm in which the dominant "firm" is the group of low-cost factory producers and the smaller plants are the artisan shops. In that model, the dominant plants set price at a level that maximizes their profits and then permit the small artisan shops to sell as many units as they wish at the prevailing price. The factories act as monopolists, the artisan shops behave like perfect competitors. See John S. Lyons, "Competitiveness or Complementarity in the Survival of Protoindustry?" (unpublished manuscript, Northwestern University, 1984), for a discussion of these scenarios in the case of British industrialization.

Another way in which the factory and handicraft shop may have been complementary is if the existence of the factory cheapened inputs for the artisanal shop, thereby making artisan shops located in close proximity to factories more efficient.

In either complementarity scenario, however large and small plants would be located in close proximity to one another, whereas when factories and handicraft production are substitutes, they will be located at a distance from one another.

VIII

The rise of the factory thus seems to have been helped by persistent, but small, productivity gains over more traditional production methods. These gains are similar to those estimated by Sokoloff for 1820 and 1850 but they are much smaller than one would have anticipated from the narrative literature. The mill and factory compared particularly favorably with sweatshops and manufactories, and competition between them led to the disappearance, or sharp economic decline, of these larger, nonmechanized producers. The smaller artisan shops also declined in relative importance, but many consumers still remained dependent upon them. They persisted despite substantial scale economies that became progressively larger as the nineteenth century progressed. These economies of large-scale production led to rapid increases in the number of large plants in every industry, consistent with Chandler's identification of the origins of modern large-scale production in the technological imperatives of new production methods.

Artisan shops, however, not only appear to have persisted but their numbers continued to increase until at least 1870, despite the emergence of more efficient and seemingly better-adapted larger plants. In many industries, this trend may have continued to the end of the century.[98] Although more work is needed on plant location, this pattern of increasing numbers of both seemingly uneconomic small plants and much more efficient large plants is consistent with a model in which competition between them is constrained by transport costs. The eventual triumph of big business and large-scale plants in many industries, then, is attributable to transportation improvements, despite scholarly efforts over the past two decades to downplay the contribution of the transportation revolution.

---

[98] This conjecture is based upon my claim that the structure of industry in 1900 does not seem to be different in most industries from the trends that were apparent in 1870. See Atack, "Industrial Structure," especially pp. 42–50.

Appendix A

TABLE 9.5
Sample Sizes by Industry and Year

Industry	1820	1850	1860	1870
Boots & shoes	25	478	471	389
Brewing	n.a.[a]	11	29	43
Clothing	n.a.	120	135	154
Cotton goods	70	41	35	14
Flour milling	44	420	627	542
Furniture	23	213	170	115
Iron	24	30	25	22
Leather	121	371	251	106
Liquor	168	30	75	24
Lumber milling	27	865	1,061	622
Meat packing	n.a.	39	67	10
Saddlery	15	181	157	159
Sheet metals	27	109	129	105
Tobacco	n.a.	66	72	84
Wagon making	20	205	270	227
Woolen goods	39	63	38	30

[a] n.a. = no observations in the sample.

Appendix B: Estimating Nonhomogeneous Production Functions

Currently the most popular form of nonhomogeneous production functions is the translog production function. In log-linear form, this may be written as [99]

$$\ln Q = b_0 + b_1 \cdot \ln L + b_2 \cdot \ln K + b_3 \cdot (\ln L)^2 + b_4 \cdot \ln K \cdot \ln L + b_5 \cdot (\ln K)^2$$

which has returns to scale,

$$b_1 + b_2 + (2b_3 + b_4) \ln L + (2b_5 + b_4) \ln K$$

This production function is both nonhomogeneous and nonhomothetic. As a result, returns to scale are usually evaluated for the different output levels generated by labor and capital in proportions equal to the mean capital-labor ratio. To the extent that different modes of production had different capital-

---

[99] See Laurits Christensen, Dale Jorgensen, and Lawrence Lau, "Transcendental Logarithmic Production Functions," *Review of Economics and Statistics* 55 (Feb. 1973), pp. 28–45; Ernst Berndt and Laurits Christensen, "The Translog Function and the Substitution of Equipment, Structures, and Labor in U.S. Manufacturing, 1929–1968," *Journal of Econometrics* 1 (March 1973), pp. 81–113, for a discussion of the methodology. See James, "Structural Change," or Louis P. Cain and Donald Paterson, "Factor Biases and Technical Change in Manufacturing: The American System, 1850–1919," *Journal of Economic History* 41 (June 1981), pp. 341–360, for examples of the application of the methodology.

The Rise of the Factory

labor ratios, they would also have had different patterns of scale economies by output. Unfortunately, the translog function is not self-dual; that is, estimation of a translog production function gives no information about the cost function. The two formulations represent different specifications of the underlying technology. This is a serious drawback for the intended use of the results.[100] Nevertheless, at times, economically nonsensical results from the preferred estimator led to the use of the translog production functions as a substitute. In general, however, the translog results were no more robust and could not be preferred on any econometric or theoretical grounds to those arrived at by other methods.

There exists a class of homothetic, nonhomogeneous production functions that are self-dual and yet have scale economies solely dependent upon the level of output. Estimators of this class will be referred to as variable scale-elasticity (VSE) estimators. These represent our preferred estimators. Functions of this class have been developed by Nerlove:[101]

$$\ln Q + c_0 (\ln Q)^2 = b_0 + b_1 \cdot \ln K + b_2 \cdot \ln L$$

which has returns to scale:

$$(b_1 + b_2)/(1 + 2c_0 \ln Q)$$

and by Zellner and Revankar:[102]

$$\ln Q + c_0 Q = b_0 + b_1 \cdot \ln K + b_2 \cdot \ln L$$

which has returns to scale:

$$(b_1 + b_2)/(1 + c_0 Q).$$

Experiments with both formulations have led to the selection of Nerlove's model because the returns-to-scale parameter is better behaved for negative values of c_0 and the estimates of c_0 are less subject to computational inaccuracy.

Regardless of formulation, the equation is estimated using a conditional maximum likelihood estimator. The parameter c_0 is free to vary within certain bounds and is chosen so as maximize the likelihood of the ordinary least-squares regression of the independent variables on the transformed dependent variable. The bounds on c_0 are such that the returns to scale are defined. That is, in Nerlove's model,

$$c_{0_{\min}} = -1(2 \ln Q_{\max})$$

---

[100] See James, "Structural Change," pp. 443–460, especially pp. 453–456, for an elaboration of this point.

[101] M. Nerlove, "Returns to Scale in Electricity Supply," in *Measurement in Economics*, Carl F. Christ, ed. (Stanford: Stanford University Press, 1963). See also V. Ringstad, "Some Empirical Evidence on the Decreasing Scale Elasticity," *Econometrica* 42 (Jan. 1974), pp. 87–102.

[102] A. Zellner and N. S. Revankar, "Generalized Production Functions," *Review of Economics Studies* 36 (April 1969), pp. 241–250.

while for the Zellner and Revankar model,

$$c_{0_{min}} = -1/Q_{max}$$

There is no upper-bound constraint on the parameter in either formulation. The cost function implied by Nerlove's model is that given by

$$\text{min.: } C = wL + rK$$

$$\text{subject to: } Qe_0^{c(\ln Q)^2} = K^{b_1}L^{b_2}$$

Solving yields

$$C_{min} = K [Qe_0^{c(\ln Q)^2}]^{(1/(b_1 + b_2))}$$

where k is a constant determined by the relative factor shares and relative factor prices. The average-cost curve corresponding to this is:

$$AC = K [Q^{-m}Qe_0^{c(\ln Q)^2}]^{(1/m)}$$

where $m = (b_1 + b_2)$.

For $m > 1$ and $c_0 > 0$, this average cost curve is U-shaped. If $m > 1$ and $c_0 < 0$, then average costs are declining at least through Q_{max}, the largest plant observed. Results for $m < 1$ (i.e., diseconomies of scale for even the very smallest plants) are inconsistent with economic theory. In the case of $m < 1$ and $c_0 < 0$, the implied average-cost curve has an inverted U-shape; that is, unit costs are increasing for very small plants, reach a maximum, and then decrease for large plants. For $c_0 > 0$, unit costs are increasing throughout but at a decreasing rate for larger plants.[103]

Where initial results were inconsistent with accepted economic theory, the data that generated them were scrutinized. Particular attention was paid to those observations with large positive residuals. By definition, plants can produce an output less than that specified by the production function (i.e., be inefficient), but they cannot produce more output, given their resources, than that defined by it.[104] That is,

$$Q_{actual} < Q_{predicted}, \quad \text{so} \quad Q_{actual} - Q_{predicted} < 0$$

Nevertheless, in some instances, $Q_{actual} > Q_{predicted}$, in which case the residuals will be positive. Plants with positive residuals were superefficient and producing beyond the production-possibilities curve, and those with the largest positive residuals were the most efficient of all.[105]

---

[103] Note that in this case, the curve is not simply the right-hand half of a "normal" U-shaped average-cost curve.

[104] See Arthur S. Goldberger, "The Interpretation and Estimation of Cobb-Douglas Functions," *Econometrica* 36 (July–Oct. 1968), pp. 464–472.

[105] See M. J. Farrell, "The Measurement of Productive Efficiency," *Journal of the Royal Statistical Society*, series A, 120, part 3 (1957), pp. 253–281; M. J. Farrell and M. Fieldhouse, "Estimating Efficient Production Under Increasing Returns to Scale,"

This result arises only because of my reliance upon least-squares estimation. By the Gauss-Markov conditions, $E(e_i) = 0$, so that except in the case of a perfect regression fit, observations will lie both above and below the predicted regression plane. There were positive residuals only because there were negative ones, and vice versa.

Of far greater concern was that the pattern of positive and negative residuals indicated a degree of autocorrelation that seemed related to the state from which the observations were drawn. A series of contiguous observations would have large positive residuals and be followed by a larger number of smaller negative residuals; that is, the production function seriously underpredicted output and productivity in some states. This raises some serious questions with respect to either the way in which the data were collected by census enumerators or the assumption of homogeneity between states.[106]

Journal of the Royal Statistical Society, series A, 125, part 2 (1962), pp. 252–257; D. J. Aigner and S. F. Chu, "Estimating the Industry Production Function," *American Economic Review* 58 (Sept. 1968), pp. 826–839. There is a technique, known as a Minimum Absolute Deviation estimator (MAD), that permits us to constrain residuals to be of one sign, but this is not widely used, is inefficient, and prevents our reliance on standard statistical tests such as hypothesis testing. See E. M. L. Beale, "On Minimizing a Convex Function Subject to Linear Inequalities," *Journal of the Royal Statistical Society,* series B, 17 (1955), pp. 173–184; George G. Judge and T. Takayama, "Inequality Restrictions in Regression Analysis," *Journal of the American Statistical Association* 61 (1966), pp. 49–61. MAD is not used here.

[106] The use of state dummy variables (both slope and intercept) and filter rules to weed out those observations exhibiting unusually high or low labor or capital productivity failed.

Appendix C

TABLE 9.6

Statistically Significant and Economically Sensible Translog Production-Function Estimates, by Year and Industry, 1820–1870

(K/L rounded to nearest $10. Value-added rounded to nearest $100)

Year and Industry	b_0	b_1	b_2	b_3	b_4	b_5	K/L	Value-added for constant returns
1820								
Flour milling	5.367	11.315	−2.138	−0.387	−1.005	0.238	1,600	4,400
(S.E.)		5.520	1.722	0.520	0.654	0.119		
Wagon making	−0.067	−0.099	1.905	−0.679	0.524	−0.186	190	2,000
(S.E.)		2.557	0.724	0.547	0.559	0.111		
1850								
Boots & shoes	1.712	1.039	0.674	−0.005	−0.051	−0.007	300	4,400
(S.E.)		0.427	0.388	0.058	0.077	0.033		
Flour milling	5.578	2.950	−0.915	−0.038	−0.257	0.117	3,250	11,100
(S.E.)		0.810	0.384	0.161	0.112	0.025		
Furniture	5.832	2.798	−0.989	0.152	−0.398	0.159	500	2,800
(S.E.)		0.812	0.622	0.125	0.151	0.053		
Lumber milling	4.517	−0.127	−0.175	−0.237	0.150	0.049	1,070	208,600
(S.E.)		0.42	0.317	0.072	0.070	0.023		
Sheet metals	6.328	6.189	−1.646	0.700	−1.001	0.245	1,150	zero
(S.E.)		1.329	0.696	0.381	0.272	0.054		
Wagon making	−3.319	−1.686	2.505	−0.179	0.339	−0.159	420	1,309,800
(S.E.)		0.927	0.947	0.188	0.183	0.070		
Woolen goods	−5.211	−2.551	2.934	−0.716	0.738	−0.226	1,220	25,700
(S.E.)		2.790	2.055	0.348	0.466	0.166		

1860							
Boots & shoes	5.426	1.333	−0.298	−0.068	−0.044	0.050	620	12,800
(S.E.)		0.608	0.464	0.082	0.112	0.042		
Flour milling	3.928	3.280	−0.402	−0.128	−0.224	0.076	2,880	12,600
(S.E.)		1.153	0.499	0.211	0.173	0.037		
Iron	13.594	6.113	−3.356	0.406	−0.793	0.308	1,590	9,900
(S.E.)		4.147	6.559	0.306	0.438	0.361		
Lumber milling	2.757	2.288	−0.080	−0.021	−0.185	0.063	1,290	11,200
(S.E.)		0.458	0.380	0.057	0.070	0.028		
Saddlery	−0.616	0.843	1.366	0.032	−0.014	−0.059	830	38,600
(S.E.)		1.168	0.681	0.193	0.212	0.061		
Woolen goods	14.859	7.055	−3.613	0.432	−0.856	0.315	1,610	33,200
(S.E.)		3.868	3.248	0.380	0.587	0.238		
1870								
Brewing	−1.511	−2.161	1.838	−0.881	0.669	−0.131	2,990	31,000
(S.E.)		2.115	1.287	0.382	0.365	0.103		
Lumber milling	−2.101	1.137	1.258	0.125	−0.122	−0.019	1,300	3,281,800
(S.E.)		0.412	0.367	0.053	0.061	0.026		
Saddlery	1.693	1.780	0.532	0.124	−0.202	0.022	900	5,000
(S.E.)		0.871	0.480	0.163	0.155	0.041		
Tobacco	0.019	−1.176	1.351	−0.319	0.292	−0.064	1,110	12,200
(S.E.)		0.975	0.810	0.184	0.173	0.062		
Woolen goods	−14.540	−2.016	4.365	−0.154	0.326	0.221	2,410	94,800
(S.E.)		3.436	2.555	0.333	0.487	0.180		

NOTE: (S.E.) = Standard error.
EQUATION: $\ln VA = b_0 + b_1 \cdot \ln L + b_2 \cdot \ln K + b_3 \cdot (\ln L)^2 + b_4 \cdot (\ln L \cdot \ln K) + b_5 \cdot (\ln K)^2$

TABLE 9.7
Statistically Significant Variable Scale-Elasticity Production-Function Estimates Implying Well-Behaved Average Costs, 1820–1870

Year and industry	c_0[a]	b_0	b_1	b_2
1820				
Cotton goods	−0.00832	2.834	0.292	0.809
(S.E.)			0.069	0.115
Flour milling	−0.01188	1.417	0.452	1.087
(S.E.)			0.137	0.348
Furniture	−0.00995	3.363	0.278	0.887
(S.E.)			0.125	0.449
Leather	−0.01096	2.935	0.308	0.811
(S.E.)			0.061	0.123
Lumber milling	−0.01161	1.441	0.504	0.764
(S.E.)			0.128	0.253
Saddlery	−0.00264	4.959	0.145	0.871
(S.E.)			0.126	0.434
Sheet metals	−0.00857	3.697	0.310	0.713
(S.E.)			0.142	0.344
Wagon making	−0.00571	3.635	0.368	0.669
(S.E.)			0.122	0.309
1850				
Flour milling	−0.00752	0.339	0.518	0.688
(S.E.)			0.109	0.041
Leather	−0.00944	2.139	0.559	0.479
(S.E.)			0.074	0.034
Liquor	−0.00788	1.363	0.352	0.656
(S.E.)			0.242	0.102
Lumber milling	−0.00639	0.288	0.302	0.779
(S.E.)			0.050	0.031
Woolen goods	−0.00720	2.390	0.845	0.379
(S.E.)			0.158	0.114
1860				
Brewing	−0.00264	0.226	0.139	0.869
(S.E.)			0.193	0.101
Flour milling	−0.00666	1.048	0.811	0.591
(S.E.)			0.098	0.038
Leather	−0.00698	2.314	0.657	0.461
(S.E.)			0.104	0.055
Lumber milling	−0.00628	1.188	0.502	0.620
(S.E.)			0.043	0.029
Saddlery	−0.00792	2.781	0.582	0.441
(S.E.)			0.100	0.057
Sheet metals	−0.00839	1.710	0.496	0.588
(S.E.)			0.126	0.066
Woolen goods	−0.00494	2.986	0.677	0.400
(S.E.)			0.198	0.159

TABLE 9.7 (*continued*)

Year and industry	$c_0{}^a$	b_0	b_1	b_2
1870				
Brewing	−0.00626	1.666	0.616	0.592
(S.E.)			0.194	0.098
Cotton goods	0.00592	2.656	0.903	0.435
(S.E.)			0.308	0.251
Flour milling	−0.00720	0.718	0.520	0.652
(S.E.)			0.104	0.032
Liquor	−0.00548	1.399	0.630	0.612
(S.E.)			0.286	0.128
Lumber milling	−0.00549	0.706	0.416	0.705
(S.E.)			0.047	0.031
Meat packing	0.01000	−2.362	0.197	1.262
(S.E.)			0.372	0.235
Woolen goods	−0.00520	0.894	0.455	0.634
(S.E.)			0.218	0.175

NOTE: (S.E.) = standard error.
EQUATION: $\ln VA + c_0 \cdot (\ln VA)^2 = b_0 + b_1 \cdot \ln K + b_2 \cdot \ln L$

[a] All estimates significantly different from zero at the five percent level. As the likelihood function is asymmetric, confidence intervals or pseudostandard errors are not reported.

10

The Structure of the Capital Stock in Economic Growth and Decline

The New Bedford Whaling Fleet in the Nineteenth Century

LANCE E. DAVIS, ROBERT E. GALLMAN,
and TERESA D. HUTCHINS

I. Introduction

In his presidential address to the International Economic History Association, Simon Kuznets characterized technical change as the "major permissive source of economic growth."[1] A few years later a then colleague of Kuznets, W. E. G. Salter, produced a path-breaking work on the connections among economic growth or decline, the structure of the capital stock, embodied production techniques, and productivity.[2] While most economists agreed with Kuznets concerning the importance of technical change, few attended closely to the questions raised by Salter. Subsequent events have shown they were pertinent questions.

The term *Rust Bowl* has become almost as much a part of the recent American patois as the descriptive label for the Great Plains from which the term was derived, and the relationship between the two terms goes beyond phonics to include patterns of deterioration and response. The erosion of the land that had darkened the skies from Kansas City to Denver made it impossible to continue dry farming on the plains and touched off the great migrations to the West Coast. The

---

Most of the research on which this paper is based was funded by the NBER research project on Productivity and Industrial Change in the World Economy. Grants to Gallman from the National Science Foundation and the Kenan Foundation helped finance the construction of the capital stock estimates. The Division of Humanities and Social Sciences at the California Institute of Technology provided funds for the collection of the labor data. Billie Norwood, Kathi Gallagher, and Karin Gleiter helped put the data in machine-readable form and helped with the programming. Paul Cyr, of the New Bedford Public Library was a kind and helpful guide through the whaling data.

[1] Simon Kuznets, Presidential Address delivered to the Third International Economic History Conference, Munich, Germany, 1965.

[2] W. E. G. Salter, *Productivity and Technical Change*, (Cambridge: Cambridge University Press, 1960).

problem created by the Dust Bowl directed scientific attention to a set of ecological problems, and the resultant research led to a series of innovations that mitigated some of the adverse effects of agricultural activity on the plains.

In a not dissimilar manner, developments of the 1950s–1970s led to an erosion of the industrial capital stock and produced a new set of socioeconomic problems (unemployment, balance-of-payments disequilibrium, and the social costs associated with the geographic displacement of economic activity, to cite only three). Just as the plight of th Joads caused a redirection of agricultural research, unemployment lines have made economists more interested in the relationships among Salter's variables. They have begun to worry about productivity, in particular, but they have not yet devoted much attention to the structure and evolution of the capital stock. To do so requires one to shift attention from the performance of the aggregate economy to that of the individual firm and its decisions about the growth of capital. At this micro level of analysis it is possible to observe the impacts of demand shifts and supply-side shocks on the level and structure of the capital stock, on embodied techniques, on productivity, and on employment opportunities.

An approach of this type to modern problems calls for new types of evidence—evidence not readily or widely available. There are as yet no studies based on modern experience; but this paper, although rooted in the past, is based on such evidence; and it has modern "Salterian" overtones. It indicates that a Salter style of analysis utilizing firm-level data can be applied to such problems. To the extent that the demonstration is successful, it is hoped that the work will stimulate more interest in the application of Salter's work to current issues. This is not to say that the substance of the historical problem is trivial—far from it. The problem is important historically and has attracted enough interest to produce a rich literature.

The New England whaling fleet rose to world dominance and then completely collapsed within the span of the century between the end of the War of 1812 and the outbreak of the First World War. The story, as it is conventionally told, certainly has a modern ring: accompanying the rise and decline of the industry—presumably producing them—were shifts in the level and structure of demand, as well as changes in the competitive vigor of West Coast whaling ports and foreign fleets; supply-side shocks, both positive (the discovery of new hunting grounds) and negative (the depletion of whale stocks); the development of new systems of organization (the rise of Honolulu as a transshipment and refitting point), a response to new hunting opportunities and problems; the search for and innovation of appropriate vessel sizes and types as well as hunting techniques, again a response, in part, to shifts in the relative importance of different hunting grounds; and finally, changes in the cost and quality of labor and capital.

Problems of this sort were precisely the type that particularly interested Salter: an industry rises rapidly and then declines; and the rise and decline carry with them important consequences for the level and structure of the capital stock and therefore changes in embodied technique, productivity, and employment. This paper focuses on these issues, but it also investigates the nature of the forces encouraging expansion and collapse, and the responsiveness of the industry to its opportunities and problems.

Section II sketches out the history of the industry and section III describes the statistical data to be analyzed, data bearing on the principal New England port, New Bedford. Section IV examines the characteristics of the New Bedford fleet in detail, while section V describes the evolution of the fleet, looking in particular at marginal changes. Section VI considers how the activities described in sections IV and V affected productivity, while section VII relates the decisions underlying the actions described in sections IV and V to questions of profitability. To put it another way, section IV is concerned with the structure of the New Bedford fleet, section V with its activities, section VI with the effects of these activities on productivity, and section VII with the extent to which these activities seem to have been guided by considerations of profit. Section VIII summarizes our conclusions.

II. Historical Background

Whales were hunted for sperm oil, whale oil, and whale bone. The occasional captive yielded a windfall—ambergris, a prized perfume fixative—but no one hunted specifically for this product, and no comprehensive records were kept on either the amount or value of ambergris brought back. During the early decades of the nineteenth century, sperm oil was primarily valued as an illuminant. In solid form, it was used for the highest-quality candles, and in its liquid state, in lighthouses and public buildings. In addition, it served as a lubricant for high-speed light machinery. The expansion of sperm oil production after 1820 matches very closely the growth of the cotton textile industry.

Whale oil obtained from baleen whales was the illuminant chosen by the average consumer from the 1820s to the 1850s. Technically inferior to sperm oil both as an illuminant and as a lubricant, it remained (because it was cheap) the most popular lamp fuel throughout most of the first half of the century. Later, as the pattern of demand changed, it became increasingly important as a lubricant for heavy machinery.

Baleen or whale bone is not true bone, but the sieve that the animal uses to screen seawater and remove food. Whale bone was used by people whenever there was a need for a strong but flexible material: in the manufacture of corset stays, hoops, whips, umbrellas, and carriage shades, to cite only a few examples.

Herman Melville notwithstanding, the whale is a mammal (order Cetacea).[3] Even among those who doubt this assertion, it is generally agreed that there are only two types of proper whales: the bone whale or baleen (suborder Mysticeti) and the toothed or cachelot (suborder Odontoceti). During the nineteenth century one type of cachelot (the sperm whale) and three types of baleens (rights, bowheads, and if all else failed, humpbacks) were actively hunted.[4] In the first half of the period, whalers concentrated on sperms and rights, and in the second, on sperms and bowheads.

The four varieties typically lived in different parts of the world's oceans. Sperms were most often found in the tropical and subtropical portions of the Atlantic, Pacific, and Indian oceans. Rights preferred cooler waters: the North Atlanic from Bermuda to Greenland, the Pacific from Japan to the northwest coast of the United States and as far north as the Arctic Ocean, and the South Atlantic and Pacific from Brazil and Chile to the Antarctic Ocean. Bowheads were usually found in the Arctic Ocean, in the Bering Strait, and in the Pacific north of the 54th parallel; but since they were able to swim from the "Pacific Arctic" to the "Atlantic Arctic," they were occasionally found in Davis Strait as well. There is some evidence that the Dutch and the English may have overhunted the bowhead in the latter ground during the eighteenth century. Such overhunting could account for the relatively small numbers subsequently found in the area. Humpbacks, like sperms and rights, preferred more temperate waters; however, they tended to swim closer to shore, particularly when breeding.

Sperm oil production increased rapidly from 1820 to 1837 and remained at a plateau until 1845 (see Table 10.1).[5] From then until the

---

[3] "Be it known that, waiving all argument, I take the good old fashioned ground that the whale is a fish, and call upon holy Jonah to back me." *Moby-Dick; or The Whale* (Berkeley: University of California Press, 1979), p. 136.

[4] The humpback was avoided because it sank when killed. Not only was there risk of losing the catch, but unless the harpooneer moved quickly, the boat and crew could go down as well.

[5] From the end of the War of 1812 until the last decade of the century the American whaling catch probably constituted more than 80 percent of the world's total. There had been a British fleet at the beginning of the period and there was some Australian activity later. The British effort, however, contracted after 1820, while the Australian

TABLE 10.1
U.S. and New Bedford Whaling Fleets: Yearly Average Catch, 1816–1905

	New Bedford					All U.S.					Real prices[e]			Price ratios	
	SPOIL[a]	WHOIL[b]	BONE[c]	Real value[d]	$ per ship ton	SPOIL[a]	WHOIL[b]	BONE[c]	Real value[d]	$ per ship ton	SPOIL	WHOIL	BONE	P_SPOIL / P_WHOIL[g]	P_WHOIL[g] / P_BONE[h]
1816–20	129	299	6	193	29.34	723	820	50	776	42.20	0.66	0.36	0.08	1.83	4.45
1821–25	575	633	15	526	40.51	2132	1606	101	1726	46.45	0.57	0.31	0.13	1.81	2.46
1826–30	919	783	117	912	38.51	2437	1781	280	2320	48.38	0.70	0.31	0.22	2.23	1.42
1831–35	1498	1616	43	1767	38.07	3660	4512	459	4557	49.13	0.84	0.31	0.18	2.76	1.71
1836–40	1885	2106	89	2292	40.72	4588	6269	1796	6229	45.72	0.85	0.32	0.18	2.66	1.75
1841–45	1848	2068	509	2889	37.33	4536	7539	2271	8490	45.72	1.00	0.41	0.38	2.41	1.10
1846–50	1445	2895	905	3411	41.77	3295	7888	2750	8597	41.26	1.24	0.45	0.35	2.73	1.31
1851–55	1363	3749	1848	5101	49.26	2694	7413	3196	9892	50.48	1.44	0.63	0.42	2.30	1.49
1856–60	1608	3302	1269	5348	50.39	2557	5931	1897	8824	45.09	1.39	0.62	0.84	2.26	0.73
1861–65	1234	1663	386	2970	40.24	1410	2805	734	4275	38.45	1.21	0.67	0.94	1.79	0.72
1866–70	966	1705	566	2616	47.80	1451	2438	827	3842	52.47	1.28	0.57	0.72	2.23	0.79
1871–75	969	935	314	1855	45.09	1283	1376	344	2465	49.50	1.16	0.50	0.84	2.33	0.59
1876–80	997	824	284	2107	60.33	1281	958	254	2407	60.80	1.07	0.48	2.27	2.21	0.21
1881–85	691	721	182	1487	63.46	828	916	356	2159	62.12	0.94	0.52	2.54	1.80	0.20
1886–90	500	146	148	1050	112.56	577	695	317	2015	75.49	0.81	0.43	3.94	1.86	0.11
1891–95	187	27	8	213	56.22	467	309	295	2092	101.82	0.84	0.52	5.22	1.62	0.10
1896–1900	173	10	6	143	57.13	429	145	232	1317	99.28	0.66	0.49	4.15	1.33	0.12
1901–05	275	11	13	247	98.52	536	85	97	871	97.82	0.64	0.42	5.07	1.53	0.08
ALL YEARS	959	1305	373	1952	46.49	1938	2971	903	4047	48.76	0.98	0.46	1.59	2.22	0.32

SOURCES: *New Bedford*, the Davis-Gallman-Hutchins tape. *All U.S.*, Walter S. Tower, *History of the American Whale Fishery* (Philadelphia: University of Pennsylvania Press, 1907). *Real Prices*, whale product prices (Tower, ibid.) divided by the wholesale price index (Warren-Pearson).

[a] Sperm oil, 000s of gallons.
[b] Whale oil, 000s of gallons.
[c] Bone, 000s of pounds.
[d] $000.
[e] 1880 = 100, price per gallon for spoil and whoil, per lb. for bone.
[f] Price of sperm oil per gallon.
[g] Price of whale oil per gallon.
[h] Price of whale bone per pound.

end of the century output followed a gentle downward trend. The growth was triggered by the domestic market; before 1845, in every year but one, exports represented less than 10 percent of total demand. After that date, exports grew in both relative and absolute terms. The British market accounted for over 85 percent of total U.S. exports, and those shipments increased by more than 50 percent between 1840 and 1852 and by an additional 130 percent between then and 1861. After the Civil War, the British market all but disappeared, and American production contracted rapidly. The real price of sperm oil tracked production fairly closely. Real prices more than doubled between 1820 and the mid-fifties, but by the end of the period, they had returned to their original level.

In the case of whale oil, the early history is similar, but the period of increasing production (an increase tied in part to a rapidly expanding export market that absorbed about half of American production) lasted until the early 1850s. From that date onward, production declined and exports absorbed an ever-smaller proportion of the shrinking total. The price profile for whale oil looks similar to that for sperm oil in the early decades (the real price doubled between 1820 and the mid-1850s); but thereafter, although prices did decline, the rate of decrease was lower. At the end of the period the real price was still about a third above its initial level. The ratio of sperm to whale oil prices was in excess of 2 to 1 over most of the period. By the beginning of the present century, it had fallen to 3 to 2.

Given the technical complementarity between whale oil and whale bone, a comparison of the time pattern of production of the two commodities produces some surprises. Although in general the pattern for bone is one of rise, stability, and decline (similar to the patterns for both sperm and whale oil), the timing is different, and the increase is

---

totals were always relatively small (in the five years 1836–1840 the British production of sperm oil was about one-fifth the American and of whale oil about one-tenth). The Norwegians and the Dutch, as well as the English, had hunted extensively in the seventeenth and eighteenth centuries (there are records of Dutch whaling off Spitzbergen as early as 1596, and the British Moscovy Company was granted a monopoly on whaling in 1613), but their fleets had all but disappeared before the Americans had more than a few ships in the North Atlantic. The Dutch never reentered. The Norwegians did not reemerge as a significant force until the last decade of the century. (In 1881, for example, Norwegian production of sperm and whale oil was about 9 percent of U.S., but in 1868 it had been less than one-half percent.) The new Norwegian thrust temporally overlapped American whaling, but the Norwegians provided, at most, marginal competition. They sought meat and margarine rather than oil and bone, and they chiefly hunted the denizens of the Antarctic grounds—the rorquals (blue, sei, fin, and minke whales), creatures that swam too fast to be successfully hunted with the American techniques.

less steady. It begins slowly but accelerates rapidly after the 1830s; total production tracks the whale oil total downward after the 1860s, but it falls much less rapidly at the end of the period. The ratio of oil to bone was, thus, not as constant as a study of biology might suggest. In the earliest decade a typical catch yielded 16 barrels of oil for each pound of bone, but by 1840 the average was 3 to 1. That proportion remained fairly constant through the middle decades, but after 1880 it began to fall again, and by the early twentieth century it was less than 1. The change in the ratio reflects, at least in part, the shift in the relative price of oil and bone. As demand (particularly export demand) increased, the real price of whale bone rose from 8 cents to $5.07 a pound; and the ratio of the price of a barrel of oil to a pound of bone fell from almost 4.5 to 1 to less than 1 to 10. The end was in sight, however, and even before World War I, the emergence of specialty steels and changes in consumer tastes drove the price of bone back to its 1820 levels.[6]

The changing ratio of oil to bone in the catch brought back by the New Bedford captains reflects both shifts in the types of whales hunted (bowheads yielded more bone per barrel of oil than did rights) and changes in production practices. Although whalers regarded the taking of a whale just for its baleen or just for its oil as sinfully wasteful, the data suggest that many whalers were prepared to sin, at least a little. At times baleen was jettisoned to make room for oil, and at other times it was the oil that was sacrificed.

Over the ninety years between 1816 and 1905, the New Bedford fleet represented about one-half of the total capacity of the American fleet (see Table 10.2). During the middle decades, when the industry was at its maximum size, the city's proportion was nearer three-fifths, but early and late it was substantially less important. Before the 1830s, Nantucket and other New England ports were important rivals; the Nantucket fleet was the largest in the United States until the early 1820s. After 1880, the New Bedford whalers lost out to vessels based in San Francisco (see Table 10.3). The western city had a substantial locational advantage for the Arctic and a somewhat lesser one for the

---

[6]The price of bone also responded to developments in the whale oil market, as J. R. McCulloch pointed out with respect to the British industry of an earlier date: "Oil, being the main product of the fishery, regulates its extent; which being diminished by the low price, the quantity of whalebone is lessened, while the demand for it continuing as great as before, the value consequently rises." *A Dictionary of Commerce* (London: Longman, Brown, Green, and Longmans, 1842), p. 1238. By the end of the nineteenth century, the "main product" of the American fishery was no longer oil. Nonetheless, the point holds.

TABLE 10.2
Annual Average Tonnage: U.S. and New Bedford Whaling Fleets,
1816–1905

	Total tons	New Bedford tons	New Bedford/total (%)
1816–20	18389	6650	36.1%
1821–25	37161	13162	35.4
1826–30	47953	23885	49.8
1831–35	92750	46659	50.3
1836–40	133897	56585	42.5
1841–45	185678	73007	39.3
1846–50	208347	82659	39.7
1851–55	195938	106105	54.2
1856–60	195692	109805	56.1
1861–65	116565	77734	66.7
1866–70	79870	60127	75.3
1871–75	55334	46020	83.2
1876–80	45655	40503	88.7
1881–85	40278	27365	67.9
1886–90	31566	11028	34.9
1891–95	25194	4648	18.6
1896–1900	17605	3185	18.1
1901–05	11017	3101	28.1
ALL YEARS	86694	44046	50.8%
1816–45	85971	36665	42.6
1846–75	144719	80408	55.6
1876–1905	27968	14972	54.1

SOURCES: See Table 10.1.

Pacific grounds. With the opening of the transcontinental railroad, its vessels had rapid access to both East Coast and European markets.

The value of the U.S. industry's total output increased from about $1.2 million per year in the first decade after the War of 1812 to almost ten times that amount in the 1850s. After that date the rising price of bone dampened the decline in the value of output, but physical production declined rapidly. In the decade that spans the end of the nineteenth and the beginning of the twentieth century, the total production of sperm oil was about 18 percent, whale bone about 6 percent, whale oil less than 2 percent, and the total value of output about 12 percent of the midcentury peak levels. By the turn of the century, total revenues had fallen to about $1.1 million a year.

The New Bedford vessels ranged over six of the seven seas. While no one searched for whales in the Mediterranean, the ships' logs listed fifty-one separate hunting grounds, including such exotic places as Patagonia, Delago Bay, the Okhotsk Sea, and Altaho. For analytical purposes the fifty-one have been grouped into five major grounds:

TABLE 10.3
Whaling Fleet Size

PANEL A: VESSELS SAILING FROM NEW ENGLAND PORTS, 1820–1860

	1820	1825	1830	1835	1840	1845	1850	1855	1860
Number of Vessels									
Nantucket	45	27	21	20	81	77	62	43	21
New Bedford	36	33	66	77	177	239	238	314	301
Other	13	14	34	49	114	132	109	109	97
Percentages									
Nantucket	47.9%	36.4%	17.4%	13.7%	21.8%	17.2%	15.2%	9.2%	5.0%
New Bedford	38.3	44.6	54.5	52.7	47.6	53.3	58.2	67.4	71.8
Other	13.8	18.9	28.1	33.6	30.6	29.5	26.7	23.4	23.2

PANEL B: WHALING VESSEL TONNAGE, NEW BEDFORD AND SAN FRANCISCO, 1871–1905

	1871–75	1876–80	1881–85	1886–90	1891–95	1896–1900	1900–05
Annual Average Tonnage							
New Bedford	46020	40503	27365	11028	4648	3185	3101
San Francisco	740	472	3837	7289	9117	6320	3696
Percentages							
New Bedford	98.4%	98.8%	87.7%	60.2%	33.8%	33.5%	45.6%
San Francisco	1.6	1.2	12.3	39.8	66.2	66.5	54.4

SOURCES: Alexander Starbuck, *History of the Whale Fishery*, and Reginald B. Hegarty, *Returns of Whaling Vessels Sailing from American Ports*.

(1) the Atlantic, (2) Hudson's Bay, (3) the Indian Ocean, (4) the Pacific Ocean, and (5) the Arctic. Although their relative importance changed from year to year, over the entire nine decades the largest number of vessels sailed to the Pacific. That ground was followed in order of importance by the Atlantic, the Indian Ocean, the Arctic, and lastly, by Hudson's Bay.

There was, over the course of the nineteenth century, little change in the basic organization of the industry or the process of catching whales, although there were important changes in detail. The vessel's owners (often including the captain) hired an agent who was responsible for signing on a crew, provisioning the vessel, and otherwise preparing it for sea. The agents, the captain, and even the owners of a vessel frequently changed from one voyage to the next. Thus, a vessel voyage is a useful unit of analysis, approximating the notion of a "firm" in economic theory.

The captain managed affairs once the vessel was at sea. While his responsibilities included any refitting, reprovisioning, and recruiting that was required, his chief task was to find whales and catch them.

Since virtually all of the New Bedford vessels traveled under sail (the four steam ships have been excluded from the data set), capture could not be managed from the vessel itself. Thus search and capture were separate and distinct operations. Once the whale had been sighted and approached, small boats were lowered; the attack was launched from them. A few men were left on the vessel to keep it under way, to watch the drama from the masts, and to signal to the boats any of the whale's movements that could not be seen from water level.

The specialists in the boats were the steersman and the harpooneer. The latter harpooned the whale and made it fast to the boat by a line attached to the harpoon and then, once the whale had tired and been reapproached by the boat, the steersman killed it with a lance. This method was not the only way to handle the task—Japanese hunters drove the whale into shallow water, leaped on its back, and stabbed it to death, while Norwegian hunters of the late nineteenth century pursued their quarry by steam vessel and shot it with a harpoon gun mounted on the deck—but harpooning from a small boat was the way usually adopted by the Americans.[7]

With the whale dead and floating, the whaling vessel hove to. A work stage was rigged, and the crewmen sliced off the blubber with cutting implements called spades. Ambergris and baleen (if any) were removed and the blubber was loaded aboard. There it was "tried" out—the oil boiled out over a fire built in a brick "try" works—and barreled. Necessarily, the crew included artisans who could build barrels, repair the try works, and perform other skilled tasks.

In its broad outlines, this procedure was unchanged in the period analyzed, but the technology employed in taking the whale did change. Charles Scammon, noted naturalist and a former whaling captain, reported in 1876: "There has been as great a revolution in the mode of killing whales during the last twenty years, as there has been in the art of naval warfare; were it not for this, but few whalers would now be afloat."[8]

Through the 1850s, most harpooneers used hand-held harpoons and lances. A harpoon gun was invented early, and it had the advantage of getting the harpoon to the whale quickly. It was, however, inaccurate, difficult to load, and its kick could knock the harpooneer overboard or even capsize the boat. Not until the breechloader or, better yet, the bomb

---

[7] The Japanese hunted little in the nineteenth century. The Norwegians, with their steamboats, were able to pursue the fast-swimming rorquals and did so, leaving the other baleens and the sperms mainly to the Americans.

[8] Charles M. Scammon, *The Marine Mammals of the Northwestern Coast of North America* (New York: Putnam, 1874), p. 206.

lance (a weapon that harpooned and killed in one act) were invented in the 1850s did harpooneers submit to mechanization. The dart gun invented in the 1860s proved even more deadly and it was widely adopted after 1880. In addition, the innovation of steam winches (again about midcentury) greatly aided the process of "flensing off" the blubber from the carcass. Our data set does not show the types of equipment used on each voyage. The general sources, however, indicate that except for the harpoon gun, the innovations were adopted quickly; and statistical analysis suggests that they did improve productivity.[9]

Through the course of the nineteenth century, the industry was subject to two types of shocks: (1) external market changes that altered the relative prices faced by the firms and (2) changes in the technical, institutional, and natural environment that affected the resource cost of production and the ability of the New England whalers to compete. In terms of the market shifts, the ninety years between 1816 and 1906 saw significant changes in both factor and output prices. The whaling literature indicates that the quality of crews declined, and this erosion of labor quality may have raised the cost of labor services relative to the cost of capital. At the same time, the industry's output prices were also, as previously noted, subject to substantial long-term movements. The real price of whale and sperm oil rose rapidly over the first forty years as the demand for illuminants and lubricants grew, but the price fell almost equally quickly as first coal oil and then petroleum and gas emerged as important competitors. The real price of whale bone, on the other hand, rose almost exponentially as its superiority as a strong but flexible material became more widely appreciated.

The industry was also confronted with a series of environmental, technical, and institutional changes. The stock of animals hunted was subject to depletion, but throughout the period new grounds were being opened to exploitation: grounds offering new supplies, but also posing new hunting and organizational problems. Entry into the Indian and Pacific Oceans meant that vessels took longer to reach the prime hunting grounds and were longer at sea. Larger vessels evolved, and efforts were made to systematize the shipment home of products during a voyage. Honolulu and Lahaina emerged as stations where a vessel long at sea could off-load product and arrange for its shipment, refit, replace crew members, and generally prepare for another period of hunting. The opening of the Bering Strait and the Okhotsk Sea—the Arctic areas north of the Pacific—called for more maneuverable ves-

[9] L. E. Davis, R. E. Gallman, T. D. Hutchins, "Productivity in American Whaling: The New Bedford Fleet in the Nineteenth Century," manuscript.

sels and hunting devices gauged to deal with the efforts of whales to seek shelter under the ice. Finally, with the completion of the transcontinental railroad, the New England firms found themselves facing new competitive threats from whalers based on the Pacific coast.

Over the ninety years covered by this study, the interaction between the New England businesses and these market, technical, environmental, and institutional changes underwrote a period of very rapid growth followed by a period of almost equally rapid decline. This study focuses on a subset of the industry—the ships sailing from New Bedford, whose vessels mirror closely the entire New England industry. The city's fleet, almost nonexistent in 1815, rose to become the largest in the world, and then retreated to a level hardly greater than its initial position. It is this history—the exogenous displacements and their effect on the capital stock, productivity, and economic returns—that is the immediate focus of this paper. More fundamentally, it is the lessons that the history of New Bedford whaling offers to modern economists, concerned with modern problems, that is of primary interest.

III. Variables, Data, and Sources

Six major characteristics of whaling vessels are identifiable in the sources from which our data are drawn. First of all, five types of sailing vessels participated in the New Bedford fleet: sloops, schooners, brigs, barks, and ships (see Table 10.4). The sloop and schooner were small, fore-and-aft–rigged vessels, the former with one mast, the latter with two. Their activities were very nearly confined to the North Atlantic, Davis Strait, and Hudson's Bay, and they composed only a tiny fraction of the New Bedford fleet. The brig, a two-masted square-rigged vessel, ventured as far as the South Atlantic and occasionally farther; but it, too, was relatively unimportant. These three types of vessels have been grouped together in the category "other" in the statistical analysis.

The bark (39 percent of the New Bedford vessels) and the ship (56 percent) were larger, three-masted vessels. The latter was square-rigged, the former, square-rigged on the fore and main masts, but fore-and-aft rigged on the mizzen. Ships, normally the larger of the two, were in the second half of the period not infrequently rerigged as barks. Rerigging gave those vessels greater maneuverability. It was a relatively simple operation (just replacing the square-rigged mizzen with a fore-and-aft rig) and could be carried out quickly. Much less time was needed to rerig than to build a new vessel. The literature asserts that maneuverability made the bark valuable in the Arctic areas,

TABLE 10.4
Vessel-Class Composition, New Bedford Fleet, 1816–1905
(Annual averages)

	Number of vessels				Tonnage					% tonnage		
	Total	Ships	Barks	Other	Unknown	Total	Ships	Barks	Other	Unknown	Ships	Barks	Other & Unknown
1816–20	24.8	18.4	1.4	5.0		6650	5505	378	767		82.8%	5.7%	11.5%
1821–25	46.6	38.0	3.4	5.2		13162	11733	658	771		89.1	5.0	5.9
1826–30	80.0	66.0	6.4	7.6		23885	21313	1513	1059		89.2	6.3	4.4
1831–35	147.0	120.2	21.0	5.8		46695	40423	5388	884		86.6	11.5	1.9
1836–40	178.2	140.4	28.8	9.0		56585	47785	7564	1236		84.4	13.4	2.1
1841–45	227.0	175.8	45.0	6.4		73007	60356	11833	818		82.7	16.2	1.1
1846–50	253.2	190.6	59.2	3.4		82659	66402	15744	513		80.3	19.0	0.6
1851–55	315.0	206.6	103.4	3.4	1.6	106105	75971	29291	517	330	71.6	27.6	0.8
1856–60	323.2	178.6	141.6	1.0	2.0	109805	66417	42833	142	413	60.5	39.0	0.5
1861–65	233.6	108.0	122.0	1.8	1.8	77734	39624	37479	261	370	51.0	48.2	0.8
1866–70	187.0	58.2	122.0	5.8	1.0	60127	21501	37617	809	200	35.8	62.6	1.7
1871–75	143.4	40.2	97.2	5.6	0.4	46020	14878	30291	771	80	32.3	65.8	1.8
1876–80	135.8	29.4	89.0	17.4		40503	10777	27293	2433		26.6	67.4	6.0
1881–85	93.6	17.6	60.2	15.8		27365	6620	18565	2180		24.2	67.8	8.0
1886–90	40.4	7.2	24.6	8.6		11028	2798	7059	1171		25.4	64.0	10.6
1891–95	19.6	2.2	12.0	5.4		4648	822	3134	692		17.7	67.4	14.9
1896–1900	15.0	0	9.2	5.8		3185	0	2380	805		0.0	74.7	25.3
1901–05	14.0	0	8.4	5.6		3101	0	2233	868		0.0	72.0	28.0
AVERAGE													
ALL YEARS	137.6	77.6	53.0	6.6	0.4	44046	27385	15625	959	77	62.2	35.5	2.4%
1816–45	117.2	93.1	17.6	6.5		36665	31186	4556	923		85.1%	12.4%	2.5%
1846–75	241.5	130.4	107.6	3.5	1.1	80408	47465	32209	502	232	59.0	40.1	0.9
1876–1905	53.1	9.4	33.9	9.8		14972	3503	10111	1358		23.4	67.5	9.1

SOURCE: Davis-Gallman-Hutchins tape.

where ice was a constant threat and where agility in a vessel was valued; twice the Arctic fleet was caught and crushed because the ocean froze so rapidly.

The other five characteristics that the data capture are crew size and skill composition; size of vessel (tonnage, a capacity measure);[10] age of vessel; length of time the vessel had served in the New Bedford fleet; and mode of entry into and exit from the fleet—some vessels were built for the fleet, some were rerigged, while others transferred in from the merchant marine or other whaling ports, and some vessels transferred out, some were lost at sea, and some were condemned.

The data were drawn from a number of sources. The most important is a manuscript (stored in the Baker Library at the Harvard Business School) prepared by Joseph Dias, a whaling captain.[11] Dias collected

---

[10] In 1865 a new method of measuring capacity was adopted. As vessels returned to port, they were officially remeasured, the process of converting the whole fleet spreading over several years. The definition of a ton of capacity was changed, from 95 cubic feet to 100 cubic feet, and a new method for measuring the cubic footage of a vessel was also adopted. Unfortunately, the precise ways in which the new and old systems of establishing cubic footage differed are unknown.

According to *Historical Statistics* (1975 Bicentennial edition, vol. 2, p. 743), the combination of changes in the definition of a ton and in the method of measuring cubic footage probably reduced the official capacities of brigs, schooners, and sloops but raised the official capacities of ships and barks. Our research in Starbuck and Dias (see below, text, and footnotes 11 and 13) demonstrates that these generalizations cannot be extended to whaling vessels. The new system of admeasurement apparently reduced the official capacities of whaling ships by about 5 percent, on average; lowered the capacities of barks by about one-sixth; and reduced the capacities of brigs, schooners, and sloops even more. In the absence of actual reports of these tonnages, we have no exact way of specifying for each vessel the capacity that would have been assigned to it under old and new admeasurement systems.

Fortunately for us, Dias, our basic text (see below, text, and footnote 11), standardized on the old capacity measurements whenever he could. That is, if he had an old capacity measurement for a vessel—and for most of them he did—he almost invariably chose to use it, in preference to the new capacity measurement for that vessel. Research in Starbuck and Dias (see below, text, and footnotes 11 and 13) permitted us to adjust the capacities of most of the remaining vessels, so that they also were expressed in terms of the old system. The capacities of the handful of vessels left were adjusted from the new to the old basis by use of formulas developed from data on vessels for which we had both old-style and new-style measurements:

$$\text{ships:} \quad \text{Old} = -85.018 + 2.11 \text{ New} - 0.00217 \text{ (New)}^2$$
$$\text{barks:} \quad \text{Old} = 13.37 + 1.44 \text{ New} - 0.00113 \text{ (New)}^2$$
$$\text{others:} \quad \text{Old} = 1.4 \text{ New}$$

Throughout this paper, then, tonnage refers to a measure of capacity established under rules and definitions in force before 1865.

[11] J. Dias "The New Bedford Whaling Fleet 1790–1906," manuscript on deposit at Baker Library, Harvard Business School. For a discussion of the manuscript see Anonymous, "The New Bedford Whaling Fleet 1790–1906" *Bulletin of the Business History Society* 6 (Dec. 1931), pp. 9–14.

data covering voyages conducted as early as 1797 and as late as 1906. Since the trade was much disturbed by the embargo and the War of 1812, the pre-1816 data have been ignored. The Dias data are comprehensive until fairly late in the period. In the later years they have been augmented with material drawn from the work of Reginald B. Hegarty.[12] For the earlier years Alexander Starbuck's account provided checks, and permitted a few gaps to be filled.[13] Additional information concerning the origin of vessels entering the fleet and the destination of those leaving it was obtained from the official ship registers of New Bedford.[14]

The labor data are drawn from the whaling-crew lists, and contain information on the number of men who shipped out on particular whaling voyages between 1807 and 1925. The lists were compiled from three separate sources, the coverage varies, and not every voyage is included.[15] Two caveats: (1) the study's coverage of labor is much better in the middle years (1840–1880) than it is for either the beginning or the end of the period; (2) the data refer primarily to crew members recruited in New Bedford. There is little information on additions made after departure, but this shortcoming appears to be of little importance for present purposes (see below, and especially footnote 39).

IV. The Technical Characteristics of the New Bedford Fleet

The British blockade had so effectively throttled the embryonic whaling industry that by 1814 the entire American fleet totaled but 562 tons, and only two vessels returned in that year (one to New Bedford) with any catch.[16] In the first full peacetime year, 1816, the New Bedford fleet comprised 12 vessels (9 ships, 1 bark, and 2 brigs); but over the ensuing four decades, growth was rapid. The number of ships rose

[12] Reginald B. Hegarty, *Returns of Whaling Vessels Sailing from American Ports* (New Bedford: Old Dartmouth Historical Society and Whaling Museum, 1959).

[13] Alexander Starbuck, "History of the Whale Fishery," *Report of the U.S. Commission on Fish and Fisheries, 1878*, part 4.

[14] The Survey of Federal Archives, Division of Professional and Service Projects, Works Project Administration. *Ship Registers of New Bedford, Massachusetts* (Boston: The National Archives Project, 1939–1942).

[15] One series is drawn from the National Archives and covers two periods: 1807–1834 and 1840–1900. These lists include considerable personal data about the crew members (birthplace, age, height, eye and skin color) and data on both outward-bound and returning crews. The second source consists of the records of the New Bedford port society, and they encompass most voyages between 1843 and 1885. This series is generally considered less reliable than the National Archives series, since it is not based on the "ship's papers." Finally, the whalemen's shipping papers provide a third source of crew lists. Fairly reliable, it is limited in scope to crewmen's names and occupations and in coverage to the years between 1841 and 1857 and the year 1866.

[16] Starbuck, "History of the Whale Fishery," pp. 121 and 663.

to a peak of 214 in 1853, and barks rose to a total of 145 seven years later. Those levels were never reached again, and by 1905 the fleet had shrunk to 10 vessels (there were no ships, 6 barks, and 4 brigs, sloops, and schooners).

Even during the prosperous 1840s and 1850s, the composition of the fleet was changing. Ships dominated in the early years and through most of the period of expansion. Gradually, however, the number of barks rose in both absolute and relative terms. Rare before 1820, they increased in number slowly over the succeeding twenty years; but in the middle decades barks emerged as the dominant technology. In the years before 1845, they accounted for one-eighth of the total tonnage; their share rose more than three times over the next thirty years; and they represented more than two thirds of the total in the last third of the period. In 1846 the ratio of ships to barks stood at 5.3 to 1, but by 1875 it had fallen to 0.53 to 1. Before the Civil War, the number of barks increased faster than the number of ships. Thereafter, their numbers declined the more slowly. Between 1855 and 1875, for example, the number of ships fell by 80 percent, but the number of barks declined by hardly more than 5 percent.

Brigs, sloops, and schooners were never important. They accounted for less than a fourth of the 1,049 voyages to the Atlantic and to Hudson's Bay and for only eleven of the 2,352 completed voyages to the other three grounds. During the period of rapid expansion, a few brigs, sloops, and schooners entered the fleet and remained until replaced by larger vessels. More importantly, the last two decades saw a significant substitution of "other" vessels for barks and ships in the Atlantic and, to a lesser extent, in Hudson's Bay. Dias records 115 voyages by "other" vessels after 1875; that figure is more than one-half of all voyages made by these vessels between 1816 and 1906.

Not only were there changes in the rigging of a typical whaler, but the average size of each class changed as well (see Table 10.5). For ships, size increased from less than 300 tons in the first decade to almost 380 by the early 1890s and averaged just over 340 tons. The pattern was, however, not uniform. The increases amounted to more than 5 percent per decade until the mid 1860s, but tonnage edged up only slowly thereafter. The increase in the early years was accompanied by a widening dispersion of tonnage. Thereafter, although size increased somewhat, the dispersion narrowed. The coefficient of variation increased until the mid-1850s (from 15 to 19), but fell thereafter (it was only 7 in the decade 1886–1895). It appears that larger ships were more productive, but that the industry's adjustment to the new technol-

TABLE 10.5
Technical Character of the New Bedford Whaling Fleet, 1816–1905

	Numbers			Average tons			Average age			Avg. yrs. in fleet	
	Stock	+	−	Stock	+	−	Stock	+	−	Stock	−
SHIPS										
1816–25	28.0	45.0	10.0	298	320	380	12.2	7.6	11.4	6.4	2.8
1826–35	93.0	109.0	23.0	320	342	295	16.6	11.7	17.1	7.0	5.2
1836–45	158.0	86.0	24.0	333	354	331	21.6	13.6	22.2	10.8	13.8
1846–55	199.0	76.0	79.0	352	407	350	25.7	10.5	27.4	14.5	14.3
1856–65	143.0	26.0	153.0	363	380	374	29.3	20.3	32.0	19.0	19.9
1866–75	49.0	4.0	31.0	364	370	364	35.6	3.0	41.6	24.4	28.4
1876–85	24.0	1.0	26.0	372	387	366	41.3	0.0	47.0	32.0	34.3
1886–95	5.0	0.0	9.0	379	—	388	38.6	—	42.8	34.2	36.5
1896–1905	0.0	0.0	0.0	—	—	—	—	—	—	—	—
AVG., ALL YRS.	78.0	3.9	3.9	343	347	358	25.1	12.3	30.6	14.7	17.9
BARKS										
1816–25	2.0	4.0	0.0	182	170	—	8.7	11.3	—	4.0	—
1826–35	14.0	20.0	2.0	249	261	150	15.0	11.7	8.0	6.0	5.0
1836–45	37.0	37.0	8.0	250	264	217	18.5	13.8	23.5	8.9	9.0
1846–55	80.0	96.0	27.0	261	293	259	19.3	10.7	25.3	8.3	13.6
1856–65	132.0	74.0	91.0	283	319	295	22.0	21.5	24.0	12.0	12.5
1866–75	109.0	51.0	70.0	298	302	305	29.9	24.8	32.0	17.6	17.7
1876–85	74.0	19.0	72.0	321	330	315	33.1	10.2	37.3	21.7	21.9
1886–95	18.0	0.0	21.0	295	—	313	37.9	—	38.2	29.5	31.6
1896–1905	9.0	2.0	4.0	276	373	259	47.2	26.5	57.2	37.0	51.2
AVG., ALL YRS.	53.0	3.4	3.3	285	296	297	25.8	16.3	31.7	14.9	18.2
OTHER										
1816–25	5.0	23.0	2.0	153	156	173	1.2	5.5	7.8	3.2	1.5
1826–35	7.0	13.0	13.0	137	127	126	12.2	9.8	9.5	6.6	4.0
1836–45	8.0	13.0	15.0	129	128	134	16.1	11.6	19.2	4.6	6.9
1846–55	4.0	6.0	8.0	128	132	141	10.6	12.0	13.3	3.3	4.8
1856–65	3.0	6.0	5.0	188	141	128	14.5	15.0	17.5	2.4	2.2
1866–75	6.0	11.0	7.0	120	139	148	19.8	17.9	15.6	2.7	3.0
1876–85	17.0	18.0	15.0	137	144	144	24.8	19.4	26.5	4.5	6.2
1886–95	7.0	1.0	6.0	130	188	150	29.6	ND	38.0	12.4	13.0
1896–1905	6.0	3.0	2.0	136	342	136	39.2	34.6	58.0	17.4	16.5
AVG., ALL YRS.	7.0	10.0	10.0	135	144	114	20.7	13.5	19.8	6.4	5.6

SOURCE: Davis-Gallman-Hutchins tape.
NOTE: + refers to additions to the fleet, − to vessels that left the fleet. Stock is an average for the period.

ogy was not instantaneous. This conclusion is supported by a comparison of the ships entering and leaving the fleet. Over the first sixty years (no ship joined the fleet after 1885), a new ship entering the fleet was almost 5 percent larger than a ship that left, and if the first decade is excluded, the difference was almost 7 percent.

Barks were on average smaller than ships. Their 285-ton average was less than 85 percent of that of ships; however, not all barks were smaller. In fact, the average size of a "rerigged" bark (344 tons) was almost identical to that of the "average" ship (343 tons). Moreover, in the decade 1866–1875, when twenty-five ships were rerigged as barks (more than half of all reriggings), the average rerigged vessel was significantly larger than the average ship in the fleet (389 tons as compared with 364).

While the time-tonnage profile for barks was similar to that for ships, the movements were much accentuated. For ships, the increase between the first decade and 1856–1865 was about 20 percent. For barks, the increase in size was 30 percent over the first decade and an additional 30 percent between then and the seventh decade (1876–1885). There were few barks in the early decades and the observed increase may be merely a small-numbers illusion. Of the later and certainly more important increase, however, almost all occurred after 1845. As with ships, even if the first decade is ignored, it appears that efficient vessel size was increasing over the middle third of the period. In every decade through 1875 barks entering the fleet were on average 3 percent larger than those they replaced. In the case of barks, however, there is no evidence of any systematic change in the size distribution of tonnages.[17]

For both classes there is evidence that the larger vessels hunted the more distant grounds and the smaller ones concentrated in the Atlantic and in Hudson's Bay (see Table 10.6). The Arctic drew the largest vessels. Barks hunting there were, for example, 20 percent above average size. The Pacific attracted ships that were about 2 percent larger than average and barks that were 8 percent larger. Vessels reporting the Indian Ocean as their destination were somewhat smaller than average; but those that hunted in Hudson's Bay were almost 10 percent below mean levels, and those in the Atlantic more than 15 percent below. While through most of the period the hunting grounds to which the largest vessels were attracted were the grounds that were becoming relatively more important, these facts do not entirely explain the in-

---

[17] The coefficients of variation for the nine decades are 38.8, 24.2, 18.1, 25.1, 25.8, 25.0, 24.7, 25.6, 26.8, respectively.

TABLE 10.6
Tonnage Relatives by Hunting Ground, Ships and Barks, 1816–1905
(All grounds = 100)

	All	Atlantic	Hudson's Bay	Indian	Pacific	Arctic
		SHIPS				
1816–25	100	102	87	—	98	—
1826–35	100	95	91	100	106	—
1836–45	100	94	—	95	104	90
1846–55	100	69	—	91	101	110
1856–65	100	92	98	95	100	103
1866–75	100	96	88	98	101	104
1876–85	100	96	—	98	103	97
1886–95	100	98	—	—	101	—
1896–1905	—	—	—	—	—	—
ALL YEARS DECADE WTS.[a]	100	93	91	96	102	101
ALL YEARS VESSEL WTS.[a]	100	80	90	95	102	110
		BARKS				
1816–25	100	86	—	—	155	—
1826–35	100	98	—	100	104	—
1836–45	100	93	—	100	105	95
1846–55	100	78	—	92	109	129
1856–65	100	76	115	91	108	122
1866–75	100	81	88	102	112	119
1876–85	100	87	53	92	116	112
1886–95	100	92	76	101	110	—
1896–1905	100	100	89	—	—	—
ALL YEARS DECADE WTS.[a]	100	89	84	97	115	115
ALL YEARS VESSEL WTS.[a]	100	85	91	92	108	120

SOURCE: Davis-Gallman-Hutchins tape.

[a] In some of the remaining tables two averages are presented. The first (decade weights) recognizes that the size of the fleet changed, and it therefore provides an estimate of the capacity of the fleet in an average decade. The second (vessel weights) provides a measure that captures the average capacity of all vessels ever in the fleet.

crease in the average size of vessels, which characterized both ships and barks in every ground.

As time passed, vessels remained in the fleet for longer periods. The average age of both ships and barks gradually increased despite the fact that in the early years substantial numbers of vessels were built for New Bedford whaling. If the first decade is excluded from the analysis for barks (there were very few) and the last for ships (all had left the fleet), both classes aged at a rate of just under five years per decade. Thus between 1816 and 1905 the average age of the fleet almost tripled. It reached more than thirty-eight years for ships in the decade 1885–1895 and more than forty-seven for barks in the decade that spanned the turn of the century.[18]

On average, older vessels left the fleet and younger ones transferred in from the merchant marine or other whaling ports (see Table 10.7). For ships, new entrants were half as old as the stock, while those that left were only about five years (20 percent) older than those that remained. Moreover, while there is considerable decade-to-decade variation, there does not appear to have been any trend in the ratio of age of entry to age of exit. For barks the story is similar, but not identical. Overall, the average age of both ships and barks was approximately equal (25.1 years for the former, 25.8 for the latter). Barks entering the fleet were about two-thirds of the average age of the stock, but the average age of exits was about the same, relative to the age of the stock, as it was for the larger class. For comparable periods, the average "whaling life" was an almost identical 14.8 years for both classes of vessels; nor was there any significant difference in their physical life (the average age of condemned ships was 36.8 years and that of barks 38.0). Thus, it seems possible to infer that the opportunity cost for both classes was similar.

The vessels that entered the fleet came from one of three sources. Some were "born" into it—that is, constructed to order—some transferred into the fleet either from other maritime activities or from another whaling port, and some were "reborn"—that is, rerigged. During the ninety years under consideration, 71 ships and 49 barks were "born," 275 ships and 207 barks transferred, and rerigging added 1 ship and 57 barks.

In solely quantitative terms transfers were the most important source, but an examination of the time pattern of the entrants' profile suggests that the other two categories are analytically the more interesting. Rerigging, a change in masts and sail that converted a vessel of one class into a vessel of another, was important only in the years between 1846 and 1875; but in that period it accounted for more than one-fifth of the entering barks. Almost all of those "new" barks had previously gone to sea as ships. Rerigging was, of course, a market response to the rapid expansion of the Arctic grounds where barks were relatively more effective. The process itself was chosen because, given the urgency of demand, it was quicker to rerig than to build, and because rerigging

---

[18] The reader may wonder how it is possible that the average age can be 38 and 47 years, when the average age of a vessel at condemnation was only 37 years. The latter figure is the overall average and is comparable with the overall average ages of twenty-five and twenty-six years.

TABLE 10.7
Technical Characteristics of Vessels in the New Bedford Whaling Fleet, 1816–1905 (Ships & Barks Only)

PANEL A: VESSELS ENTERING THE FLEET

	Addition to the fleet							Relatives (fleet = 100)							The fleet	
	Tons				Age			Tons				Age				
	All additions	Born	XFR[a]	Rerig	All additions	Born	XFR	Rerig	All additions	Born	XFR	Rerig	All additions	Born	XFR	Rerig	Tons	Age

SHIPS

1816–25	320	325	322	247	7.6	0.0	8.9	4.0	107	109	108	82	62	0	73	33	298	12.2
1826–35	342	370	338	—	11.7	0.0	13.6	—	107	116	106	—	70	0	82	—	320	16.6
1836–45	354	375	357	—	13.6	0.0	16.2	—	106	113	105	—	63	0	75	—	333	21.6
1846–55	407	395	416	—	10.5	0.0	17.8	—	116	112	118	—	40.8	0	70	—	352	25.7
1856–65	380	429	364	—	20.3	0.0	26.7	—	105	118	100	—	70	0	91	—	363	29.3
1866–75	370	441	347	—	3.0	0.0	45.0	—	102	121	95	—	84	0	126	—	364	35.6
1876–85	387	387	—	—	0.0	0.0	—	—	104	104	—	—	0	0	—	—	372	41.3
1886–95	—	—	—	—	—	—	—	—	—	—	—	—	—	—	—	—	379	38.6
1896–1905	—	—	—	—	—	—	—	—	—	—	—	—	—	—	—	—	—	—
ALL YEARS	347	381	352	247	12.3	0.0	15.0	4.0	101	110	103	72	49	0	60	16	343	25.1

BARKS

1816–25	170	—	170	—	11.3	—	11.3	—	93	—	93	—	130	—	130	—	182	8.7
1826–35	261	377	246	288	11.7	0.0	11.1	16.5	105	151	99	116	78	0	74	110	249	15.0
1836–45	264	336	260	241	13.8	0.0	13.8	24.0	106	134	104	96	75	0	75	130	250	18.5
1846–55	293	342	277	294	10.7	0.0	13.0	23.5	112	131	106	113	55	0	67	121	261	19.3
1856–65	319	377	274	350	21.5	0.0	19.6	32.2	113	133	97	124	98	0	90	146	283	22.0
1866–75	302	372	276	378	24.8	0.0	23.2	31.6	101	125	93	127	83	0	78	106	298	29.9
1876–85	330	349	284	380	10.2	0.0	32.0	25.0	103	109	88	118	31	0	97	76	321	33.1
1886–95	—	—	—	—	—	—	—	—	—	—	—	—	—	—	—	—	295	37.9
1896–1905	373	—	373	—	26.5	0.0	26.5	—	135	—	135	—	56	0	56	—	276	47.2
ALL YEARS	296	853	272	338	16.3	0.0	17.3	29.2	104	124	95	119	63	0	67	113	285	25.8

356

PANEL B: VESSELS LEAVING THE FLEET

	Tons					Age				Years in fleet				The fleet		
	All	Rerig	XFR	Lost	Cond.	All	Rerig	XFR	Lost	Cond.	All	Rerig	XFR	Lost	Cond.	Tons	Age	Years

SHIPS

1816–25	380	—	—	ND	ND	11.4	—	—	9.0	ND	2.8	—	—	0.0	3.0	298	12.2	6.4
1826–35	295	265	339	273	—	17.1	16.0	22.2	15.0	—	5.2	11.2	3.5	2.1	—	320	16.6	7.0
1836–45	331	368	381	326	321	22.2	20.0	18.3	19.7	25.6	13.8	13.0	6.0	14.0	17.6	333	21.6	10.8
1846–55	350	329	380	357	302	27.4	24.5	22.3	28.2	31.8	14.3	18.1	9.5	13.6	18.4	352	25.7	14.5
1856–65	374	362	385	372	363	32.0	31.7	31.0	33.2	30.9	19.9	18.8	19.6	21.4	20.1	363	29.3	19.0
1866–75	364	394	396	359	312	41.6	33.4	48.0	38.7	54.0	28.4	24.3	35.7	23.2	39.5	364	35.6	24.4
1876–85	366	—	374	366	343	47.0	—	40.6	47.5	52.4	34.3	—	37.5	29.1	38.8	372	41.3	32.0
1886–95	388	—	383	—	—	42.8	—	41.0	—	—	36.5	—	37.5	—	—	379	38.6	34.2
1896–1905	—	—	—	—	—	—	—	—	—	—	—	—	—	—	—			
ALL YEARS	358	354	379	358	328	30.6	29.1	30.5	31.3	36.8	17.9	18.9	19.7	17.7	24.3	343	25.1	14.7

BARKS

1816–25	—	—	—	—	—	—	—	—	—	—	—	—	—	—	—	182	8.7	4.0
1826–35	150	—	—	—	217	28.0	—	—	18.1	8.0	5.0	—	—	—	4.0	249	15.0	6.0
1836–45	217	—	205	253	216	23.5	—	24.6	21.6	—	9.0	—	10.1	5.5	—	250	18.5	8.9
1846–55	259	—	319	231	269	25.3	—	28.6	25.5	31.2	13.6	—	14.5	10.7	20.5	261	19.3	8.3
1856–65	245	—	305	296	241	24.0	—	21.5	32.0	28.2	12.5	—	10.9	14.2	13.3	283	22.0	12.0
1866–75	305	—	316	324	252	32.0	—	30.0	32.0	40.0	17.7	—	15.8	20.8	14.4	298	29.9	17.6
1876–85	315	—	342	323	197	37.3	—	32.0	38.7	45.8	21.9	—	19.7	23.5	26.4	321	33.1	21.7
1886–95	313	—	340	—	—	38.2	—	31.0	—	48.0	31.6	—	30.4	—	38.0	295	37.9	29.5
1896–1905	259	—	305	243	—	57.2	—	50.0	59.6	—	51.2	—	50.0	51.6	—	276	47.2	51.2
ALL YEARS	297	—	316	301	250	31.7	—	28.7	31.5	38.0	18.2	—	17.1	19.0	19.3	285	25.8	14.9

PANEL C: EXITS—ALL YEAR RELATIVES (FLEET = 100)

	All	Rerig	XFR	Lost	Cond.	All	Rerig	XFR	Lost	Cond.	All	Rerig	XFR	Lost	Cond.	Tons	Age	Years
Ships	104	103	110	104	96	122	116	122	125	147	122	129	134	120	165	100	100	100
Barks	104	—	111	106	88	123	—	111	122	147	122	—	115	128	130	100	100	100

SOURCE: Davis-Gallman-Hutchins tape.
[a] XFR refers to vessels that transferred into or out of the fleet.

357

extended the life of ships that were beginning to become unprofitable as returns from their favored grounds in the Pacific and Indian oceans suffered a temporary decline. In the two decades between 1856 and 1875 one-third of entering barks and ships were "built to order," while in the succeeding decade the only ship and three-fourths of the new barks came directly from the shipyards. As Chapelle has noted, and as the rigging and tonnage data indicate, in the second half of the century, vessels (particularly barks) that were specifically designed for whaling came to replace the transfers of the earlier era.[19] In the early decades, almost any ship or bark could earn above normal profits, and merchant ships were redirected to whaling. As early as 1863, however, the editor of the "Annual Review of the Whale Fishery" reported that the explanations for the better performance of the industry included

> the greater number of suitable vessels that have this year been fitted for the fishery, compared with those that have been fitted during the last few years; the growing determination in the minds of merchants not to introduce into service any more such expensive vessels as new clippers and bulky ships that were never meant for whalers, but introduce the vessels of proper size, and only such as may be built expressly for the business, and that can sell at comparatively low figures.[20]

Slightly more than a decade later the same source, emphasizing the change in relative costs, reported, "some vessels may possibly be added to the fleet from the merchant service; but as such ventures are attended with so heavy an outlay for repairs, alterations and whaling inventories, it is not possible that many such additions will be made."[21] Previously, such merchant ships required little alteration to hunt profitably. Two years later, in 1877, the point is again underscored: "Shipbuilding has revived, and twelve whalers were built during the year, it being now apparent that at present prices new vessels can be built cheaper than merchantmen can be altered into whaleships."[22] "The building of ships for the whaling service marks a new era in the business."[23] Moreover, the shift in the composition of entrants from transfers to rerigs to new construction indicates that the new barks were more cost effective.

[19] Howard I. Chapelle, *The History of American Sailing Ships* (New York: Norton, 1935), p. 288.
[20] "Annual Review of the Whale Fishery" in *Merchant Transcript*, Jan. 13, 1863, for 1862.
[21] Ibid., Jan. 11, 1876, for 1875.
[22] Ibid., Jan. 15, 1878, for 1877.
[23] Ibid., Jan. 16, 1877, for 1876. The author clearly is using the word "ship" in the nontechnical sense, i.e., a vessel.

Some vessels transferred out of the fleet to other whaling ports or other maritime activities (including, during the Civil War, to the Stone Fleet—ships sunk in an attempt to blockade southern ports). Some were rerigged and reborn as "new" vessels. Paralleling "birth," death came at times by condemnation and at times by loss at sea. In all, 108 ships and 125 barks transferred, 48 ships but no barks were rerigged, 125 ships and 101 barks sank, and 41 ships and 45 barks were condemned.

For ships, those that were born into the fleet were on average almost 10 percent larger than the group that they joined, but there was little difference between the size of the ships that transferred into the fleet and that of those that were already there. In the case of barks, both those that came from the shipyards and those that were rerigged were about 15 percent larger than the average member of the class. Transfers were, if anything, somewhat smaller than the fleet and, again, they were barely more than half as old. Contemporaries believed that the whaling fleet had become a dumping ground for vessels too old to be profitably employed in other pursuits. The evidence drawn from the experience of the New Bedford fleet, however, indicates a somewhat different story. The "Annual Review" for 1876, for example, lamented that "the character of the fleet has suffered of late by the adding of worn out merchant vessels, which obtain insurance at the same rate as new ships just from the stocks."[24] Two years later the same source reports that while "the new vessels added recently have improved the general character and average quality of whaleships . . . it is to be regretted that so many vessels in an unseaworthy condition are sent out upon whaling voyages."[25] The record of the New Bedford fleet suggests that these observations may have been partially correct for ships in the decade 1866–1875 and probably were correct for barks in the next decade, but they were not an accurate account of the history of the fleet. In the years before 1866, ships that transferred in were on average only 70 percent as old as the average ship in the fleet, and the figure for barks in the decade before 1876 was only 65 percent. The three ships that transferred in in the immediate postbellum decade, however, were on average, equal in age to the vessels of that class in the fleet, and the sixteen barks that transferred in in the next decade were more than 60 percent *older* than those they joined. Thereafter, however, no ships and only two barks entered by transfer (and those two, while no younger, were no older than the group they joined). Thus it appears that only in the case of barks in the decade 1876–1885 is there strong evidence that the New Bedford fleet had become a dumping ground.

[24] Ibid.
[25] Ibid., Jan. 14, 1879, for 1878.

Of the vessels that withdrew from the fleet, those that were condemned were older (by about 40 percent) and more experienced (i.e., had put in more time in the whaling fleet) than those that remained. For barks, and perhaps ships as well, the condemned vessels were somewhat smaller than the average; but since standardization for age removes most of that difference, it probably reflects nothing but the gradual increase in average vessel size.

Perhaps more surprisingly, the vessels lost at sea were somewhat larger (almost 10 percent so in the case of barks) than the average in their class; less surprisingly, they tended to be somewhat older. The correlation between size and loss rate is partly (but only partly) explained by the experience in the Arctic. That ground drew the largest vessels, and it was also the most dangerous. As already noted, for example, almost all vessels hunting there were lost to a sudden freeze on two occasions. The fact that vessels lost at sea were almost 20 percent more experienced reflects not only the correlation between age and experience, but also the fact that whaling was more risky than alternative maritime pursuits.

Vessels that left to assume other maritime pursuits tended to be older, larger, and more experienced than either vessels that remained or that through loss or condemnation left the fleet. No barks were re-rigged, but almost half a hundred ships reemerged as barks. Those vessels were slightly larger than the stock from which they were drawn, but the difference is small. Rerigged ships were, however, 10 percent older and almost a third more experienced. Both of these observations on age provide additional support for the belief that the Arctic expansion provided employment for a number of ships that had begun to appear unprofitable and that barks had emerged as the technique of choice.

Transfers dominated the new entrants to the fleet, and they played an important but less dominant role in accounting for the vessels that left. They constituted almost three-quarters of the entrants and almost 40 percent of those that exited. The relative importance of transfers underscores the unique nature of major capital investment in the whaling industry: capital was highly malleable. Owners could quickly shift their capital stock from activity to activity in response to shifts in relative profit rates. For this industry, the long run was very short indeed.

For ships, the average crew size was about twenty-seven, but that figure rose from twenty-one to thirty-three (see Table 10.8). In the case of barks, crew size tended to be smaller (in any given decade the average size of a bark's crew was about six-sevenths that of a ship), but the aver-

TABLE 10.8
Crew and Crew per Ton, New Bedford Whalers, 1816–1905

	Average crew size			Men per ton		
	Ships	Barks	Other	Ships	Barks	Other
1816–25	20.88	14.33	14.0	.071	.074	.101
1826–35	23.37	19.79	14.75	.074	.080	.117
1836–45	26.31	22.74	16.79	.081	.092	.114
1846–55	29.10	25.77	18.30	.084	.099	.157
1856–65	29.98	27.14	20.10	.082	.095	.125
1866–75	30.87	27.35	17.71	.084	.096	.153
1876–85	31.10	27.79	18.33	.084	.099	.142
1886–95	33.00	28.36	16.53	.083	.103	.138
1896–1905[a]	—	26.40	17.00	—	.119	ND
ALL YEARS	27.49	26.48	17.06	.080	.096	.138

SOURCE: Davis-Gallman-Hutchins tape.
[a] 1895–1900 only.

age also rose over the period. For brigs, sloops, and schooners, typical crew size was substantially less than for the larger classes, but even for these smaller vessels, the average size of crew tended to increase. For all three rigging classes, however, the increase was concentrated in the first half of the period.[26]

If attention is shifted from average crew size to a measure adjusted for differences in vessel size (men per vessel ton, for example), there is still evidence of an increase in labor-intensity, but it is not so heavily concentrated in the early years.[27] Moreover, the measure indicates that ships used proportionately less labor than did barks, and barks in turn used less than brigs, sloops, and schooners. Over the entire period, the men-per-ton figure was 0.08 for ships, 0.10 for barks, and 0.14 for "others."

The increasing labor-capital ratio might provide support for the general view that the quality of crews was deteriorating, a view that also receives some support from indirect evidence. Between 1850 and

[26] Percent change in average crew size:

	1816/25–1856/65	1856/65–1886/95
Ships	44	10
Barks	89	5
Other	44	−6

[27] Percent change in labor-capital ratio:

	1816/25–1856/65	1856/65–1886/95
Ships	15	1
Barks	28	8
Other	24	10

1880, for example, the proportion of U.S. citizens in the typical crew declined from three in four to one in two. The shift suggests pressure in the labor market that operated to reduce crew quality. It is obvious from any reading of the qualitative sources that while life on a whale ship may never have been good, its attractions relative to other pursuits got progressively worse as incomes in other sectors rose. Melville twice deserted. While comments on labor turnover were infrequent in the early years, they later became common. In 1850 the "Annual Review" noted, with some surprise, "that accounts from the Sperm whalers in the Pacific are not at all encouraging. This arises . . . in no small part from desertions and want of discipline among the crew, arising from the discovery of the gold regions and the other attractions of California."[28] Three decades later when there was no gold rush to blame, the criticism was repeated:

> The business, however, is subject to many serious drawbacks, some of which, if not corrected, bid fair to impair its success. Chief among these are the influences of those ports where officers and crew are constantly leaving vessels, causing a large expense in replacing them.[29]

Although observations on skill level are limited to the years between 1842 and 1879, the data indicate that about one-third of the members could be classified as skilled (a classification that includes officers, boat steerers, harpooneers, and artificers), while the remainder appear to have been largely unskilled (seamen, green hands, boys, etc.). There does not appear to have been any trend in the skilled-unskilled ratio, nor is there any significant difference between the ratio for ships and that for barks.[30]

V. Whaling: The New Bedford Fleet at Work

With the Treaty of Ghent, New England whaling entered a half-century period of rapid expansion. At first most activity was in grounds close

---

[28] "Annual Review," Jan. 7, 1851, for 1850.
[29] Ibid., Jan. 6, 1880, for 1879.
[30] Ratio of skilled to unskilled workers:

	Ships	Barks	Other
Skilled	36.8	35.9	31.1
Unskilled	63.2	64.1	68.9

The traditional labor contract in the industry called for crewmen to receive a fraction of the catch rather than a set wage. The total labor share averaged about one-third (34 percent), with agents, owners, and costs absorbing the remainder. Although there was considerable year-to-year and voyage-to-voyage fluctuation, there is little evidence of trend or of systematic redistribution within labor's share. The captain received between

The New Bedford Whaling Fleet

to home: the Atlantic drew over one-half the fleet's capacity between 1815 and the mid-thirties (see Table 10.9). Thereafter, the number of vessels hunting in that ground declined (they accounted for a mere 4 percent of capacity at the peak of New Bedford's dominance). With increasing competition from San Francisco and with the general contraction of the industry after 1875, the Atlantic reassumed its former dominant role. There were, in fact, more voyages returning from that ground in the decade 1876–1885 than there had been in any previous decade save one.

Even during the early years, the New Bedford captains had begun to range into the Pacific, a ground that provided Melville with the basis for his novels and the city and industry with the basis for their prosperity. Although no vessels returned from that ground after 1895, over the entire ninety-year period the Pacific drew more than one-half of all tonnage; during the prosperous midcentury decades, its share was almost two-thirds. The whalers did not limit their lengthy voyages to the Pacific. Between the mid-thirties and the mid-fifties the Indian Ocean attracted almost a fifth of the city's tonnage. Later, as whale bone became more valuable, the captains discovered the bowheads of the Arctic. That ground, as defined in this paper, had been visited by two vessels before 1845, although the first voyage to the "true" Arctic probably occurred in 1848. In that year Captain Roys, in the bark Superior, worked his vessel through the Bering Strait and into the Arctic Ocean; there he found, "whales innumerable, some of which yielded two hundred and eighty barrels of oil."[31] From 1850 until West Coast competition prevailed, the Arctic was the destination of more than 15 percent of the tonnage that returned and, given the very high loss rates, an even higher fraction of the tonnage that left New Bedford.

Because the sperms, rights, bowheads, and humpbacks lived in separate areas, the composition of the catch differed from ground to ground. Even within a ground, to the extent possible, captains adjusted their catch to favor those animals that were the most valuable. Whales were not always easy to find, however, and most captains were unwilling to pass up the chance to catch one, even if that decision might prove *ex post* not to have been the most profitable. Over half of the vessels returned with some mix of baleen and sperm oil, and some with a few barrels of walrus or black fish oil to boot. There was, however, a distinct tendency toward voyage specialization (see Table 10.10). Not

---

$1/12$ and $1/16$ of the catch, other officers from $1/20$ to $1/25$, boat steerers from $1/75$ to $1/90$, artificers from $1/150$ to $1/160$, and seamen from the latter figure to $1/200$.

[31] Scammon, *Mammals*, p. 58.

TABLE 10.9
Voyages by Vessel Class by Grounds, 3,369 Returning Voyages, 1816–1905

	Total voyages	Tons[a]	% voyages				% tonnage returning			
			Atl.	Hud.	Ind.	Pac.	Arc.	Atl.	Hud.	Ind.	Pac.	Arc.

	Total voyages	Tons[a]	Atl.	Hud.	Ind.	Pac.	Arc.	Atl.	Hud.	Ind.	Pac.	Arc.
1816–25	136	37.5	56.6%	2.2%	0.7%	40.4%	0.0%	54.4%	2.1%	0.3%	43.2%	0.0%
1826–35	380	114.8	54.5	0.3	2.6	42.6	0.0	51.4	0.3	2.4	46.0	0.0
1836–45	583	178.6	27.4	0.0	21.1	51.1	0.3	22.7	0.0	20.9	56.0	0.2
1846–55	697	228.7	6.2	0.0	23.8	61.7	8.3	3.6	0.0	21.6	65.0	9.7
1856–65	665	222.4	11.9	2.3	14.0	55.8	16.1	8.5	2.3	13.1	58.2	18.0
1866–75	373	116.0	31.1	3.8	18.2	33.5	13.4	24.1	3.2	19.0	37.7	15.9
1876–85	377	109.8	48.9	3.7	6.1	28.4	12.7	39.0	2.3	6.6	36.5	15.7
1886–95	109	28.7	53.2	3.6	5.5	37.6	0.0	42.4	3.1	6.3	48.3	0.0
1896–1905	52	10.7	84.6	15.4	0.0	0.0	0.0	84.0	16.0	0.0	0.0	0.0
ALL YEARS	3369	1047.2	28.7%	1.8%	14.5%	47.1%	7.9%	22.8%	1.4%	14.3%	52.0%	9.4%

SOURCE: Davis-Gallman-Hutchins tape.
[a] Tons 000's.

TABLE 10.10
Specialization in Catch, New Bedford Whalers, All Voyages, By Ground, 1816–1905
(Percent)

	All Grounds			Atlantic			Hudson's Bay		
	Baleen[a]	Mixed[b]	Sperm[c]	Baleen	Mixed	Sperm	Baleen	Mixed	Sperm
1816–25	54.5%	1.4%	44.1%	73.8%	1.2%	25.0%	100.0%	0.0%	0.0%
1826–35	46.6	4.9	48.5	79.3	5.7	14.9	100.0	0.0	0.0
1836–45	39.7	18.2	42.1	55.7	11.4	32.9	—	—	—
1846–55	45.5	17.3	37.2	16.7	23.8	59.5	0.0	0.0	100.0
1856–65	43.7	14.8	41.6	6.3	26.6	67.1	100.0	0.0	0.0
1866–75	32.2	25.2	42.6	7.8	26.1	66.1	92.9	7.1	0.0
1876–85	31.4	20.4	48.3	6.0	24.7	69.2	100.0	0.0	0.0
1886–95	15.1	20.8	64.2	1.8	15.8	82.5	100.0	0.0	0.0
1896–1905	9.6	5.8	84.6	0.0	0.0	100.0	62.5	37.5	0.0
ALL YEARS	39.8%	16.2%	44.0%	34.3%	15.4%	50.3%	91.1%	7.1%	1.8%

	Indian Ocean			Pacific			Arctic		
	Baleen	Mixed	Sperm	Baleen	Mixed	Sperm	Baleen	Mixed	Sperm
1816–25	—	—	—	24.6%	1.8%	73.7%	—	—	—
1826–35	20.0%	60.0%	20.0%	7.6	2.1	90.3	—	—	—
1836–45	62.3	28.7	9.0	21.7	17.3	61.3	50.0%	50.0%	0.0%
1846–55	47.9	20.6	31.5	39.0	17.7	43.3	96.7	1.7	1.7
1856–65	16.1	32.3	51.6	41.2	10.8	48.0	92.3	6.8	6.8
1866–75	11.6	34.8	53.6	37.6	26.4	36.0	86.0	12.0	2.0
1876–85	0.0	21.7	78.3	45.3	20.8	34.0	91.7	8.3	0.0
1886–95	0.0	16.7	83.3	30.0	30.0	40.0	—	—	—
1896–1905	—	—	—	—	—	—	—	—	—
ALL YEARS	37.1%	21.3%	35.6%	32.5%	15.0%	52.5%	91.6%	7.3%	1.1%

SOURCE: Davis-Gallman-Hutchins tape.
[a] [sperm oil / (sperm oil + whale oil)] ≤ 20%.
[b] [sperm oil / (sperm oil + whale oil)] > 20% but < 80%.
[c] [sperm oil / (sperm oil + whale oil)] ≥ 80%.

only was there a tendency for individual captains to specialize, but specialization characterized the entire fleets of many of the American whaling ports. Nantucket was noted for its sperm whalers, Provincetown for her "plum-pu-dn-rs" employed on short voyages in the Atlantic, New Londoners for the right whale fishery and for contending with the ice and snow of the Davis Strait and Hudson's Bay in search of bowheads, and Sag Harbor and Stonington for the northern and southern right whale fishery. New Bedford, on the other hand, "in the course of her absorption of the greater portion of the whaling commerce of the United States, prosecuted the enterprise in its various branches all over the world."[32] Given the ubiquity of the New Bedford fleet, the total of all New Bedford vessels probably provides a reasonable index to the composition of the total U.S. whaling catch. From those figures it appears that on average, about 40 percent of the vessels specialized in baleens and a slightly larger fraction in sperm whales. The remainder of the returning vessels brought back a more thoroughly mixed cargo. If the last decades are excluded (the fleet had largely withdrawn to the Atlantic), those ratios (4 to 4 to 2) appear to have been fairly stable.

In the Arctic, with its large bowhead population, and in Hudson's Bay, with its rights, baleens constituted the vast majority of the catch. Over nine-tenths of the voyages returned with more than 80 percent whale oil; another 8 percent returned with mixed cargoes. In the Pacific, sperm oil was most important but as the price of bone rose, the proportion of mixed and "baleen" voyages increased. By the time the New Bedford vessels withdrew, they were returning with as much whale bone as sperm oil. The impact of rising bone prices in the Arctic and Pacific grounds is underscored by the increasing importance of bone to the whalers' income. In the early years many captains who brought back sizable quantities of whale oil often returned with no bone. In the last quinquennium that New Bedford whalers ventured as far as the Arctic, however, two-thirds of their revenue came from bone; and in the Pacific in the years 1891–1895, bone produced 4 in every 5 dollars of total revenue. This trend had, in fact, been noted as early as the mid-1860s. At that time, the "Annual Review" reported that "the sperm whale fishery, which in former years was prosecuted with success on the Pacific Coast, seems to have been exhausted of late in those grounds; and ships with few exceptions have done little or nothing within the past few years."[33] Estimates of the stock of sperm whales in

[32] Ibid., p. 240.
[33] "Annual Review," Feb. 2, 1864 for 1863.

the ocean, however, do not indicate any serious depletion. Given the prices of whale bone, the shift may have been triggered solely by the larger potential profits from baleens.

In the Indian Ocean, an early concentration on baleens gave way after the mid-1850s to voyages specializing in sperms. In the last decade that any significant number of New Bedford vessels hunted that ground (1876–1885), no vessel returned with less than 20-percent sperm oil; and eighteen of the twenty-three reported a figure of 80 percent or more. This experience in the face of rising bone prices provides some indirect evidence that in that ground, baleen stocks may have been declining.

The Atlantic ground displays what at first glance appears to be a puzzling pattern. In the early decades when bone went begging, the fleet specialized in baleens; but as the value of right whales increased, the captains reported an ever declining fraction of whale oil. Between 1816 and 1845, 70 percent of the returning captains reported that sperm oil constituted less than one-fifth of their catch. In the years after 1855, only 5 percent of the voyages reported a sperm oil to whale oil ratio of less than 0.2, and almost fifteen times that number returned with more than 80 percent sperm oil. Moreover, for the baleen whales caught in those years of high bone prices, the ratio of bone to whale oil was substantially less than it had been earlier. Unless one believes that the captains were irrational, it appears not only that the ratio of baleens to sperm whales in the ground had shifted dramatically, but also that the baleens that remained yielded less bone. Since older and larger whales have a higher ratio of baleen to oil, the Atlantic decline may provide some indirect evidence of depletion in that ground. If there was some depletion and the baleens remaining were in fact "bone light," then the shift to sperm whales may have been a normal response to changing market conditions for bone and whale oil. In the words of the editor of the "Annual Review" for 1872, however, "nearly three-fourths of the [Atlantic] fleet is sperm whaling, . . . and that branch of our business promises to survive, as substitutes are not as readily found as for whale oil."[34]

---

[34] Ibid., Jan. 17, 1872 for 1873, p. 240. The same issue offers a different explanation of the "bone lightness" of baleens captured in the Atlantic after 1860: "with few exceptions the sperm whale fleet has been largely engaged in Humpbacking between seasons," presumably because of the propinquity of the sperm and humpback migration routes. Humpback whales are poor sources of bone. The decline in the ratio of bone to oil may, therefore, reflect a shift from hunting rights to hunting humpbacks, a shift resulting from concentration of the Atlantic fleet on sperm whales during the sperm whale season.

TABLE 10.11
Vessel Class by Grounds, New Bedford Whalers, 3,369 Returning Voyages, 1816–1905

PANEL A: NUMBER OF VOYAGES

	Atlantic			Hudson's Bay and Davis Strait			Indian Ocean			Pacific			Arctic		
	Ships	Barks	Other	Ships	Barks	Other	Ships	Barks	Other	Ships	Barks	Other	Ships	Barks	Other
1816–25	54	8	15	3	0	0	0	0	1	51	2	2	0	0	0
1826–35	165	28	13	1	0	0	6	2	2	139	19	3	0	0	0
1836–45	84	40	36	0	0	0	95	27	1	250	48	0	1	1	0
1846–55	7	25	11	0	0	0	100	66	0	346	84	0	47	11	0
1856–65	11	67	1	5	9	1	40	53	0	210	161	0	71	36	0
1866–75	18	80	18	3	8	3	21	47	0	47	78	0	21	29	0
1876–85	18	101	65	0	3	11	6	17	0	36	71	0	9	39	0
1886–95	5	25	28	0	4	0	0	6	0	12	29	0	0	0	0
1896–1905	0	23	21	0	4	4	0	0	0	0	0	0	0	0	0
ALL YEARS	362	397	208	12	28	19	268	218	4	1091	492	5	149	116	0

PANEL B: PERCENTAGE DISTRIBUTION, NEW BEDFORD WHALERS GROUND AND DECADE RELATIVES, 1816–1905 (SHIPS AND BARKS ONLY)

	All grounds		Atlantic		Hudson's Bay		Indian Ocean		Pacific		Arctic	
	Ships	Barks	Ships	Barks	Ships	Barks	Ships	Barks	Ships	Barks	Ships	Barks
					% Distribution								
1816–25	92.1%	7.9%	87.7%	12.3%	100.0%	0.0%	—	—	96.6%	3.3%	—	—
1826–35	86.4	13.6	85.4	14.5	100.0	0.0	75.0%	25.0%	88.1	11.9	—	—
1836–45	78.8	21.2	67.7	32.3	—	—	77.9	22.1	83.9	16.1	50.0%	50.0%
1846–55	72.9	27.1	21.9	78.1	—	—	60.2	39.8	80.5	19.5	81.0	19.0
1856–65	50.8	49.2	14.1	85.9	35.7	64.2	43.0	57.0	56.6	43.4	66.4	33.6
1866–75	31.2	68.8	18.4	81.6	27.3	72.7	30.9	69.1	37.6	62.4	42.0	58.0
1876–85	23.0	77.0	15.1	84.9	0.0	100.0	26.1	73.9	33.6	66.4	18.8	81.2
1886–95	21.0	79.0	16.7	83.3	0.0	100.0	0.0	100.0	29.3	70.7	—	—
1896–1905	0.0	100.0	0.0	100.0	0.0	100.0	—	—	—	—	—	—
ALL YEARS	60.2%	39.8%	47.9%	52.1%	30.0%	70.0%	55.1%	449.0%	69.1%	30.9%	52.2%	43.8%

PANEL B: *(continued)*

Ground Relatives (All grounds = 100)

1816–25	100	100	95	156	109	—	105	43	—	—
1826–35	100	100	99	107	116	184	102	88	—	—
1836–45	100	100	86	152	—	104	107	76	63	235
1846–55	100	100	30	288	—	147	110	72	111	70
1856–65	100	100	28	175	70	116	111	88	130	68
1866–75	100	100	59	119	87	101	120	91	134	84
1876–85	100	100	66	110	0	96	146	86	82	106
1886–95	100	100	79	105	0	127	139	90	—	—
1896–1905	100	100	—	—	—	—	—	—	—	—
ALL YEARS	100	100	80	131	50	113	114	78	100	110

Decade Relatives (All years in ground = 100)

1816–25	153	20	183	24	333	—	140	11	—	—
1826–35	143	34	122	29	333	56	127	39	—	—
1836–45	131	53	141	62	—	49	121	52	96	114
1846–55	121	68	46	150	—	89	116	63	155	43
1856–65	84	123	29	165	119	127	82	140	127	77
1866–75	52	173	38	157	91	154	54	202	80	132
1876–85	38	193	32	163	0	165	49	215	36	185
1886–95	35	198	35	160	0	223	42	229	—	—
1896–1905	0	251	0	192	0	—	—	—	—	—
ALL YEARS	100	100	100	100	100	100	100	100	100	100

SOURCE: Davis-Gallman-Hutchins tape.

Like the bowheads, rights, sperms, and humpbacks, the brigs, sloops, barks, ships, and schooners displayed preferences for particular hunting grounds (see Table 10.11). The first few years aside, the smallest class (the brigs, sloops, and schooners) never ventured further afield than the Atlantic and Hudson's Bay. They were three times as likely to hunt the former and four times as likely to work the latter as their total numbers would suggest. Of the two larger and numerically more important vessel classes, barks appeared (in relative terms) much more frequently in Hudson's Bay and in the Atlantic, somewhat more frequently in the Arctic and the Indian Ocean, and substantially less often in the Pacific. Ships, of course, dominated hunting in the latter and most important ground, and they appeared almost as frequently as chance would suggest in the Indian and Arctic oceans.

Over time, there was a general substitution of barks for ships; and, at the very end of the period, a substitution of brigs, sloops, and schooners for both barks and ships in the Atlantic, and perhaps in Hudson's Bay as well. The substitution of barks for ships did not, however, proceed at a uniform pace. Although the process was gradual, if the average proportion of a class over the entire period is taken as a base, barks began to appear in higher than expected proportions in the Atlantic about 1845, in the Indian Ocean about five years later, and in the Pacific in yet another five years. That benchmark was not reached in the Arctic until 1865—the war may have slowed the process—and, although the pattern is somewhat blurred, perhaps not until the early 1880s in Hudson's Bay. By the third quarter of the nineteenth century, barks had proven themselves technically superior to ships, but dominance was not achieved at a uniform rate and seems to have depended in part, at least, on the ground hunted and the whales sought.

It has been shown that average vessel size was increasing. The data also indicate that barks built after 1855 were larger not only in absolute terms but also relative to ships. In the decade 1876–1885, for example, twelve barks were built, and they were almost nine-tenths as large as the ships then in the fleet. In the decade 1846–1855, twenty-one barks had been built, and they were only 80 percent as large as the ships they joined. Between 1856 and 1875, forty-one ships were re-rigged as barks, and they were, on average, 3 percent *larger* than the ships that remained. The increased size reduced the relative disadvantage in labor costs that barks had suffered. And as noted already, the bark, with its fore-and-aft rig, had special advantages in maneuvering as compared with the square rigged ships.[35] "Where economy of han-

---

[35] Michael J. Maran, *The Decline of the American Whaling Industry* (Ph.D. diss., University of Pennsylvania, 1974).

dling was of special importance, the bark rig was used."[36] Thus, as opportunities increased in the Arctic and Hudson's Bay grounds, with their great premium on maneuverability, the new, larger bark was innovated rapidly. It took time to build new vessels, as has been described, but the process of technical improvement was hastened by rerigging ships in the more productive configuration.

The examination of the average size of vessels in the fleet indicated that carrying capacity was increasing (see Table 10.5); however, disaggregation by ground suggests that for both ships and barks there was little further increase in the size of vessels hunting in a particular ground after the early 1880s (see Table 10.12). There were also very substantial interground differences in average tonnage, and there was a close correlation between size and the distance between New Bedford and the ground hunted. In the case of both ships and barks, the smallest vessels were used in the Atlantic. The average size increased as the destination shifted from that ground to Hudson's Bay, to the Indian Ocean, to the Pacific, and finally, to the Arctic, where the largest vessels were employed. Moreover, although the size of all vessels changed, this geographic profile was largely invariant with the passage of time.

Since new technologies alter the productivity of the capital stock through the addition of new vessels and the retirement of old ones, "marginal" changes are important. An analysis of those "marginals" provides some additional insight into the relationship between the capital stock and the industry's economic history (see Table 10.13).

Over time, the average size of vessels increased, and the nature of that increase can be seen in a comparison of the "marginal" vessels with the remainder of the fleet. Taking a rig-decade-ground as the unit of observation, the data indicate that in the years before 1865 (years that saw the vast majority of the increase in vessel size), vessels making their first voyage were on average 5 percent larger than the group they joined.[37] Moreover, of the thirty-five bark and ship observations, new vessels were larger in twenty-seven, equal in two, and smaller in only six cases. As for leavers, over the entire ninety-year period, the vessels that left were on average 4 percent smaller than their "peers." There were in the period fifty-three observations on barks and ships; leavers were smaller in thirty, equal in two, and larger in twenty-

[36] John G. B. Hutchins, *The American Maritime Industries and Public Policy 1789–1914: An Economic History,* Harvard Economic Studies, vol. 71 (Cambridge, Mass.: Harvard University Press, 1941), pp. 218–219.

[37] Hudson's Bay is an exception, but only seven vessels made their first voyages to that ground, and only three their last.

TABLE 10.12
Average Tonnage, by Class and Ground, New Bedford Whalers, 1816–1905

		Barks						Ships						Other			
	All	Atl.	Hud.	Ind.	Pac.	Arc.	All	Atl.	Hud.	Ind.	Pac.	Arc.	All	Atl.	Hud.	Ind.	Pac.	Arc.
1816–25	193	166	—	—	300	—	302	309	264	—	297	—	158	157	—	99	182	—
1826–35	249	244	—	248	259	—	321	305	291	321	390	—	138	133	—	137	159	—
1836–45	255	237	—	256	269	243	335	316	—	319	347	302	129	129	—	159	—	—
1846–55	266	208	—	246	291	344	353	244	—	322	358	390	128	128	—	—	—	—
1856–65	293	223	336	268	315	358	369	339	361	352	371	380	210	158	262	—	—	—
1866–75	300	243	265	307	335	357	369	355	325	362	372	385	120	107	197	—	—	—
1876–85	319	276	168	293	369	358	372	358	—	364	383	359	137	137	184	—	—	—
1886–95	289	265	221	293	317	—	385	377	—	—	389	—	129	129	188	—	—	—
1896–1905	275	274	245	—	—	—	346	—	—	—	—	—	136	126	—	139	172	—
Avg. (vessel)[a]		248	265	270	315	349		313	310	329	352	381		126	178	139	172	—
Avg. (decade)[b]		237	247	273	307	332		325	310	340	353	363		133	208	132	171	—

GROUND RELATIVES (BARKS IN ATLANTIC = 100)

1816–25	100	—	—	181	—	186	159	—	179	—		95	—	50	110	—
1826–35	100	—	102	106	—	125	119	132	139	—		55	—	56	65	—
1836–45	100	—	108	114	103	133	—	135	146	127		54	—	67	—	—
1846–55	100	—	118	140	165	117	—	155	172	188		62	—	—	—	—
1856–65	100	151	120	141	160	152	161	158	166	170		70	117	—	—	—
1866–75	100	109	126	138	147	146	134	149	153	158		44	81	—	—	—
1876–85	100	61	106	134	130	130	—	132	139	130		46	67	—	—	—
1886–95	100	83	111	120	—	142	—	—	147	—		49	69	—	—	—
1896–1905	100	89	—	—	—	—	—	—	—	—		46	—	—	—	—
Avg. (vessel)	100	107	109	127	141	126	125	133	142	154		51	72	56	69	—
Avg. (decade)	100	104	115	130	140	137	130	143	151	153		56	88	56	72	—

DECADE RELATIVES (DECADAL AVERAGE OF RIG IN GROUND = 100)

1816–25	70	—	—	98	—	95	85	—	83	—		118	—	75	106	—
1826–35	103	—	91	84	—	94	94	94	95	—		100	—	104	93	—
1836–45	100	—	94	88	73	97	—	94	97	83		97	—	120	—	—
1846–55	88	—	90	95	104	75	—	95	100	107		96	—	—	—	—
1856–65	94	136	98	103	108	104	116	104	104	105		119	126	—	—	—
1866–75	103	107	112	109	108	109	105	106	104	106		80	95	—	—	—
1876–85	116	68	107	120	108	110	—	107	107	99		96	88	—	—	—
1886–95	112	89	107	103	—	116	—	—	109	—		97	90	—	—	—
1896–1905	116	99	—	—	—	—	—	—	—	—		95	—	—	—	—
Avg. (vessel)	105	107	99	103	105	96	100	97	99	105		95	86	105	100	—
Avg. (decade)	100	100	100	100	100	100	100	100	100	100		100	100	100	100	—

SOURCE: Davis-Gallman-Hutchins tape.
[a] Average of all vessels ever in the fleet (each vessel has weight of one).
[b] Average of the decadal averages (each decade average carries weight of one).

TABLE 10.13
First and Last Voyages by Class and Ground, All Years and Vessel Weights, Time at Sea (interval), and Vessel Characteristics (tonnage, age, years in fleet), New Bedford Whalers, 1816–1905

	Barks						Ships						Other				
	All	Atl.	Hud.	Ind.	Pac.	Arc.	All	Atl.	Hud.	Ind.	Pac.	Arc.	All	Atl.	Hud.	Ind.	Pac.	Arc.

(A1) INTERVAL (MONTHS)

Fleet	34.3	27.0	17.1	36.1	39.8	36.2	32.9	17.0	15.8	32.1	37.6	38.9	16.3	16.0	17.0	16.0	22.0	—
New	36.4	27.8	24.0	37.7	41.9	41.7	32.0	19.6	13.0	30.3	37.2	37.7	15.1	15.2	14.0	none	15.5	—
Old	36.1	27.1	43.0	38.5	35.6	38.2	39.3	23.4	12.5	39.2	41.7	40.8	16.6	16.5	15.8	15.5	21.0	—

(A2) INTERVAL RELATIVES [FLEET IN GROUND = 100]

Fleet	100	100	100	100	100	100	100	100	100	100	100	100	100	100	100	100	100	—
New	106	103	140	104	105	115	97	115	82	94	99	97	93	95	82	—	70	—
Old	105	100	251	107	89	106	119	138	79	122	111	105	102	103	93	95	95	—

(B1) TONNAGE

Fleet	288	249	265	272	315	349	346	314	320	332	355	382	142	135	178	139	228	—
New	289	234	388	290	308	387	360	318	305	354	366	438	150	136	218	99	385	—
Old	296	255	340	277	314	360	363	324	407	336	365	392	147	140	211	135	169	—

(B2) TONNAGE RELATIVES [FLEET IN GROUND = 100]

Fleet	100	100	100	100	100	100	100	100	100	100	100	100	100	100	100	100	100	—
New	100	94	146	107	98	111	104	101	95	107	103	115	106	101	122	71	168	—
Old	103	102	128	102	100	103	105	103	127	101	103	103	104	104	119	97	74	—

(C) AGE (DEPARTURE AGE)

Fleet	25.1	29.1	32.4	23.3	22.1	24.3	23.1	20.3	24.0	25.8	22.5	28.4	21.7	22.5	21.0	7.5	9.8	—
Old	28.3	33.8	14.0	26.2	26.4	24.4	28.2	38.1	38.5	28.9	25.6	31.2	20.1	20.8	23.2	9.0	7.7	—
Relative	113	116	43	112	119	100	122	188	160	112	114	110	93	92	110	120	79	

(D) YEARS IN FLEET (AT DEPARTURE)

Fleet	13.9	16.0	21.1	11.9	12.5	14.3	12.4	9.2	8.6	13.8	12.3	17.7	6.3	6.5	4.3	3.3	5.7	—
Old	15.1	16.7	1.0	12.7	15.9	12.4	16.4	19.9	21.0	16.8	14.7	19.7	5.9	6.2	4.7	1.0	6.0	—
Relative	109	104	5	107	127	87	132	216	244	122	120	111	94	95	109	30	105	

(E) YEARS/AGE

Fleet	55	55	65	51	57	59	54	45	36	53	55	62	29	29	20	44	58	
Old	53	49	7	48	60	51	58	52	55	58	57	63	29	30	20	11	78	
Relative	96	89	11	94	105	86	107	116	153	109	104	102	100	103	100	25	134	

SOURCE: Davis-Gallman-Hutchins tape.

one. It is interesting to note that of that latter total, the Arctic contributed six (out of a total of only ten observations on that ground).

These data provide strong support for the conclusion that the increase in size was the result of two quite different phenomena—one an adjustment to new economic environments and the other pure technical change. On the one hand, larger size proved relatively more productive in the distant grounds, and a part of the increase reflects nothing but an adjustment to the changing geographic character of whaling. On the other hand, within each ground there is also evidence of increasing capacity. That change is not "environment induced," and it indicates that the entire fleet was moving toward a more efficient technical configuration.

Ships were larger than barks, but they were less maneuverable. Other differences between rigging classes could also have affected productivity. In every decade the average voyage length for ships was greater than the average time at sea for barks; however, when correction is made for grounds hunted, an overall difference of almost two months vanishes (see Table 10.14). Ships were at sea less time in five grounds, and an almost identical period in the sixth; but none of the differences, except in the Arctic, is large. Hudson's Bay aside, however, both ships and barks remained at sea longer by any measure, than the smaller brigs, sloops, and schooners. In that ground there were so few voyages that comparisons are doubtful, but the smaller vessels may have been at sea as long as the larger ones. It appears, then, that hunting ground was important in the explanation of time at sea, rig type, and size of vessel; however, in the case of size and rig type, changes also occurred within grounds. The technological shifts were not therefore merely a response to changes in the geographic distribution of economic activity.

In terms of the performance of new and old vessels relative to "standard practice," barks making their first voyage and those that were making their last tended to be at sea longer than the "average" (see Table 10.13). For ships, new entrants spent slightly shorter periods hunting, while those on their last voyage were away longer than average. In general, it appears that the vessels that left the fleet tended to be less efficient than the average. On the other hand, there is no evidence that new entrants were more or less efficient than the other vessels in the fleet they entered. It may be that the favorable effects of the technical superiority of the new vessels (if, in fact, they were superior) were canceled by the costs of breaking in the new vessel to the New Bedford whaling routine (but see below the discussion on learning by doing in section VIII).

TABLE 10.14
Voyage Length, by Vessel Class and Ground, New Bedford Whalers, 1816–1905
(Months)

	Barks					Ships					Other			
	Atl.	Hud.	Ind.	Pac.	Arc.	Atl.	Hud.	Ind.	Pac.	Arc.	Atl.	Hud.	Ind.	Pac.	Arc.
1816–25	12.3	—	—	32.0	—	12.4	10.3	—	24.2	—	11.5	11.0	12.0	17.0	—
1826–35	12.6	—	23.0	30.3	—	12.2	11.0	14.8	33.3	—	16.9	—	15.5	28.7	—
1836–45	18.7	—	22.4	35.5	21.0	20.7	—	24.9	36.3	21.0	12.7	—	22.0	—	—
1846–55	24.0	—	34.0	37.6	38.1	26.9	—	33.7	37.5	34.8	16.3	—	—	—	—
1856–65	24.5	17.8	40.3	44.4	42.4	20.9	19.6	41.0	43.6	41.3	22.0	17.0	—	—	—
1866–75	28.7	16.8	41.4	45.2	52.1	22.9	16.3	42.0	47.2	46.8	14.3	16.7	—	—	—
1876–85	33.4	16.0	38.8	33.4	16.2	36.3	—	40.5	37.6	8.8	16.9	14.8	—	—	—
1886–95	34.9	18.3	35.7	34.7	—	37.6	—	—	33.6	—	22.3	24.8	—	—	—
1896–1905	28.3	16.0	—	—	—	—	—	—	—	—	19.0	—	—	22.0	—
AVG.[a]	27.0	17.1	36.1	39.8	36.2	17.0	15.8	32.1	37.6	38.9	16.3	17.0	16.3	—	—
AVG.[b]	17.0	33.6	36.7	32.0	23.7	14.3	32.8	36.7	30.5	16.9	16.9	16.5	22.8	—	—

GROUND RELATIVES [BARKS IN ATLANTIC = 100]

1816–25	100	—	—	260	—	101	84	—	197	—	93	89	98	138	—
1826–35	100	—	183	240	—	97	87	117	264	—	134	—	123	228	—
1836–45	100	—	120	190	112	107	—	133	194	112	68	—	118	—	—
1846–55	100	—	142	157	159	112	—	140	156	145	68	—	—	—	—
1856–65	100	73	164	181	173	85	80	167	178	169	90	69	—	—	—
1866–75	100	59	144	157	182	80	57	146	164	163	50	58	—	—	—
1876–85	100	48	116	100	49	108	—	121	113	26	50	44	—	—	—
1886–95	100	52	102	99	—	108	—	—	96	—	64	—	—	—	—
1896–1905	100	57	—	—	—	—	—	—	—	—	67	88	—	81	—
AVG.[a]	100	63	134	147	134	122	59	119	139	144	60	63	60	94	—
AVG.[b]	100	70	139	152	132	98	59	136	152	126	70	70	68	—	—

DECADE RELATIVES [ALL YEARS VESSEL AVERAGES = 100]

1816–25	51	—	—	87	—	52	72	—	66	—	68	65	73	75	—
1826–35	52	—	68	83	—	51	77	45	91	—	100	—	94	126	—
1836–45	77	—	67	96	66	87	—	76	99	69	75	—	133	—	—
1846–55	99	—	101	102	119	114	—	103	102	114	96	—	—	—	—
1856–65	101	105	120	121	132	88	137	125	119	135	130	101	—	—	—
1866–75	119	99	123	123	163	97	114	128	129	153	85	99	—	—	—
1876–85	138	94	115	91	51	153	—	123	102	29	100	88	—	—	—
1886–95	144	108	106	95	—	159	—	—	92	—	132	147	—	—	—
1896–1905	117	94	—	—	—	—	—	—	—	—	112	—	—	96	—
AVG.[a]	112	101	107	108	113	152	110	98	102	127	96	101	99	96	—
AVG.[b]	100	100	100	100	100	100	100	100	100	100	100	100	100	100	—

SOURCE: Davis-Gallman-Hutchins tape.
[a] Vessel weights.
[b] Decade weights.

TABLE 10.15
Average Vessel Age, at Departure, by Year and Ground, New Bedford Whalers, 1816–1905
(Years)

	Barks					Ships					Other			
	Atl.	Hud.	Ind.	Pac.	Arc.	Atl.	Hud.	Ind.	Pac.	Arc.	Atl.	Hud.	Ind.	Pac.	Arc.
1816–25	7.0	—	—	14.0	—	12.1	8.3	—	11.5	—	8.1	4.0	—	3.4	—
1826–35	14.5	—	12.3	14.8	—	15.9	11.0	19.8	16.3	—	10.9	—	6.0	12.0	—
1836–45	15.1	—	20.9	17.1	19.5	22.1	—	23.7	20.3	18.5	15.2	—	7.0	—	—
1846–55	11.0	—	20.0	17.5	17.5	31.0	—	26.2	24.4	24.2	16.0	—	11.0	—	—
1856–65	24.3	20.8	25.4	20.9	21.3	32.8	28.7	28.4	27.5	31.3	19.0	0.0	—	—	—
1866–75	32.2	31.9	26.3	28.6	29.9	46.0	34.7	36.7	32.4	28.4	24.2	9.3	—	—	—
1876–85	34.2	28.6	27.6	27.8	27.8	45.0	—	24.0	42.0	53.8	25.1	25.4	—	—	—
1886–95	41.4	48.0	38.3	30.5	—	39.0	—	—	39.5	—	30.0	26.0	—	—	—
1896–1905	47.9	55.0	—	—	—	—	—	—	—	—	36.8	—	—	33.0	—
AVG.[a]	29.1	32.4	23.3	22.1	24.3	20.3	24.0	25.8	22.5	28.4	22.5	21.0	7.5	9.8	—
AVG.[b]	25.3	36.9	24.4	21.4	23.2	30.4	20.7	26.5	26.7	31.2	20.6	12.9	8.0	16.1	—

GROUND RELATIVES [BARKS IN ATLANTIC = 100]

1816–25	100	—	—	200	—	173	119	—	164	—	116	57	—	49	—
1826–35	100	—	85	102	—	110	76	137	112	—	75	—	41	83	—
1836–45	100	—	138	113	129	146	—	157	134	123	101	—	46	—	—
1846–55	100	—	182	159	159	282	—	238	222	220	145	—	100	—	—
1856–65	100	86	105	86	88	135	118	117	113	129	78	0	—	—	—
1866–75	100	99	99	82	93	143	108	114	101	88	75	29	—	—	—
1876–85	100	84	81	81	81	132	—	70	123	157	73	74	—	—	—
1886–95	100	116	93	74	—	94	—	—	95	—	72	63	—	—	—
1896–1905	100	115	—	—	—	—	—	—	—	—	77	—	—	69	—
AVG.[a]	100	111	80	76	84	70	82	89	77	98	77	72	26	34	—
AVG.[b]	100	146	96	85	92	120	82	105	106	123	81	51	32	64	—

DECADE RELATIVES [DECADE AVERAGE FOR GROUND = 100]

1816–25	28	—	—	65	—	40	40	—	43	—	39	31	—	21	—
1826–35	57	—	50	69	—	52	53	75	61	—	53	—	75	75	—
1836–45	60	—	86	80	84	73	—	89	76	59	74	—	88	—	—
1846–55	43	—	82	82	75	102	—	99	91	78	78	—	138	—	—
1856–65	96	56	104	98	92	108	139	107	103	100	92	0	—	—	—
1866–75	127	86	108	134	129	151	168	138	121	91	117	72	—	—	—
1876–85	135	78	113	130	120	148	—	91	157	172	122	197	—	—	—
1886–95	164	130	157	143	—	128	—	—	148	—	146	202	—	—	—
1896–1905	189	149	—	—	—	—	—	—	—	—	179	—	—	205	—
AVG.[a]	115	88	95	103	105	67	116	97	84	91	109	163	94	61	—
AVG.[b]	100	100	100	100	100	100	100	100	100	100	100	100	100	100	—

SOURCE: Davis-Gallman-Hutchins tape.
[a] Vessel weights.
[b] Decade weights.

The fleet gradually aged (see Table 10.15). On average ships were older than barks, but the relationships among grounds were more complex. In the early decades it was the older barks that sailed for the distant grounds, but the ships that ventured into the Indian, Pacific, and Arctic oceans were somewhat younger than those in the Atlantic. In the middle decades all grounds appear to have drawn from about the same age pool. In the later decades it was the older vessels that hunted in the Atlantic, while the younger ones searched for whales in the Indian, Pacific, and Arctic. A glance at the entering and exiting vessels (Table 10.13) suggests that on average it was the older units that left and that there were few ground-to-ground differences. Only ships in the Atlantic show any substantial deviation, so perhaps the Atlantic had become something of a dumping ground for the oldest ships in the fleet. That conclusion finds additional support in an examination of the average age of vessels hunting the Atlantic. After 1845 they were significantly older than ships hunting in other grounds.[38] It is possible, therefore, that the contemporaries' view of the fleet as a dumping ground, discussed earlier, was colored by their knowledge of ships hunting close to home.

It is difficult to disentangle the effects of age and experience. The figures for "years in fleet" mirror closely the pattern displayed by the age calculations (see Table 10.16). Standardizing years in fleet by average ages indicates that there were few important interground differences. The same lack of interground pattern holds for vessels that left the fleet—for any given rig type, those that exited were neither more nor less experienced than those that remained.

In almost every ground for all three classes of vessels, the ratio of labor to tonnage rose over the first four decades (see Table 10.17), but the disaggregated figures show no further increase. Thus there is no evidence of any increase in the ratio of labor to capital after the middle of the century. It is interesting to note that the non-ground-related component of size increase was accompanied by an increase in the labor-capital ratio until midcentury, but the increases associated with the new technical configurations in the years after the Civil War did not demand any disproportionate increase in labor. In fact, given the evidence of declining crew quality, they may have used less quality-adjusted labor per ton.

For barks, there were substantial interground differences in the relative proportions of the two factors of production. The labor-capital ratios were highest in the Atlantic and successively smaller in the In-

---

[38] By decade, beginning in 1846–1855, the ratios of the age of ships in that ground to all ships was 1.26, 1.14, 1.30, and 1.01.

TABLE 10.16
Years in Fleet, by Decade and Ground, New Bedford Whalers, 1816–1905
(Departure year)

	Barks					Ships					Other			
	Atl.	Hud.	Ind.	Pac.	Arc.	Atl.	Hud.	Ind.	Pac.	Arc.	Atl.	Hud.	Ind.	Pac.	Arc.

YEARS

1816–25	2.9	—	—	5.5	—	4.3	7.0	—	5.7	—	2.0	0.0	1.0	1.8	—
1826–35	3.8	—	2.0	7.0	—	4.7	11.0	1.8	7.0	—	3.4	—	5.0	12.0	—
1836–45	6.1	—	9.6	7.7	1.5	11.3	—	9.9	9.7	9.0	2.9	—	4.0	—	—
1846–55	3.1	—	8.1	7.5	7.1	22.1	—	16.0	13.2	13.3	3.2	—	0.0	—	—
1856–65	12.0	13.3	15.0	10.8	12.9	20.0	9.0	19.4	17.0	20.3	2.3	0.0	—	—	—
1866–75	17.6	21.9	15.0	17.5	15.0	24.6	8.7	29.1	22.6	20.8	3.0	1.3	—	—	—
1876–85	21.4	6.4	11.6	21.5	20.5	32.8	—	24.0	31.2	42.1	4.1	4.5	—	—	—
1886–95	31.4	38.8	30.5	27.8	—	39.0	—	—	34.9	—	12.4	7.0	—	0.0	—
1896–1905	39.2	38.7	—	—	—	—	—	—	—	—	19.3	5.0	3.3	5.7	—
AVG.[a]	16.0	21.1	11.9	12.5	14.3	9.2	8.6	13.8	12.3	17.7	6.5	4.3	3.3	5.7	—
AVG.[b]	15.3	23.8	13.1	13.2	11.4	20.0	8.9	16.7	17.6	21.1	5.8	0.8	2.5	4.6	—

GROUND RELATIVES [BARKS IN ATLANTIC = 100]

1816–25	100	—	—	190	—	148	241	—	197	—	69	0	34	62	—
1826–35	100	—	53	184	—	124	289	47	184	—	89	—	132	316	—
1836–45	100	—	157	126	25	185	—	162	159	148	48	—	66	—	—
1846–55	100	—	261	242	229	713	—	516	426	429	103	—	0	—	—
1856–65	100	111	125	90	108	167	75	162	142	169	19	0	—	—	—
1866–75	100	124	85	99	85	140	49	165	128	118	17	7	—	—	—
1876–85	100	30	54	100	96	152	—	112	146	197	19	21	—	—	—
1886–95	100	124	97	89	—	124	—	—	111	—	39	22	—	36	—
1896–1905	100	99	—	—	—	—	—	—	—	—	49	13	—	—	—
AVG.[a]	100	132	74	78	89	58	54	86	77	111	41	27	21	36	—
AVG.[b]	100	156	86	86	75	131	58	109	115	138	38	5	16	30	—

AVERAGE YEARS/AVERAGE AGE × 100

1816–25	41	—	39	—	—	36	84	—	50	—	25	0	—	53	—
1826–35	26	—	16	47	—	30	100	9	43	49	31	—	83	100	—
1836–45	40	—	46	45	8	51	—	42	48	55	19	—	57	—	—
1846–55	28	—	41	43	41	71	—	61	54	65	20	0	0	—	—
1856–65	49	64	59	52	61	61	31	68	62	73	12	14	—	—	—
1866–75	55	69	57	61	50	53	25	79	70	78	12	18	—	—	—
1876–85	63	22	42	77	74	73	—	—	74	—	16	27	—	—	—
1886–95	76	81	80	91	—	100	—	100	88	—	41	—	—	0	—
1896–1905	82	70	—	—	—	—	—	—	—	—	52	—	—	—	—
AVG.[a]	55	65	51	57	59	45	36	53	55	62	29	20	44	58	—
AVG.[b]	60	65	54	62	49	66	43	63	66	68	28	6	31	29	—

SOURCE: Davis-Gallman-Hutchins tape.
[a] Vessel weights.
[b] Decade weights.

378

TABLE 10.17
Men per Ton, by Decade and Ground, New Bedford Whalers, 1816–1905

	Barks					Ships					Other			
	Atl.	Hud.	Ind.	Pac.	Arc.	Atl.	Hud.	Ind.	Pac.	Arc.	Atl.	Hud.	Ind.	Pac.	Arc.
1816–25	.0735	—	.0833	.0802	—	.0698	.0799	.0769	.0723	—	.1029	—	—	.0907	—
1826–35	.0803	—	.0999	.0883	.0947	.0780	.0722	.0852	.0695	.0795	.1173	—	.1262	.1007	—
1836–45	.0912	—	.1003	.0931	.0871	.0818	—	.0882	.0779	.0794	.1438	—	—	—	—
1846–55	.1185	—	.0950	.0910	.0850	.0952	—	.0842	.0832	.0832	.1574	.1107	—	—	—
1856–65	.1110	.0911	.0950	.0885	.0854	.0773	.0944	.0839	.0817	.0803	.1392	.1264	—	—	—
1866–75	.1072	.1098	.0900	.0885	.0867	.0831	.0955	.0831	.0850	.0918	.1601	.1179	—	.0941	—
1876–85	.1044	.1251	.0975	.0875	—	.0870	—	—	.0808	—	.1493	—	—	—	—
1886–95	.1102	.1061	.0990	.0972	—	.0809	—	—	.0835	—	.1383	—	—	—	—
1896–1905	.1193	.1193	—	—	—	—	—	—	—	—	.1408	—	—	—	—
AVG.[a]	.1049	.1056	.0967	.0905	.0858	.0785	.0887	.0860	.0798	.0815	.1421	.1191	.1262	.0941	—
AVG.[b]	.1017	.1103	.0950	.0890	.0878	.0816	.0855	.0836	.0792	.0828	.1388	.1183	.1262	.0957	—

GROUND RELATIVES [BARKS IN ATLANTIC = 100]

1816–25	100	—	104	100	—	95	109	96	98	—	140	—	—	123	—
1826–35	100	—	110	97	—	97	90	93	87	87	146	—	157	125	—
1836–45	100	—	85	97	104	90	—	74	85	67	158	—	—	—	—
1846–55	100	—	86	79	74	80	—	74	70	75	133	100	—	—	—
1856–65	100	82	84	82	77	70	85	76	74	75	125	118	—	—	—
1866–75	100	102	93	83	80	78	89	78	79	88	149	113	—	—	—
1876–85	100	120	93	84	83	83	—	80	77	—	143	—	—	90	—
1886–95	100	96	90	88	—	73	—	—	76	—	125	—	120	94	—
1896–1905	100	100	—	—	—	—	—	—	—	—	118	—	—	—	—
AVG.[a]	100	101	92	86	82	75	85	82	76	78	135	114	120	90	—
AVG.[b]	100	108	93	88	86	80	84	82	79	81	136	116	124	94	—

DECADE RELATIVES [DECADE WEIGHT AVERAGE IN GROUND = 100]

1816–25	72	—	88	90	—	86	93	92	91	—	74	—	—	95	—
1826–35	79	—	105	99	—	96	84	102	88	—	85	—	100	105	—
1836–45	90	—	106	105	108	100	—	106	98	96	104	—	—	—	—
1846–55	117	—	100	102	99	117	—	101	105	96	113	94	—	—	—
1856–65	109	83	95	99	97	95	110	101	103	100	100	107	—	—	—
1866–75	105	100	103	98	97	102	112	99	107	97	115	100	—	94	—
1876–85	103	113	104	109	99	107	—	—	102	—	108	—	—	—	—
1886–95	108	96	—	—	—	99	—	—	105	111	100	—	100	98	—
1896–1905	117	108	—	—	—	—	—	—	—	—	101	—	—	—	—
AVG.[a]	103	96	102	102	98	96	104	103	101	98	102	100	100	98	—
AVG.[b]	100	100	100	100	100	100	100	100	100	100	100	100	100	100	—

SOURCE: Davis-Gallman-Hutchins tape.
[a] Vessel weights.
[b] Decade weights.

dian Ocean, the Pacific, and the Arctic hunting grounds. At the height of Arctic whaling, for example, the labor-capital ratio was only about three-quarters of that found in the Atlantic. The smaller figure is in part a reflection of the use of larger vessels in the more distant grounds.[39] Between 1846 and 1885, a bark hunting in the Arctic was on average half again as large as one assigned to the Atlantic. For ships, in contrast, there were few interground differences, although those that hunted the Indian Ocean used slightly more labor. Brigs, sloops, and schooners employed relatively much more labor, but they appeared so infrequently in any but the Atlantic ground that there is no basis for an estimate of interground differences.

Between classes, even if the focus is narrowed to the Atlantic, brigs, sloops, and schooners continued to display much higher labor-capital ratios than either barks or ships. By the end of the period, however, the differential between the smaller vessels and barks had narrowed from close to 40 to less than 20 percent. That narrowing, when combined with their greater maneuverability, provides at least a partial explanation for their increasing popularity.

Overall, barks used more labor per ton than ships, but that conclusion does not hold equally for all grounds, and the differential diminishes as the distance from New Bedford increases. In the Atlantic, barks employed one-fourth more labor per ton of vessel; in the Indian and Pacific, the difference was less than half that great; in the Arctic, the difference was less than 5 percent. In part, this phenomenon can be explained by the relationship between vessel size and ground hunted, but the complete explanation appears to be more complex. A simple regression using the ratio of the tonnage of barks to the tonnage of ships and the square of that ratio "explains" about one-half of the variance in the ratio of men per ton in the two vessel classes.[40] The remain-

---

[39] There is a possibility that there is some measurement bias: vessels hunting the Atlantic may have been less likely to take on additional crew after leaving New Bedford than vessels hunting the Pacific and Arctic oceans. But if this explanation were correct, one would expect to find the same interground variations in the labor-ton ratio for ships. In fact there are none.

[40] The regression equation is:

$$\text{relmen} = 0.1308 + 3.3269 \text{ relton} - 2.6098 \text{ relton}^2$$
$$(1.789) \quad (-2.244)$$
$$r^2 = 0.4937, \text{ the } F\text{-ratio } 12.189, \text{ and the Durbin Watson } 2.0103,$$

where relmen is the ratio of men per ton in barks to men per ton in ships, relton the ratio of the average tonnage of barks to that of ships, and relton^2 is that latter ratio squared. The argument is that the New Bedford fleet represented a sample of all whaling vessels. It is in that spirit that the T- and F-statistics are offered. To the extent that we are talking only about New Bedford vessels, those statistics, of course, have no meaning.

ing variance is not easily explained. It is not related to the passage of time, nor is it directly related to the species of whales caught. It is, however, in part at least, related to some interaction between the technological variables and the species hunted. The addition of neither time nor a variable directly reflecting the proportion of baleens caught improves the explanatory power of the model. Inclusion of "cross terms" between the proportion of baleens and the two tonnage variables does.[41] While there is still a substantial unexplained residual (perhaps reflecting the possible undercounting of crews in vessels fishing distant grounds), it does appear that there is some relationship between the capital stock and productivity, and that the simple analysis has not captured that relationship.

The average difference was not great, but barks hunting in the Arctic used somewhat less labor after 1876. In the last decade of Arctic hunting, barks appeared in larger than normal proportions, and the reversal of the labor-cost differential could go a long way toward explaining the increase in their popularity.

It was not difficult for an owner to recognize when a vessel reached an age where it could no longer perform or when it might pay to transfer it to another port or another activity. It was more difficult to determine *ex ante* which vessels were due to be lost at sea, but vessels were lost in large numbers. Of the 750 vessels in the fleet, 231 failed to return from a voyage. There was, however, considerable year-to-year variation in the rate of loss (see Table 10.18). Confederate raiders sank many vessels during the first years of the war, and again, there were two years when a large fraction of the Arctic fleet was caught by a sudden freeze and lost in the ice. There was no long-term trend in losses for ships and barks, although for brigs, sloops, and schooners, there is some evidence of improvement over the last few decades.[42] There were, however, marked differences between the classes and, for a given class, among the grounds hunted. The crude loss rate per voyage for all vessels was 8.7, but the figure for ships was 7.2, for barks 10.4, and for

---

[41] The regression equation is:

$$\text{relmen} = 0.1229 + 4.2197 \text{ relton} - 3.6452 \text{ relton}^2 - 0.0056 (\text{rel} - \text{bone})^2$$
$$(2.420) \qquad (-3.015) \qquad (-3.445)$$
$$+ 0.00068 \text{ rel}^2 \times \text{bone})$$
$$r^2 = 0.6929, F = 12.974, \text{DW} = 2.0084, \qquad (3.192)$$

where (rel × bone) and (rel² × bone) are the cross terms between the proportion of baleen and the tonnage variables.

[42] The average loss rate for "other" vessels between 1816 and 1845 was 10.6. Between 1846 and 1875 it rose to 14.4, but between that date and 1905 it fell to 5.6 percent.

TABLE 10.18
Annual Loss Rate,[a] by Ground and Class, New Bedford Whalers, 1816–1905

	Barks					Ships					Other			
	All	Atl. & Hud.	Ind.	Pac.	Arc.	All	Atl. & Hud.	Ind.	Pac.	Arc.	All	Atl. & Hud.	Ind.	Pac.	Arc.
1816–25	0.0	0.0	—	0.0	—	3.0	3.3	—	2.7	—	3.1	43.0	0.0	—	—
1826–35	2.7	3.6	0.0	2.4	—	1.5	4.0	0.0	0.5	—	17.4	19.2	25.8	20.9	—
1836–45	0.8	0.0	3.7	0.0	0.0	1.7	1.4	2.4	1.6	0.0	11.2	11.8	0.0	—	—
1846–55	3.8	1.9	2.0	4.5	5.2	3.0	5.6	2.3	2.7	5.9	7.9	6.1	54.5	—	—
1856–65	3.7	4.0	2.3	3.4	12.8	2.7	6.5	4.4	2.1	3.4	27.3	27.3	—	—	—
1866–75	5.4	3.3	3.3	5.9	8.4	3.5	6.1	1.3	3.6	3.6	8.0	8.0	—	—	—
1876–85	2.6	3.8	1.7	1.9	7.7	3.8	1.7	0.0	4.6	11.5	10.4	10.4	—	—	—
1886–95	0.5	1.2	0.0	0.0	—	0.0	0.0	—	0.0	—	1.8	1.8	—	—	—
1896–1905	3.1	3.1	—	—	—	—	—	—	—	—	4.7	4.7	—	—	—
AVG.[b]	3.5	3.2	2.7	3.5	7.0	2.6	3.2	2.7	2.2	4.3	8.7	8.6	34.5	20.9	—
AVG.[c]	2.8	2.3	1.9	2.3	6.8	2.4	3.6	1.7	2.2	4.9	10.2	10.4	40.1	20.9	—

SOURCE: Davis-Gallman-Hutchins tape.
[a] Number of vessels lost per vessel year of voyaging × 100.
[b] Vessel weights.
[c] Decade weights.

"others" 12.0.[43] Brigs, sloops and schooners were at sea for relatively short periods of time, and adjustment for exposure makes the contrast even sharper. The loss rate per year at sea for all classes averaged 3.2. It was 2.6 for ships, 3.5 for barks, and an astounding 8.7 for "other" vessels."[44]

Although the few brigs and schooners that ventured that far did not fare well, the Indian Ocean was the safest ground for barks and was reasonably safe for ships as well.[45] The Pacific was kindest to ships but was the second most dangerous ground for barks. The Atlantic and Hudson's Bay display the same time-adjusted loss rate for ships and barks, but the figure is relatively low for the latter and moderately high for the former. For both ships and barks, the Arctic proved most treacherous. In that ground the rate was more than 7.0 for barks and 4.0 for ships. For ships, the loss rate was a third higher in the Arctic than in the Atlantic and Hudson's Bay (the next most dangerous ground). For barks it was twice as high as the Pacific rate—a distant second on the danger list for that class of vessel.

On average ships sank less frequently than barks, but the difference was not independent of the ground hunted. Nowhere was the experience of barks better than that of ships, but it was almost identical in the Atlantic and Hudson's Bay and in the Indian Ocean. In the Pacific and the Arctic, however, the loss rate was more than 60 percent higher. These were profitable grounds and barks were innovated in both, but they were also dangerous grounds for the new entrants.

VI. Productivity Changes

The productivity of the New Bedford whaling fleet rose in the first few years of the industry's expansion, then fell fairly dramatically in the period of rapid growth. As the industry reached its maximum size, the decline slowed, and through the rest of the industry's history, its productivity followed the profile of a shallow plate, declining moderately, stabilizing, and then rising moderately. Through the period in which the fleet was most active, it would be fair to say that productivity changed little, except for the year-to-year variations exhibited by any industry in which a struggle with nature is central to the production process. There is no evidence to support the conventional view that a

[43] The crude loss rate per voyage is the number of vessels lost divided by the number of voyages × 100.
[44] The rate measures number of vessels lost per number of vessel years at sea × 100.
[45] Because of the small numbers in Hudson's Bay, that ground has been combined with the Atlantic for this analysis.

decline in productivity was one factor leading to the destruction of the New Bedford fleet.[46] It is, however, certainly true that during a period of dynamic productivity change in U.S. agriculture, transportation, and manufacturing, the whaling industry was a laggard. While it was not destroyed by declining productivity, its viability may very well have been affected by its failure to keep up.

The dip in productivity, the subsequent rise when U.S. whaling went into decline, and the replication of that overall dip and rise for each hunting ground, as the intensity of hunting increased and then diminished, could be interpreted to mean that productivity was responding to changes in the supplies of whales, a view widely held by contemporaries. The quantitative evidence, however, suggests otherwise. For example, best estimates indicate that there were between 1.8 and 2.4 million sperm whales in the early nineteenth century. According to Alexander Starbuck, American whalers captured about 188,000 and killed but lost another 38,000 in the seventy-three-year period 1804–1876. Those figures certainly do not suggest an ecological disaster. Adding the whales killed by the hunters of other nations is unlikely to increase the number by as much as one-sixth of the original total.

Baleens were less abundant. Leaving out of account the North Atlantic (there are no good data), bowheads, rights, grays, and humpbacks numbered between a quarter and a half a million, while American

---

[46] This section is based on Lance E. Davis, Robert S. Gallman, and Teresa Hutchins, "Productivity in American Whaling." Total-factor productivity was measured by four separate indexes, all of which yielded similar patterns, patterns not unlike those appearing in Table 10.20, below. Index (a) is the translog multilateral productivity index (hereafter, the "superlative index"),

$$\ln_{kn} = \tfrac{1}{2}\left[\sum (R_i^k + \overline{R}_i)(\ln Y_i^k - \overline{\ln Y_i})\right] - \tfrac{1}{2}\left[\sum_n (w_n^k + \overline{w}_n)(\ln X_n^k - \overline{\ln X_n})\right]$$

where: the R's are the shares of total revenue earned by individual outputs;
the Y's are quantities of individual outputs;
the w's are factor shares in income;
the X's are quantities of factor inputs;
the $\bar{R}$'s, $\bar{Y}$'s $\bar{X}$'s and $\bar{w}$'s are average values across all observations.

Indexes (b), (c), and (d) are the ratio of constant price outputs divided by constant price inputs, the price bases selected being early in the period [index (b)], toward the middle of the period [index (c)], and late in the period [index (d)]. Labor and capital inputs were measured in man-months and ton-months. Land (stocks of whales) was not entered into the index, but was used in the form of lagged independent variables in regression analysis of productivity, productivity in this instance being measured at the level of the voyage. See Douglas W. Caves, Laurits R. Christensen, and W. Erwin Diewert, "Multilateral Comparisons of Output, Input, and Productivity Using Superlative Index Numbers," *The Economic Journal* 92 (March 1982), pp. 73–86, for a discussion of the superlative index.

whalers probably killed about 194,000 in the years 1804–1876. Initially, the bowheads, rights, and grays, on whom most of this destruction was visited, probably numbered fewer than 200,000. Even so, one may doubt that nineteenth century hunting drastically thinned baleen stocks, except in the North Atlantic and, perhaps, among the California grays. Baleen reproductive capacity is formidable; the rate of natural increase runs close to 5 percent per year when the herds are below the carrying capacity of the available feed. There were few nineteenth century years when the Americans killed more than this figure, the maximum sustainable yield of baleens.

The Americans almost certainly did not permanently damage whale populations (again the North Atlantic rights and California grays are exceptions). Intense hunting in a specific, narrowly defined ground might nonetheless have led to reduced productivity for vessels in that ground until the whalers learned to look elsewhere. The source of the decline might not have been a reduction in the number of whales in an area, but the distribution of the given supply among a larger number of vessels. It is difficult to examine this problem rigorously since, except in the cases of sperm whales and bowheads, there are no whale-stock data specific to narrowly defined hunting grounds; and, in the case of the sperm whales, there are no good aggregate hunting data specific to those narrowly defined grounds. Furthermore, since a whaler on any given voyage could and did hunt through more than one ground and took both baleens and sperms, intense hunting of one species in one part of one ocean might not necessarily be followed by dramatic declines in productivity; it might instead be followed by a moderate shift in hunting ground and an increase in the catch of the other species.

Given these problems, no firm and final conclusion can be reached with respect to the effects of hunting on subsequent productivity. Nonetheless, regression analysis, taking into account levels of hunting and maximum sustainable yields, suggests that intense hunting of baleens in the Atlantic and baleens and sperms in the Indian may have been followed by declining productivity.[47] Further analysis, including an attempt to extract the effect of the number of vessels hunting in a ground, may provide additional insights into the source of this problem.

Similarly, in confirmation of the conventional view, increases in real wages ashore of both skilled and unskilled workers were followed by declines in whaling productivity. There is some hint, then, in the statistical analysis, of a whaling labor problem—a problem widely discussed in the literature.

[47] Davis, Gallman, and Hutchins, "Productivity in American Whaling."

The steps taken by the owners and captains of whalers that appear directed toward enhancing productivity are shown, by the statistical analysis, to have been on the whole effective. Thus the shift to larger vessels definitely raised productivity, as did the shift toward barks and toward hunting in the Arctic. The first burst of technical improvement, centered on the 1850s, does not show up in enhanced productivity, but the second, in the late 1870s, clearly does.

The statistical analysis, then, suggests the following story of productivity change in the New Bedford fleet: as the industry expanded rapidly in the years after the War of 1812, productivity initially declined, at least in part due to intense hunting of the grounds first exploited. The whalers adjusted by shifting to new, richer grounds; adopting bigger, more effective vessels with rigs appropriate to the new hunting problems; and adopting new hunting techniques. These changes counteracted the forces tending to push productivity down and, taken together with the subsequently diminished intensity of hunting, they eventually raised productivity above the levels attained at the apogee of the American and New Bedford industries.

They did not, however, save New Bedford. The diminution of hunting brought, as its counterpart, growing unemployment and other economic losses in the town. These impacts were muted by the ability of whaling-vessel owners—and their crews—to shift to the West Coast and continue whaling there or to convert to the merchant marine. Such shifts were less easy, however, for the onshore workers and for the capital tied up in the processing of whale products. This congeries of effects resembles those experienced in recent decades in the U.S., as investment has shifted from the old industrial heartland to the Sun Belt or overseas. Mobile workers and capital escape; immobile workers and capital face long-term unemployment. What distinguishes the New Bedford whaling industry from most modern, troubled industries is that such a large fraction of the capital and labor was mobile. "Fixed capital" consisted importantly of vessels, which could move both geographically and sectorally. But these options were temporary. As the U.S. merchant marine continued its long decline and as West Coast whaling petered out, these doors were closed. But by then the fleet was small and old. It expired quietly.

VII. Revenues and Profits

Over the ninety years 1816–1905 the revenue-generating ability of ships declined, that of barks increased somewhat, and that of the brigs, sloops, and schooners, although subject to substantial short-term varia-

TABLE 10.19
Average Yearly Earnings per Ton, New Bedford Whalers, by Ground and Class, 1816–1905
(Constant dollars of 1910–1914)[a]

	Barks						Ships						Other				
	All	Atl.	Hud.	Ind.	Pac.	Arc.	All	Atl.	Hud.	Ind.	Pac.	Arc.	All	Atl.	Hud.	Ind.	Pac.	Arc.
1816–25	46	63	—	—	28	—	65	68	91	—	61	—	69	69	—	43	68	—
1826–35	64	68	—	35	62	47	64	67	70	83	60	94	55	53	—	48	66	—
1836–45	53	51	—	63	50	53	56	50	—	65	55	61	51	52	—	8	—	—
1846–55	54	44	150	57	54	60	58	69	51	60	57	55	23	23	37	—	—	—
1856–65	54	59	49	48	47	52	51	41	61	45	51	55	86	135	47	—	—	—
1866–75	49	53	40	48	44	52	54	55	—	45	55	62	64	67	52	—	—	—
1876–85	81	68	71	44	87	139	65	58	—	53	67	96	64	67	—	—	—	—
1886–95	51	51	28	59	47	—	45	38	—	—	48	—	79	79	111	—	—	—
1896–1905	80	89	—	—	—	—	—	—	—	—	—	—	89	85	—	—	—	—
AVG.[b]	58	60	81	52	54	79	57	60	63	58	56	59	64	65	63	33	67	—
AVG.[c]	59	61	68	51	52	70	57	50	68	59	57	74	64	70	62	33	67	—

GROUND RELATIVES (ALL BARKS = 100)

1816–25	100	137	—	—	61	—	141	148	198	—	133	—	150	150	—	93	148	—
1826–35	100	106	—	55	97	—	100	105	109	130	94	—	86	83	—	75	103	—
1836–45	100	96	—	119	94	89	106	94	—	123	104	177	96	98	—	15	—	—
1846–55	100	81	277	106	100	98	107	128	94	111	106	113	43	43	69	—	—	—
1856–65	100	109	100	89	87	111	94	76	94	83	94	102	159	250	96	—	—	—
1866–75	100	108	49	98	90	106	110	112	124	92	112	127	131	137	64	—	—	—
1876–85	100	84	139	54	107	172	80	72	—	65	83	119	79	83	—	—	—	—
1886–95	100	100	35	116	92	—	88	75	—	—	94	—	155	155	139	—	—	—
1896–1905	100	111	—	—	—	—	—	—	—	—	—	—	111	106	—	—	—	—
AVG.[b]	100	103	140	90	93	136	98	103	109	100	97	102	110	112	109	57	116	—
AVG.[c]	100	103	115	86	88	119	97	85	115	100	97	125	108	119	105	56	114	—

YEAR RELATIVES (AVERAGE DECADE IN GROUND = 100)

1816–25	78	103	—	—	54	—	114	136	134	141	107	—	108	99	—	130	101	—
1826–35	108	111	—	69	119	—	112	134	103	110	105	165	86	76	—	145	99	—
1836–45	90	84	—	124	96	67	98	100	—	102	96	107	80	74	—	24	—	—
1846–55	92	72	221	112	104	76	102	138	75	76	100	96	36	33	60	—	—	—
1856–65	92	97	72	94	90	86	89	82	90	76	89	109	134	193	75	—	—	—
1866–75	83	87	59	94	85	74	95	110	—	76	96	168	100	96	84	—	—	—
1876–85	137	111	104	86	167	199	114	116	—	90	118	—	123	113	179	—	—	—
1886–95	86	84	41	116	90	—	79	76	—	—	84	—	139	121	—	—	—	—
1896–1905	136	146	—	—	—	—	—	—	—	—	—	—	—	—	—	—	—	—
AVG.[b]	98	98	119	102	104	113	100	120	93	98	98	104	100	93	102	106	100	—
AVG.[c]	100	100	100	100	100	100	100	100	100	100	100	100	100	100	100	100	100	—

SOURCE: Davis-Gallman-Hutchins tape.
[a] Current price earnings deflated by Warren-Pearson price index.
[b] Vessel weights.
[c] Decade weights.

tion, displayed little trend (see Table 10.19). Between classes, "others" produced slightly more revenue per ton year than barks or ships, but the difference was not great. Moreover, that finding rests almost entirely on the performance in the last third of the period, when ships disappeared and hunting contracted into the Atlantic and Hudson's Bay. In those grounds the smaller vessels enjoyed a relative advantage.[48] Revenue is not profit, however, and the smaller vessels displayed higher labor-to-capital ratios and experienced a much higher loss rate.

Not all of these "aggregate" changes can be traced to technical shifts. Interground differentials provide a substantial part of the explanation. The ability of the average bark to generate revenue was pushed upward by its very substantial earning power in the northern (Arctic and Hudson's Bay) grounds, and also, marginally, by its performance in the Atlantic. For ships, the Atlantic provided a rich reward in the first decade, but not thereafter. The Indian Ocean, a ground that produced above average returns, did so only because of the catch in the years before 1855; and the Pacific, the largest and most important ground, while yielding marginally less revenue than the average, kept the level of total revenues up. For ships, like barks, it was the Arctic that produced the greatest monetary rewards; however, that ground was not as relatively rich for them as it was for their fore-and-aft-rigged peers.

There is little trend in the revenue series from the Pacific and the Arctic, although there is a suggestion of a small decline for ships in the former. If it exists, it is substantial only in the last decade. In general, ships returned more revenue than did barks in the Pacific, but that result was reversed after 1875. The Atlantic, the earliest ground hunted, proved rewarding in the early decades, suffering a precipitous decline in the late 1830s, but remained relatively stable thereafter. For both classes of vessel, the Indian Ocean generated substantial returns over the first half of the period, but steadily smaller returns thereafter.

Two facts stand out. The Arctic provided an important source of revenue after 1855, and the industry suffered when those revenues stopped. It was not, however, a decline in the ground's resources that caused the withdrawal of the New Bedford fleet. Revenues from the Arctic for both ships and barks were higher in the last decade than they had ever been in any other decade in any other ground. Neither declining resources nor falling prices were the cause of the relocation; instead, it was competition from the San Francisco fleet that forced the Yankees to change their ways. This conclusion is supported by the

---

[48] For the Atlantic in the years between 1826 and 1875, the earnings rates were $4.54 for barks, $4.92 for ships, and $3.89 for brigs, sloops, and schooners.

activities of the vessels that withdrew from the Arctic. Most did not turn to other maritime pursuits but transferred their base to the West Coast port.

The dramatic relocation of the New Bedford fleet raises doubts about the validity of interground comparisons after 1885. There were no ships in the fleet and no voyages by barks to any of the three western grounds in the last decade, and there were only forty-seven voyages by ships and barks to those grounds in the previous ten years. That period aside, however, there is little evidence that whaling was becoming more difficult, although there were some exceptions (see Tables 10.20 and 10.21). The New Bedford captains appear to have had somewhat more difficulty finding sperm whales during the middle decades—following a period of intense hunting—but later the catch recovered. Of more lasting importance, there is evidence that the catch of baleens was declining in the Atlantic and in the Indian Ocean, although the decline was offset in part by increasing catches of sperm whales. The exceptions, therefore, do not appear to have been important enough to account for the demise of the New Bedford fleet. Instead, the decline appears rooted in changes in market demand, coupled with an increasingly noncompetitive position in the one ground where the value of the catch was still increasing.

That conclusion receives additional support from the data on rates of return (see Table 10.22). Potential profits provided the signals that led the owners to substitute barks for ships (and later brigs, sloops, and schooners for barks); to redirect activities from the Atlantic to the Indian and Pacific, to the Arctic, and then back again to the Atlantic; and to first expand and then contract the size of the fleet. That it was profits rather than revenues that triggered these changes would surprise no economist, but it does appear to have raised some questions in the eyes of contemporaries. In 1877, for example, the editor of the "Review" wrote,

> The present tendency being to cruise those grounds nearest to home, so that the catchings may be shipped at the earliest moment. We find in the North and South Atlantic Oceans a fleet of 100 vessels, while the more fruitful grounds of the Pacific Ocean, Japan, New Zealand and Sooloo Sea, are almost neglected.[49]

There are innumerable problems inherent in any attempt to measure returns to enterprise. The data on the New Bedford fleet do provide the basis for a set of estimates, but they are tentative, and the reader should

---

[49] *Merchant Transcript*, Jan. 15, 1878.

TABLE 10.20
Average Catch per Vessel Ton per Year of Hunting, New Bedford Whalers, 1816–1905

	Barks						Ships						Other			
	All	Atl.	Hud.	Ind.	Pac.	Arc.	All	Atl.	Hud.	Ind.	Pac.	Arc.	All	Atl.	Hud.	Ind.	Pac.	Arc.

(A) SPERM OIL (BBLS)

1816–25	2.12	2.90	—	—	1.34	—	1.24	0.39	—	—	2.19	—	1.66	1.58	—	0.00	2.92	—
1826–35	1.81	1.40	—	0.57	2.43	—	1.36	0.54	0.04	1.84	2.31	0.05	2.20	2.10	—	2.14	2.78	—
1836–45	1.06	0.93	—	0.84	1.31	0.99	1.11	0.66	—	0.74	1.42	0.15	1.69	1.73	—	0.24	—	—
1846–55	0.81	0.84	0.31	0.97	0.76	0.14	0.57	1.49	—	0.57	0.60	0.20	0.53	0.53	0.05	—	—	—
1856–65	0.80	1.29	0.04	1.01	0.70	0.16	0.46	0.76	0.20	0.64	0.50	0.19	1.29	2.53	0.11	—	—	—
1866–75	0.89	1.21	0.00	1.02	0.81	0.30	0.71	1.24	0.00	0.81	0.75	0.11	1.31	1.52	0.00	—	—	—
1876–85	0.98	1.33	0.00	1.19	0.79	0.16	0.91	1.04	—	1.10	0.91	—	1.63	1.92	—	—	—	—
1886–95	1.36	1.77	0.12	2.10	0.99	—	1.04	1.32	—	—	0.92	—	2.85	2.85	0.00	—	—	—
1896–1905	3.57	4.17	—	—	—	—	—	—	—	—	—	—	3.49	4.16	0.02	0.79	2.28	—
AVG.[a]	1.02	1.48	0.13	1.02	0.88	0.20	0.85	0.66	0.09	0.69	1.05	0.18	1.97	2.16	0.04	0.79	2.85	—
AVG.[b]	1.49	1.76	0.09	1.10	1.14	0.35	0.93	0.93	0.08	0.95	1.20	0.14	1.85	2.10				

(B) WHALE OIL (BBLS)

1816–25	0.08	0.12	—	—	0.03	—	4.26	6.22	—	—	2.01	—	3.24	3.81	—	4.04	0.00	—
1826–35	1.81	3.06	—	1.57	0.19	—	3.16	5.58	8.51	3.66	0.33	4.73	0.42	0.53	0.88	0.00	0.00	—
1836–45	1.69	2.11	—	2.55	0.85	1.51	1.84	3.01	6.75	3.00	0.98	2.34	0.28	0.29	—	0.00	—	—
1846–55	0.98	0.43	—	0.91	1.08	1.88	1.70	0.19	—	2.04	1.54	1.67	0.17	0.17	—	—	—	—
1856–65	0.73	0.36	3.77	0.24	0.62	1.83	1.12	0.51	1.07	0.76	1.04	2.01	0.68	0.48	0.88	—	—	—
1866–75	0.65	0.34	1.54	0.41	0.64	1.66	1.05	0.37	2.06	0.62	1.00	1.77	0.69	0.37	2.48	—	—	—
1876–85	1.49	0.98	0.58	0.21	1.94	3.02	0.96	0.99	—	0.28	0.98	—	0.41	0.27	1.20	—	—	—
1886–95	0.39	0.31	0.50	0.26	0.46	—	0.33	0.34	—	—	0.33	—	0.22	0.22	—	—	—	—
1896–1905	0.09	0.08	0.17	—	—	—	—	—	—	—	—	—	0.09	0.01	0.50	1.35	0.00	—
AVG.[a]	0.97	0.83	1.86	0.77	0.88	2.11	1.89	4.08	3.21	2.05	1.14	1.97	0.52	0.43	1.24	1.35	0.00	—
AVG.[b]	0.88	0.87	1.31	0.88	0.73	1.98	1.80	2.15	4.60	1.73	1.03	2.50	0.69	0.68	1.27		0.00	

(C) WHALE BONE (LBS)

1816–25	0.51	1.02	—	—	0.00	—	3.01	5.51	—	—	0.53	—	0.00	0.00	—	0.00	0.00	—
1826–35	10.34	18.71	—	0.00	0.00	—	6.34	11.19	0.00	15.67	0.18	41.63	0.86	1.10	—	0.00	0.00	—
1836–45	6.36	4.34	—	14.34	3.63	28.40	7.03	6.14	46.09	13.24	4.80	28.45	0.91	0.93	—	0.00	—	—
1846–55	9.83	3.61	64.94	7.13	11.41	20.30	16.95	0.70	—	15.47	16.14	16.42	0.00	0.00	—	—	—	—
1856–65	7.82	1.27	23.98	1.74	6.51	19.79	11.34	3.40	18.32	5.08	11.21	25.85	7.97	1.13	14.82	—	—	—
1866–75	5.82	1.40	12.87	2.53	5.19	40.86	11.75	1.30	33.32	4.10	11.47	21.33	2.64	2.43	3.80	—	—	—
1876–85	13.40	6.38	12.09	0.40	15.35	—	9.05	4.31	—	4.94	11.01	—	3.24	1.35	13.93	—	—	—
1886–95	2.37	0.33	4.16	0.38	3.51	—	2.75	0.20	—	—	3.82	—	0.60	0.60	—	—	—	—
1896–1905	0.62	0.00	—	—	—	—	—	—	—	—	—	—	3.89	0.00	24.32	0.00	0.00	—
AVG.[a]	8.17	3.89	32.14	4.97	7.67	26.43	10.63	7.50	21.61	11.95	9.57	22.04	2.07	0.95	14.57	0.00	0.00	—
AVG.[b]	6.34	4.12	23.60	3.79	5.70	21.87	8.53	4.09	24.43	9.75	7.40	26.74	2.23	0.84	14.22	0.00	0.00	—

SOURCE: Davis-Gallman-Hutchins tape.
[a] Vessel weights.
[b] Decade weights.

TABLE 10.21
Physical Index of Catch per Ton per Year of Hunt, by Product, Ground, and Decade, New Bedford Whalers, 1816–1905
(For each product, Barks in 1856–1865 = 100)

	Barks						Ships						Other					
	All	Atl.	Hud.	Ind.	Pac.	Arc.	All	Atl.	Hud.	Ind.	Pac.	Arc.	All	Atl.	Hud.	Ind.	Pac.	Arc.

(A) SPERM OIL

1816–25	265	363	—	—	168	—	155	49	—	—	274	—	208	198	—	0	365	—
1826–35	226	175	—	71	304	—	170	68	5	230	289	—	275	263	—	268	348	—
1836–45	133	116	—	105	164	124	139	83	—	93	178	6	211	216	—	30	—	—
1846–55	101	105	—	121	95	18	71	186	—	71	75	19	66	66	6	—	—	—
1856–65	100	161	39	126	88	20	58	95	25	80	63	25	161	316	14	—	—	—
1866–75	111	151	5	128	101	38	89	155	0	101	94	24	164	190	0	—	—	—
1876–85	123	166	0	149	99	20	114	130	—	138	14	14	204	240	—	—	—	—
1886–95	170	221	0	263	124	—	130	165	—	—	115	—	356	356	—	—	—	—
1896–1905	446	521	15	—	—	—	—	—	—	—	—	—	436	520	0	—	—	—
AVG.[a]	128	185	16	128	110	25	106	83	11	86	131	23	246	270	3	99	285	—
AVG.[b]	186	220	11	138	143	44	116	116	10	119	150	18	231	263	5	99	356	—

(B) WHALE OIL

1816–25	11	16	—	—	4	—	584	852	1166	—	275	—	444	582	—	553	0	—
1826–35	248	419	—	215	26	—	433	764	925	501	45	648	56	73	—	0	0	—
1836–45	232	289	—	349	116	207	252	412	—	411	134	321	38	40	121	0	—	—
1846–55	134	59	—	125	148	258	323	26	—	279	211	229	23	23	340	—	—	—
1856–65	100	49	516	33	85	251	153	70	147	104	142	275	93	66	164	—	—	—
1866–75	89	47	211	56	88	227	144	51	282	85	137	242	94	51	0	—	—	—
1876–85	204	134	79	29	266	414	132	136	—	38	134	—	56	37	69	—	—	—
1886–95	53	42	68	36	63	—	45	47	—	—	45	—	30	30	170	185	—	—
1896–1905	12	11	23	—	—	—	—	—	—	—	—	—	12	1	174	185	—	—
AVG.[a]	133	114	255	105	121	289	259	559	440	281	156	270	71	59	—	—	—	—
AVG.[b]	121	119	179	121	100	271	247	295	630	237	141	342	95	93	—	—	—	—

(C) WHALE BONE

1816–25	7	13	—	—	0	—	38	70	—	—	7	—	0	0	—	0	0	—
1826–35	132	239	—	0	0	—	81	143	589	200	2	—	11	14	—	0	0	—
1836–45	81	55	—	183	46	363	90	79	—	169	61	532	12	12	190	—	—	—
1846–55	126	46	—	91	146	260	217	9	—	198	206	364	0	0	49	—	—	—
1856–65	100	16	830	22	83	253	145	43	234	65	143	210	102	14	178	—	—	—
1866–75	74	18	307	32	66	523	150	17	426	52	147	331	34	31	311	—	—	—
1876–85	171	82	165	5	196	—	116	55	—	63	141	273	41	17	186	—	0	—
1886–95	30	4	155	5	45	—	35	3	—	—	49	—	7	7	182	0	0	—
1896–1905	8	0	53	—	—	—	—	—	—	—	—	—	50	12	—	—	—	—
AVG.[a]	104	50	411	64	98	338	136	96	276	153	122	282	26	12	—	0	0	—
AVG.[b]	81	53	302	48	73	280	109	52	312	125	95	342	29	11	—	0	0	—

SOURCE: Table 10.20.
[a] Vessel weights.
[b] Decade weights.

TABLE 10.22

Index of Profit Rates, by Rig and Ground, New Bedford Whalers, 1816–1905[a]

(All Barks 1846–1855 = 100)

	All	Atl.	Hud.	Ind.	Pac.	Arc.
			BARKS			
1816–25	70	123	—	—	18	—
1826–35	169	182	—	57	159	—
1836–45	107	99	—	139	100	84
1846–55	100	62	—	110	104	99
1856–65	146	159	568	121	118	177
1866–75	131	145	125	130	112	150
1876–85	288	225	83	118	322	559
1886–95	119	118	193	150	107	—
1896–1905	221	256	27	—	—	—
AVG.[b]	152	148	276	125	140	274
AVG.[c]	145	132	247	122	125	203
			SHIPS			
1816–25	126	139	203	—	117	—
1826–35	171	180	193	240	157	—
1836–45	124	102	—	153	120	253
1846–55	119	149	—	124	116	130
1856–65	141	101	136	115	142	157
1866–75	162	165	185	118	162	197
1876–85	219	185	—	167	231	274
1886–95	104	80	—	—	114	—
1896–1905	—	—	—	—	—	—
AVG.[b]	136	151	165	135	131	160
AVG.[c]	143	137	180	152	142	218
			OTHER			
1816–25	95	101	—	loss	91	—
1826–35	90	85	—	39	124	—
1836–45	66	58	—	loss	—	—
1846–55	loss	loss	—	—	—	—
1856–65	214	397	loss	—	—	—
1866–75	139	148	84	—	—	—
1876–85	156	162	120	—	—	—
1886–95	179	179	—	—	—	—
1896–1905	216	201	293	—	—	—
AVG.[b]	123	139	155	loss	110	—
AVG.[c]	117	141	137	loss	105	—

SOURCE: See text.
[a] No adjustment for time in port and crew maintenance at $60.
[b] Vessel weights.
[c] Decade weights.

approach them with caution.[50] The interdecadal, interrig, and interground rate-of-return differentials are robust to a wide variety of specifications; but the *level* of profit seems very high, even given the riskiness of whaling (see Table 10.22).[51] The analysis is confined to indices of

[50] In the profit calculation the revenue figures are those from Table 10.19. They have been adjusted for (1) the mariners' lay, estimated at 34 percent of the value of the catch, (2) seamen's maintenance (estimated at $60 per man per year in dollars of 1844 or $83 in 1910–1914 dollars), (3) depreciation and losses at sea, and (4) forgone interest on capital investment. The average age of a ship or bark when it was condemned was 33 years and that of a brig, sloop or schooner 22½. Therefore depreciation was calculated at the rate of 3 percent per year for the first two classes and 4½ percent for the last. Losses were calculated from losses per year by ground, adjusted for the difference between the average age of a vessel when lost and the average age when condemned. For ships, this difference amounted to 16 percent, for barks 20 percent, and for "others" 26 percent of their value. Forgone interest was estimated on the basis of a linked interest series of railroad and government bonds. By decades, beginning in 1816–1825, the rates were 7.9, 7.1, 7.6, 6.9, 8.1, 7.7, 5.7, 4.5, and 4.0 percent. The cost of a vessel was derived from a compilation of a number of literary sources. In dollars of 1910–1914, by decade, they averaged $70, $55, $60, $64, $47, $46, $45, $55, and $55 for ships and barks, and $79, $62, $68, $72, $53, $50, $49, $59, and $59 for brigs, sloops, and schooners. In each case the prices used were the costs of a new vessel in that decade (thus the estimated "profits" reflect returns in whaling alone and are net of capital gain or loss). Although a variety of sources were used, most of the ship-cost estimates are from John G. B. Hutchins, *The American Maritime Industry* (Chapters 6, 7, 9, and 17), Howard I. Chapelle, *The Search for Speed Under Sail 1700–1855* (New York: Norton, 1967); Henry Hall, "Report of the Shipbuilding Industry in the United States," *U.S. Tenth Census,* part 8 (printed as Miscellaneous Document of the House of Representatives, 47th Congr., 2nd sess. 1882–83, vol. 13, no. 42, part 8.

After the calculations were made, Peter Coclanis drew to our attention the following table from *Debow's Review,* which achieves results remarkably like ours [from *Debow's Review* 26 (May 1859), 590].

WHALING INTERESTS OF THE UNITED STATES

ESTIMATED VALUE OF THE 661 WHALE VESSELS SAILING FROM THE UNITED STATES, INCLUDING THEIR OUTFITS, PROVISIONS, AND THE ADVANCES MADE TO SEAMEN ON THE DAY OF SAILING, AT THE RATE OF $25,000 EACH	$16,525,000
SBx PER CENT, PER ANNUM, INTEREST ON THE SAME	991,500
TEN PER CENT, PER ANNUM, ALLOWED FOR WEAR AND TEAR	1,600,000
TWO AND A HALF PER CENT, INSURANCE	413,125
FRESH SUPPLIES PURCHASED BY THE MASTERS, EQUAL TO ABOUT $1,200 PER ANNUM EACH	793,000
AMOUNT OF MONEY PAID TO MASTERS, OFFICERS, AND CREW, BEING THEIR SHARES OF THE OIL TAKEN, EQUAL TO ONE THIRD OF THE GROSS VALUE OF THE PRODUCTS	4,013,601
TOTAL AMOUNT OF MONEY INVESTED, INCLUDING INTEREST, &C	24,336,226
VALUE OF THE ANNUAL AMOUNT OF OIL TAKEN, SHOWING A CLEAR YEARLY PROFIT OF 46 PER CENT	12,040,805
DIFFERENCE BETWEEN THE WHOLE CAPITAL INVESTED AND THE YEARLY PROFITS	12,295,421

[51] Costs incurred between voyages have not been included, and that omission certainly means that the levels are too high. There do not, however, appear to be any differences in the length of the refit between classes of vessels or between most grounds. The only exception is the Arctic. Both barks and ships returning from that ground appear to have taken more than one month longer to refit. As to level, our results are not unlike Debow's (footnote 50, above).

the rate of return on whaling investment by decade, ground and rig type, although the profit estimates are easily derived.[52]

The time profile of returns is similar for all rig classes and for all grounds, except for barks in the Arctic. Early, the number of vessels in any ground was small and the returns high. Entry was easy and profits attracted a swarm of new entrants. As any student of introductory economics would predict, profits declined; and the decline continued until the first postbellum decade. Then, however, as vessels withdrew and the price of bone began to soar, the average rate of return moved up again. The pattern of decline followed by an increase in returns tracks productivity and reflects in some measure the same underlying phenomena. Late nineteenth century profit recovery was substantial, even if the size of the total catch did not rebound to its previous peak. The barks that remained earned as much over the last three decades as they had in their most profitable years; and brigs, sloops, and schooners, although their numbers were small, earned more than they ever had. It was only ships that found the new environment wanting, and it is their sluggish profit performance that explains the observed substitution of barks for ships.

Disaggregation suggests that the scenario was similar for vessels in almost every ground. Increased activity in the Atlantic pushed returns down, entry into the Indian and Pacific oceans was triggered both by that decline and by substantial returns in those grounds. Profits there, however, also soon fell. Finally, the opening of the Arctic provided the greatest profit opportunities.

The profit estimates also tend to confirm that the new technical configurations altered the profitability of ships relative to barks and barks relative to brigs, sloops, and schooners. In the sixty years prior to 1875, ships were more profitable than barks (and much more profitable than "others") in almost every ground. The only exception was the Indian Ocean, where the fiscal dominance of ships disappeared in the mid-fifties. After 1875, however, ships were less profitable than barks in every ground.

Even in the Arctic there is indirect evidence of the decline in the relative profitability of ships. Despite the record high profits, only nine ships returned between 1876 and 1885 (and only four after 1880)—a sharp reduction from the twenty-one of the previous decade. For barks,

---

[52] The average profit rate for "All Barks" 1846–1855 was 32.1 percent. Multiplying this figure by the index number of any other category (and dividing by 100) will yield its actual profit rate. Thus the profit rate for Ships in the Pacific during 1876–1885 was $231/100 \times 32.1 = 74.2$ percent; for Barks in the Atlantic during 1896–1905 it was $256/100 \times 32.1 = 82.2$ percent.

the story is reversed. The number returning was thirty-nine, up more than a third from the total in the decade 1866–1875. It appears that economic pressure had, by 1880, driven all but the most profitable ships from the Arctic ground. Between 1876 and 1895, even if the Arctic is included, the almost 30-percent profit advantage that had characterized ships' voyages was reversed and turned into an almost equally large differential in favor of barks. So unprofitable had the larger vessels become that by 1895 every one had left the fleet. For brigs, sloops, and schooners the story is different. Barely two-thirds as profitable as barks for most of the period, over the last two decades (as hunting concentrated in the Atlantic and Hudson's Bay), they proved themselves almost one-fifth more profitable.

If the study is narrowed to the years 1835–1885 and to the Atlantic, Indian, Pacific, and Arctic (decades and grounds where ships and barks hunted side by side—or perhaps stern by bow), ships proved more profitable in every ground except at times in the Arctic. The Arctic proved the most lucrative for both classes, but ships found the Indian and Pacific oceans reasonably profitable and the Atlantic the least rewarding. In the case of barks, the Pacific and Atlantic grounds stood in the intermediate position, and the Indian Ocean produced the lowest monetary rewards.

Entry—and in whaling entry was easy—undercut profits, but technical accommodation left the surviving "firms" with substantial earnings. Moreover, while the Arctic continued to be profitable to New Bedford vessels, their owners found that profits were even higher when they transferred their base of operations to the West Coast. Thus New Bedford did not abandon the Arctic; New Bedford vessels abandoned New Bedford. Finally, the technical restructure that accompanied the late nineteenth century shift from western to eastern grounds, produced a fleet of barks, brigs, sloops, and schooners that, while much smaller in number, was as profitable as the fleet of barks and ships had been fifty years before.

VIII. Conclusions

The history of New Bedford whaling provides useful insights into the functioning of a competitive market blessed with a very malleable capital stock. Increases in the prices of the industry's final products drew waves of new entrants, while later declines in consumer demand pushed firms out of the industry almost as rapidly. Early entrants reaped supernormal profits. By the end of the Civil War, however, entry had caused average profit rates to decline by almost 40 percent, and in response owners redirected their vessels to other, more profitable, maritime opportunities. Fewer and more efficient "firms" caused profits to rebound. By the turn of the century, although the fleet was very different in both size and composition, profit rates were not substantially lower than they had been during the prosperous 1830s and 1840s. It might also be noted that, while profits declined as more vessels entered the hunt, there is little evidence of the kind of overhunting that might have endangered the stock of whales.

Of more interest, given the central focus of this paper, are questions relating to the evolution of the capital stock and the connections between the stock and industrial productivity. The industry is unique, because the capital stock, far from being specialized, was very malleable indeed. Thus both entry and exit were easy; and owners could quickly respond to changes in the profitability of whaling relative to other maritime pursuits. Such ease suggests that new entrants did not always reflect the most modern technical configuration.

It is clear that as the area of search widened, average vessel size increased; and the recorded increase is not a statistical artifact related to rig type or ground hunted. Between 1815 and 1845 the average size of all vessels increased. For ships, at least, the increase was accompanied by a reduction in variance. In the twenty-five years after 1845, ship size continued to edge upwards, albeit slowly; but the average size of barks increased even more rapidly. By the 1870s, however, it appears that an optimal size configuration had been achieved; thereafter there is no evidence of further change.

While the size composition of each class changed not at all after the 1870s, the class composition of the fleet continued to change over the entire period. The relative proportion of ships to barks had been decreasing since the 1820s, but in 1845 there were still five times as many ships as barks. Over the next thirty years, however, the ratio shifted from 5.30 to 1 to 0.53 to 1. Although the substitution of barks for ships occurred in every ground, it happened first in the Atlantic, and then in the Indian, the Pacific, and finally the Arctic. Barks were more maneuverable than ships and maneuverability is a virtue; but because of their relatively small size, they were at a disadvantage in the more distant grounds. As designs changed and their relative size increased, the disadvantage receded; and the benefits derived from their greater maneuverability made it possible for them to compete ever more effectively. In the North Pacific and the Arctic, where the ability to maneuver commanded a very large premium, the now-larger barks proved more productive than their square-rigged competitors.

Finally, the postbellum years saw an increasing fraction of vessels built to order. As competition squeezed out excess profits, only the most efficient vessels could survive. By 1870, economic survival implied both large size and bark rigging. In the short run the demand for more efficient vessels was met, at least in part, by rerigging ships as barks; in the longer run, newly built vessels moved into the fleet. In the case of barks, over the first thirty years of the study, 80 percent of the entrants were transfers from other activities, 13 percent were rerigged ships, and a mere 7 percent were built explicitly for whaling. Over the next thirty years, transfers represented about 60 percent, rerigged vessels accounted for almost a quarter, and the nation's shipyards accounted for 15 percent of the new entrants. Finally, in the years after 1875 (and that means the decade 1876–1885, since there are only two subsequent entrants) transfers had fallen to less than 60 percent, rerigs had all but disappeared (they accounted for only 3 percent), but custom-made vessels accounted for about 40 percent of barks entering the fleet. The 40-percent figure—a figure that represents twelve new vessels—was achieved in a period when the total tonnage of the New Bedford fleet was declining by about 60 percent.

Technology was improving, and apparently improving quickly enough to overcome, in part at least, the effects of a collapsing market for whale oil. Still, some questions remain. For one thing, there is little evidence of learning by doing.[53] Voyage length was long, and the length of time at sea had two effects. First, there may well have been some learning by doing within a voyage; such productivity increases would not be captured by our data. Second, crews objected to returning immediately to sea, and owners almost certainly did not want their vessels out of service. As a result crews turned over almost every voyage. It would be interesting to discover if crews tended to stay together as they moved from vessel to vessel and if experienced crews displayed greater productivity. For another thing, there is strong evidence that vessels about to exit the fleet tended to perform less well than the average, but there is no evidence ones newly added caught more than their share of whales. The "first voyage" figures may be biased downward by transfers, and it would be useful to know if vessels that were rerigged or custom built did better than the average. The data set should provide some evidence on this question. Finally, although there is evidence of increasing productivity associated first with larger vessels and

---

[53] A regression of "catch" per ton on age, $(age)^2$, time, and ground does have a positive coefficient on age and a negative one on $(age)^2$, but the size of the coefficients suggests that it would take 35 years to achieve all the "benefits" of age. The average age of a condemned ship was less than that.

then with large bark-rigged vessels, there is no evidence that the new technology made it any less risky to go to sea in search of whales. Since the nineteenth century saw maritime activities in general becoming much less risky, it is hard to understand why Ahab's part of the industry was not affected.

Selected Publications of Stanley Lebergott

1944

"Chance and Circumstance: Are Laws of History Possible?" *Journal of Philosophy* 41 (June), 393–411.

1945

"Shall We Guarantee Full Employment?" *Harper's* 190 (Feb.), 193–202.
"Forecasting the National Product." *American Economic Review* 35 (March), 59–80.

1946

Comment on "Discrimination Against Older Workers in Industry." *American Journal of Sociology* 51 (Jan.), 332–325.

1947

"Wage Structures." *Review of Economics and Statistics* 29 (Nov.): 274–285.

1948

"Earnings of Non-Farm Employees in the United States, 1890–1946." *Journal of the American Statistical Association* 43 (March), 74–93.
"Labor Force, Employment and Unemployment, 1929–1939: Estimating Methods," *Monthly Labor Review* 67 (July), 50–53.
Employment, Unemployment and Labour Force Statistics: A Study of Methods. Montreal: International Labour Office.

1949

"Surveys of Income and Expenditure: the Validity of Interview." Presented at the Annual Meeting of the American Statistical Association.

1950

"Labor Force Statistics, the Task Ahead." Presented at the Annual Meeting of the American Statistical Association.

1952

"Family Incomes and Child Welfare." Presented at the White House Conference on Children and Youth.

1953

Review of *Wages and Salaries in the United Kingdom, 1920–1938*, by Chapman. *American Economic Review* 43 (March), 192–195.

"Has Monopoly Increased?" *Review of Economics and Statistics* 35 (Nov.), 349–351.

1954

"Measurement for Economic Models." *Journal of the American Statistical Association* 49 (June), 209–226.

"Measuring Unemployment." *Review of Economics and Statistics* 36 (Nov.), 390–400.

"Those Unemployment Figures." *Illinois Business Review* (Nov.), 8–9.

1955

"Economic Research and Public Policy: Discussion." *American Economic Review* 45 (May), 318–320.

Review of *An International Comparison of National Products and the Purchasing Power of Currencies*, by Gilbert and Kravis, *American Economic Review* 45 (June), 438–445.

1956

Review of *Income of the American People*, by Herman P. Miller. *Monthly Labor Review* 79 (Jan.), 85–86.

Comment on "Trends in the Participation of Women in the Working Force," by A. J. Jaffee. *Monthly Labor Review* 79 (May), 597.

1957

"Consolidated Budget for Economic Statistical Programs." *Business and Economic Section Proceedings*. American Statistical Association.

"Annual Estimates of Unemployment in the United States 1900–1954." In *The Measurement and Behavior of Unemployment*. Universities National Bureau of Economic Research. Princeton, N.J.: Princeton University Press.

"Diffusion Indices and the NBER Indicators." *Business and Economic Section Proceedings*. American Statistical Association.

"Entrepreneurial Income." In Conference on Research in Income and Wealth. *A Critique of the United States Income and Product Accounts*. Princeton, N.J.: Princeton University Press.

1959

"The Shape of the Income Distribution." *American Economic Review* 49 (June), 328–347.
"Long Term Factors in Labor Mobility and Unemployment." In *Employment, Growth and Price Levels: Hearings*. Joint Economic Committee, 86th Congr., 1st sess., 582–585. Reprint. *Monthly Labor Review* 82 (Aug.), 876–881. Reprint. Ralph Andreano, ed. *New Views on American Economic Development*. Cambridge, Mass.: Schenkman Publishing Company, 328–347.
"The Government as a Source of Anticipatory Data." *Business and Economic Section Proceedings* 7 (Dec.), 160–164. American Statistical Association.

1960

"Population Change and the Supply of Labor." In Universities National Bureau of Economic Research. *Demographic and Economic Change in Developed Countries*, 377–341.
"Wage Trends, 1800–1900." In Conference on Research in Income and Wealth. *Trends in the American Economy in the Nineteenth Century*, 449–498.
"The Shape of Income Distribution: A Reply." *American Economic Review* 50 (June), 443–444.

1961

"Patterns of Employment Since 1800." In *American Economic History*. Edited by Seymour E. Harris. New York: McGraw-Hill, pp. 281–310.
Review of *Wages and Earning in the U.S.*, by Clarence D. Long. *Journal of Economic History* 21 (June), 262–264.
"Statistics, EDP and the Tax Administrator." *National Tax Journal* 14 (Sept.), 234–247.
Review of *Real Wages in Manufacturing, 1890–1914*, by Albert Rees and Donald P. Jacobs. *American Economic Review* 51 (Sept.), 773–774.

1963

"The Labor Force as an Endogenous Factor in Econometric Analysis." *Business and Economic Section Proceedings* (Sept.): 322–332. American Statistical Association.

1964

Manpower in Economic Growth: The American Record Since 1800. New York: McGraw Hill.
Men Without Work, The Economics of Unemployment. Englewood Cliffs, N.J.: Prentice Hall.

"Factor Shares in the Long Term: Some Theoretical and Statistical Aspects." In Conference on Research in Income and Wealth. *The Behavior of Income Shares*. Princeton, N.J.: Princeton University Press.

Review of *Beyond the Melting Pot: The Negroes, Puerto Ricans, Jews, Italians and Irish of New York City*, by Nathan Glazer and Daniel P. Moynihan. *Monthly Labor Review* 87 (April), 447.

1965

"The Labor Force and Marriages as Endogenous Factors." In *The Brookings Quarterly Econometric Model of the United States*. Edited by James Duesenberry et al., New York: Rand McNally.

1966

"Labor Force and Employment, 1800–1960." In Conference on Research in Income and Wealth. *Output, Employment, and Productivity in the United States after 1800*. pp. 117–204.

"Productivity: Eight Questions and Seven Answers." *New York Times*, April 24, sec. 11.

"Directions for the Future of the Employment Act." Statement in "Twentieth Annual Employment Act of 1946. An Economic Symposium (suppl.)" *Journal of Economic History* 26, 96.

"The Accounts and the Computer." *Review of Income and Wealth* 12 (Dec.), 335–347.

"United States Transport Advance and Externalities." *Journal of Economic History* 26 (Dec.): 437–461.

"Tomorrow's Workers: The Prospects for the Urban Labor Force." In *Planning for a Nation of Cities*. Edited by Sam Warner. Cambridge, Mass.: MIT Press, pp. 124–140.

"The Behavior of Employment, 1961–1965: Discussion." In *Prosperity and Unemployment*. Edited by R. A. Gordon and M. S. Gordon. New York: Wiley, pp. 184–188.

"Standard Occupational Classification." In *Social Statistics Section Proceedings*, 207–208. American Statistical Association.

Statement in Joint Economic Committee, *Twentieth Anniversary of the Employment Act of 1946*.

1967

Review of *Occupation and Pay in Great Britain*, by Routh. *American Economic Review* 52 (March), 322–324.

"Three Aspects of Labor Supply since 1900." In *American Statistical Association. Proceedings of the Social Statistics Section*, 172–178.

1968

"Income Distribution: Size." *International Encyclopedia of the Social Sciences*. 1968 ed., vol. 7, 145–154.

"Issues in Federal Aid to Manpower Training." In United States Task Force on Occupational Training in Industry. *A Government Commitment to Occupational Training in Industry,* vol. 2. Washington, D.C.: Washington Publishing Office, 18–58.

"U.S. Transport Advance and Externalities: A Reply." *Journal of Economic History* 28 (Dec.), 635.

1969

"The Service Industries in the Nineteenth Century." In *Production and Productivity in the Service Industries.* Edited by Victor R. Fuchs. National Bureau of Economic Research. Princeton, N.J.: Princeton University Press, 365–368.

1970

"Slum Housing: A Proposal." *Journal of Political Economy* 78 (Dec.), 1362–1366.

"Migration Within the U.S., 1800–1960: Some New Estimates." *Journal of Economic History* 30 (Dec.), 839–847.

1971

"Better Bricks for OBE Structures." *Survey of Current Business* 51 part 2, (July), 117–118.

1972

"Measuring Agricultural Change." *Agricultural History* 46 (Jan.), 227–233.

"Slum Housing: A Further Word." *Journal of Political Economy* 7, (Sept.–Oct.), 28–37.

Sections on Labor. In *American Economic Growth: An Economist's History of the U.S.* Edited by Lance E. Davis, Richard A. Easterlin, William N. Parker, New York: Harper and Row.

1973

"A New Technique for Time Series: Comment." *Review of Economics and Statistics* 55 (Nov.): 525–527.

"Commentary." In *The National Archives and Statistical Research.* Edited by M. H. Fishbein. Columbus, Ohio: Ohio University Press, 205–208.

Sections in *Chemicals and Health; Report.* U.S. President's Science Advisory Committee: Panel on Chemicals and Health. Washington, D.C.: U.S. Government Printing Office.

"New Drug Development." Statement for Senate Small Business Committee.
Sections in NAS-NRC. *ROMMITTEE*
Sections in NAS-NRC. *Report on Manpower Training Evaluation.*

1975

Wealth and Want. Princeton, N.J.: Princeton University Press.

Review of *Time on the Cross,* by Robert William Fogel and Stanley L. Engerman. *American Political Science Review* 69 (June), 697–700.

"Industrialization, Regional Change and the Sectoral Distribution of the U.S. Labor Force, 1850–1880: Reply." *Economic Development and Cultural Change* 23 (July), 749–750.

"Factors Affecting Drug Industry Structure and Costs: Comments." In *Drug, Development and Marketing.* Edited by R. B. Helms. American Enterprise Institute for Policy Research, 285–287.

"How to Increase Poverty." *Commentary* (Oct.) 59–63. Reprinted. *Issues Past and Present* 2, by Phillip Paludon (1978); *Social Problems, A Policy Perspective,* by Howard E. Freeman (1979); *Economic Decisions for Consumers,* by Don Leet and Joanne Driggs, (1983).

1976

The American Economy; Income Wealth and Want. Princeton, N.J.: Princeton University Press.

"Are the Rich Getting Richer?: Trends in U.S. Wealth Concentration." *Journal of Economic History* 36 (March): 147–162.

"The Drug Regulatory Process: Choice and Consequence," and "Who Shall Die?: Costs and Benefits of Drug Regulation." Reports to Science and Technology Policy Office, National Science Foundation.

1977

Review of *Parameters and Policies in the U.S. Economy,* by Otto Eckstein. *Economic Journal* 87 (Dec.), 804–806.

1980

"The Returns to U.S. Imperialism, 1890–1929." *Journal of Economic History* 40 (June), 229–252.

1981

"Through the Blockade: The Profitability and Extent of Cotton Smuggling, 1861–1865." *Journal of Economic History* 41 (Dec.), 867–888.

1982

Review of *Personal Income Distribution: A Multicapability Theory,* by Joop Hartog. *Journal of Economic Literature* (June), 625–626.

Review of *Working Class Life: The "American Standard" in Comparative Perspective, 1899–1913,* by Peter Shergold. *Journal of Economic History* 42 (Dec.): 959–961.

1983

"Why the South Lost: Commercial Purpose in the Confederacy." *Journal of American History* 70 (June), 58–74.

1984

The Americans. An Economic Record. New York: Norton.

1985

"The Demand for Land: The United States 1820–1860." *Journal of Economic History* 45 (June), 181–212.

Contributors

PETER KILBY is Professor of Economics at Wesleyan University.

JEREMY ATACK is Associate Professor of Economics, University of Illinois.

PAUL A. DAVID is Coe Professor of American Economic History at Stanford University.

LANCE E. DAVIS is Mary Stillman Harkness Professor of Social Science at California Institute of Technology.

ROBERT E. GALLMAN is Kenan Professor of Economics and Economic History at the University of North Carolina.

CLAUDIA GOLDIN is Professor of Economics at the University of Pennsylvania.

TERESA D. HUTCHINS is Assistant Professor of Economics at the University of North Carolina.

JOHN A. JAMES is Associate Professor of Economics at the University of Virginia.

WILLIAM N. PARKER is Philip Goldin Bartlett Professor of Economics and Economic History at Yale.

JONATHAN SKINNER is Assistant Professor of Economics at the University of Virginia.

THOMAS WEISS is Professor of Economics at the University of Kansas.

GAVIN WRIGHT is Professor of Economics at Stanford University.

Index

Abbott, Edith, 139
acquisition-cost value, 219
  of capital stock, 218-222
  census data and, 219-220
  of producers' durables, 222
  reproduction costs and, 220-223
  savings behavior and, 218
Adams, Henry, 17
Adams County, Miss., 102
agricultural ladder, 116-122
  asset ownership and, 118-120
  blacks on, 122
  credit markets and, 118-119
  mobility and, 116-118
  shares vs. fixed rents on, 120-121
agricultural workers:
  blacks as, 109, 114-115, 117, 127, 128
  census data on, 272
  children as, 77, 145-146
  as common laborers, 116
  depression-formed reserve of, 78-79
  gender gap ratio for, 143-146
  hiring standards for, 78
  industrial wages and, 108
  land ownership as goal of, 110
  male, 113, 127
  manufacturing competition for, 74-77
  as plantation wage laborers, 111-112
  shortages of, 77-79
  unmarried men as, 114
  unskilled, 76-77
  wages for, 75-76, 108, 127
  women as, 136
agriculture:
  capital accumulation in, 274-275
  capital stock estimates in, 215
  direct labor requirements in, 20
  entry into, 76
  in Great Plains states, 336-337
  home production and, 136
  Indians displaced by, xxi-xxii
  land erosion and, 336-337
  land improvement in, 255-256
  land requirements in, 20
  land use efficiency in, xxi
  New England decline in, 38
  in Puritan society, 24-27
agriculture, subsistence:
  direct labor requirements in, 17-21
  in New England, 19-21
  regional comparisons of, 17-21
  in Southern colonies, 17, 19, 21

Alston, Lee J., 118-119
Althusserians, 17
ambergris, 338, 345
American Economic Growth: An Economist's History of the United States, 9, 10
American Revolution, 22
Americans: An Economic Record, The (Lebergott), xviii, xxi, 15, 99
Anderson, Abel, 120
"Annual Review of Whale Fishery," 358, 359, 362, 366, 367, 389
armories, 33
Arthur, Brian, 131
artisan shops, 286
  in boot and shoe industry, 298
  factories distinguished from, 287, 288-289
  labor supervision in, 307
  in Puritan society, 25, 27
  relative efficiency of, 303-309
  value-added in, 299
asset ownership, 118-120
Atack, Jeremy, 14, 286-335
Austrian immigrants, 60

baleens, 338-339, 345, 366-367, 384
banks, 265
Bardhan, P. K., 111
Barger, Harold, 239, 240
Baring Brothers, 55
barks, 347-348, 351, 353-354, 360-361
  labor-capital ratios in, 377-381
  loss rate for, 381-383
  new, 355-359, 374
  profitability of, 394-395
  revenue-generating ability of, 386-388
  ships replaced by, 370, 396-397
  size of, 370-371, 374, 396-397
Barnum, P. T., 34
Bateman, Fred, 291
Beney, M. Ada, 148, 149
Berlin, Ira, 175
Berlin, Isaiah, 11
Berry, Thomas Senior, 217-218, 245
Bidwell, Percy W., 145
Billings, Dwight, 100-101, 132
Birds of Passage (Piore), 106
blacks:
  as agricultural laborers, 109, 114-115, 117, 127, 128
  on agricultural ladder, 122
  in census labor force statistics, 174
  gender gap ratio for, 153

409

blacks (*continued*)
  labor mobility of, 109, 128
  as sharecroppers, 114–115
  in urban populations, 196–201
  see also free blacks; racial discrimination; slavery; slavery, urban
blacksmiths, 89, 90, 94, 95, 97
boat building industry, 25
Bogue, Allan, 4, 11
bond holdings, state, 271
bonds, federal, 270–271
boom towns, 52, 53
boot and shoe industry, 294–295, 320
  artisan shops in, 298
  efficiency of, 302
  expansion of small mills in, 321–322
  factory production in, 297–298
  labor-capital ratios in, 308
  mechanization in, 297–298, 305
  productivity in, 308
  sweatshops in, 305
  value-added in, 305, 309
Brady, Dorothy, 8, 239, 240
branch banking, 110
brass works, 31
brewing industry, 295
brigs, 347, 351, 360–361, 370, 380, 381, 383
Brown and Sharpe, 37
Bryan, Nelson, 120
building trades, 71
Bureau of the Budget, U.S., xvi
business cycles, 214
business histories, 9
butchers, 89

Cambridge Economic History, 8
Campbell, George, 103
Canada, 37
canal projects, 72–73
capital:
  human vs. physical, 256
  in migrations, 271
  NNP in ratio to, 255
capital accounting, 220
capital accumulation:
  in agriculture, 274–275
  census data on, 257, 269–272
  deflator for real worth series and, 284–285
  demographic changes and, 278–279, 282
  dramatic rise in, 255
  empirical model of, 266–281
  family size and, 267, 274–275, 278–279, 282
  for farm families, 274–275
  financial intermediation and, 282
  foreign holdings and, 271
  government debt and, 270
  income levels and, 279–280
  inequality and, 257, 264–265, 279, 282
  investment goods prices and, 256, 268
  occupational shifts and, 258
  occupational variations in, 274–276, 279–281, 282
  as real and personal wealth, 269
  real estate and, 269
  real income increases from, 280
  real net savings rates and, 266
  regional variations and, 267–268, 276
  slaves in, 268–269
  year-specific proportional variables in, 268, 276–277, 279–280
  see also savings rate
capital consumption, *see* depreciation
capital-income ratios, 256
Capital in the American Economy (Kuznets), 237
capitalism:
  mercantile, 26–27
  patronage, 110
capital-output ratios, 259
capital stock:
  acquisition-cost value of, 218–222
  census-style estimates of, 230, 235, 238, 241
  construction-flow series in, 239
  controversy over measurement of, 11–12
  demographic changes and, 258
  discount rates and, 224
  distribution of earnings and, 223
  economic growth and, 336–338
  erosion of, 337
  GNP in ratio to, 269–270
  gross and net estimates of, 230, 235, 238, 241
  in industrial rise and decline, 336–338
  long-term evolution of, 244
  market value of, 219–220, 224
  micro level analysis of, 337
  new investment and, 226
  "other construction" component of, 221, 222
  productivity and, 336, 337, 396
  reproduction cost value of, 219–222
  Salterian analysis of, 336–338
  structure of, 336, 337
  "true value" of, 220
  value measurements for, 219
  in whaling industry, 360
capital stock values, census-style estimates of:
  casualty losses and, 227–228
  Civil War losses in, 223–234
  comprehensive quality of, 215
  consistency tests for, 244
  conventional components in, 240
  Goldsmith and Gallman estimates on, 237–238
  interest rates and, 224–226
  measurement error in, 231–232

Index

as net reproduction-cost estimates, 232
perpetual inventory estimates compared with, 165–168, 214–215, 225–227, 231–237
sources of, 219–220
types of capital included in, 232–233
capital stock values, perpetual inventory estimates of:
adaptability of, 215
analytical requirements of, 216
Berry series on, 217–218
casualty losses and, 227–228
census-style estimates compared with, 214–215, 225–227, 231–237
Civil War losses in, 233–234
consistency tests for, 244
conventional components in, 240
falling interest rates and, 225
Gallman series on, 217
gross vs. net measurements in, 245
highways omitted in, 233
internal improvements in, 233
investment flows and, 214, 215, 217
Kuznets's series on, 217
measurement error in, 231–232
"other construction" element in, 216
producers' durables element in, 216
reproduction-cost value and, 226–227
service lives in, 227–228
types of capital included in, 233
Carey, Henry, 147
carpenters, 89, 94, 95
Carr, Julian, 133
Case, Harriet, 67
Cash, W. J., 100
casualty losses, 227–228
census data:
acquisition-cost value in, 219–220
on agricultural workers, 272
black population in, 174
capital accumulation data in, 257, 269–272
capital stock valuation in, 219–220
depreciation estimates in, 227, 229
of 1800, 191–202
of 1810, 202
of 1820, 202–203
of 1830, 203–204
of 1840, 55, 204
of 1850, 76–77
of 1870, 67
of 1890, 114
of 1910, 103, 116
of 1940, 140–141
of 1950, 190
Indians in, 284
labor force population statistics in, 172–174
occupation data in, 272
present estimates and, 205

real and personal property valuations in, 281–282
service life in, 227–228
wealth data in, 257, 269–271
Census of Manufacturing, 148, 149
comparability of data in, 292
of 1820, 292–293
motive power sources reported in, 293
Census of Occuapations, 56, 64, 65
Chandler, Alfred D., 9, 14, 41, 288, 327
Chapelle, Howard I., 358
Chicago, Ill.:
in "boom town" phase, 52, 53
Eastern migrations to, 58–59
foreign immigrants in, 59–63
Germans in, 59–61
internal improvement projects in, 74
Irish in, 59–63
land speculation in, 54–56
local labor supply in, 53–56
lumber industry in, 80–85
population increases in, 63
production workers in, 55–56
skill margins in, 87–89
Chicago Board of Trade and Mercantile Association, 91, 93–94
Chicago Common Council, 91
Chicago Daily Democratic Press, 86–87
Chicago Democrat, 54
Chicago Mechanics' Institute, 58
Chicago Premium Glove Factory, 57
Chicago Times, 62–63, 67
Chicago Tribune, 63
Chicago Vise and Tool Company, 64
children:
as investments, 256, 264, 267
in life-cycle capital accumulation pattern, 274–275, 282
rearing costs of, 264–265, 267
child workers, 48
in agriculture, 77, 145–146
in manufacturing, 145–146
productivity of, 48, 146
cigar making, 294–295
Civil War, U.S.:
capital destruction in, 233–234
cotton production in, xxiii
demobilization after, 93–94
draft in, 92
Lebergott's study of, xxii–xxiii
price expectations and, 259
productivity during, 68–69
skilled labor shortages in, 68–70, 91–93
whaling ships sunk in, 381
Clark, Colin, 50
Clark, Victor, 288
clerical trades, 137–138, 156
educational preparation for, 162
female participation rates and, 138
mechanization and, 137
relative earnings in, 161–162

Cliometric Society, 6
clock and watch industry, 31–34
Cobb-Douglas production function, 261, 301
Coelho, Philip, 285
Colt, Samuel, 33–34
commercial banks, 265
commercial-paper houses, 265
community organization, 41
comparable worth, 138, 139
Conference on Research in Income and Wealth (1963), 8
Conrad, Alfred, 4
construction industry:
  as demand-localized, 71
  unskilled labor requirements of, 71–72
construction price index, 222, 223
consumer durables, 269
consumer goods:
  lower-quality, 70
  western markets for, 70
consumption, household, xviii–xix
consumption, life-cycle, 274
convict-lease system, 126
Cook County Board of Supervisors, 91
corporate organization, 41
cotton production:
  during Civil War, xxiii
  exports from, xxiii
  manpower absorbed in, xxiii
cotton textiles:
  factory production in, 298
  inanimate power in, 293
Coxe, Tench, 54
craftsmen, 288
Crane, Richard T., 64
Crane Company, 64
credit markets, 118–119
creditworthiness, 110–111, 118
Crogman, W. H., 109
culture, 40
Current Population Reports, 135
Curry, Leonard, 174, 182

David, Paul, 13, 47–97, 131, 256, 258, 261, 263, 264, 265, 290
David-Solar retail price index, 269–270
Davis, Lance E., 8, 15, 227, 257, 258–259, 262, 263, 269, 280, 284, 336–398
debt peonage, 120
DeCanio, Stephen, 99, 121
demographic changes:
  capital accumulation and, 278–279, 282
  per capita income and, xx–xxi
Dent, John, 111
depreciation:
  census estimates of, 227, 229
  retirement schedules for, 229
depressions, 78–79
Dewey Report, 148

Dias, Joseph, 349–350, 351
discount rates, 224
division of labor:
  in clerical sector, 138
  in rural neighborhoods, 19
Doggett, Bassett & Hills, 56–57
domestic servants, 272
door and sash factories, 69
Du Bois, W. E. B., 114
Durand, John, 138
Dust Bowl, 336–337

ecological problems, 336–337
economic growth:
  capital-intensity and, 255
  capital stock and, 336–338
  Lebergott on, xvii–xviii
  national income approach to, 9
  technical change and, 336
  traditional explanations of, xvii
economic history, new:
  beginnings of, 3–5
  division of labor in, 7
  economists vs. historians in, 4–6, 10–11
  first generation of scholars in, 3, 9
  Fogel and Engerman in, 9–11
  Income and Wealth conferences on, 4, 6, 7–8
  industrial organization problems in, 8
  institutionalization of, 6–7
  interrelatedness in, 6
  Lebergott and, xvii, 3
  market studied in, 5
  national income studies in, 9
  second group of scholars in, 3
  social group in, 7
Economic History Association, 4
economists:
  historians vs., 4–6, 10–11
  on institutions vs. market forces, 115
Edgeworth, F. Y., 135, 138, 139
education, 59
  gender gap ratio and, 135–136, 138, 162
  periods of rapid increase in, 162
Ely, Richard T., 139
Engerman, Stanley E., 10–11
entrepreneurial labor, 309
equal pay, 139
Explorations in Economic History, 6
Explorations in Entrepreneurial History, 6

Fabricant, Solomon, 4
factory production:
  artisan production vs., 287, 288–289
  in boot and shoe industry, 297–298
  defined, 287–288
  industries dominated by, 297–299
  labor supervision in, 307
  Marxist view of, 288
  productivity gains in, 286–287, 292

quantitative studies of, 288–289
relative efficiency in, 302–309
specialized labor in, 288
steam and water power in, 287
in tobacco industry, 298–299
value-added in, 293–294, 297–299
year-round production incentive in, 292
Falconer, John F., 145
family consumption allowances, 20
family size, 256, 264, 267
federal lands, 284
fertility rates, 256, 263–264, 282
Field, Alexander, 47
financial markets, 257, 265
firm histories, 39–40
Fishlow, Albert, 239, 240
Fite, Gilbert, 100, 109
flour mills, 293, 297
  expansion of small establishments in, 321–322
  plant size determinacy in, 316–317
  relative efficiency estimates for, 303
  value-added in, 313–314, 316–317, 320
Fogel, Robert W., 10–11
Fordism, 41
France, 27
free blacks:
  mobility of, 101–104
  in urban labor force, 182–183, 207–209
Free West, 67
furniture industry, 122, 295, 297

Galena and Chicago Union railroad, 66
Gallatin, Albert, 54, 146
Gallman, Robert, 8, 12, 15, 69, 214–254, 255, 256, 257, 258–259, 262, 263, 269, 270, 280, 284–285, 336–398
Gates, Paul, 4, 77
gender-based discrimination, 138–139, 158–161
gender distinction, 23
gender gap ratio:
  in agriculture, 143–146
  Beney data on, 148, 149
  for black population, 153
  Brissenden data on, 148, 149
  British studies on, 138–139
  clerical jobs and, 137
  current literature on, 139
  discrimination and, 138, 158–160
  earnings equations and, 159
  economic progress and, 136
  education and, 135–136, 138, 162
  for 1815–1970, 143–149
  explanations of, 158–167
  in geometric means, 156
  hours worked differences in, 149, 151, 158
  human capital measures in, 158–160
  industry sex segregation and, 148–149
  job experience in, 158
  life-cycle labor market experience and, 141–143, 164–165
  in manufacturing, 143–149
  occupational distribution and, 153–158
  within occupations, 157
  participation rates and, 136, 141–143, 164–166
  piece-rate wages and, 163
  regression equations on, 149
  size and strength factor in, 162–164
  stability of, 135–136, 167–168
  technology and, 135–137
  at various dates, 158–161
  wage-rate comparisons in, 155
  as World War I era issue, 139
German immigrants, 59–62
Gerschenkron, Alexander, 4
Gertz and Loder's Brush Factory, 57
Goldin, Claudia, 13, 48, 135–170, 174, 176, 179–180, 182, 233, 234
Goldsmith, Raymond, 214, 228, 237–238, 283
Gooding, William, 72
Goodyear, Charles, 31
government debt, 270
Grand Traverse, 258, 281, 282
Great Awakening, 27
Great Britain:
  gender gap studies in, 138–139
  Industrial Revolution in, 39
  sperm oil exports to, 341
Great Plains, 336
Greece, 27
gross national product (GNP):
  capital stock in ratio to, 269–270
  per capita, *see* income, per capita
gun industry, 33–34
Gutman, Herbert, 175

Hairston Farm plantation, 102
Hall, John, 33
Hamilton, Alexander, 146, 325
hand tools, 287
harpooners, 345–346
Hegarty, Reginald B., 350
Hermitage plantation, 119–121
Hibbard, Benjamin, 119
Higgs, Robert, 101, 119, 121, 122, 125
highways, 233
hiring standards, 49–50, 65–67
  for agricultural workers, 78
  downward revision of, 65–67, 86
  productivity and, 69
  unskilled labor scarcity and, 72
  upward revisions in, 87
  women workers and, 66–67
historians:
  economists vs., 4–6, 10–11
  statistical methodology adopted by, 10–11

Holland, 27
Holmes, George K., 114
Holmes, Oliver Wendell, Jr., xxiii
home manufactures, 286
home production, 136
Hong Kong, 29
household consumption, xviii–xix
house painters, 89, 94, 97
housing, public, xxii
hunting and fishing societies, xxi
Hunt's Merchants' Magazine, 80
Hutchins, Teresa, 15, 227, 336–398

Illinois and Michigan Canal, 55, 72, 91
Illinois Central railroad, 73, 74
Illinois Staats-Zeitung, 62
illiteracy, 59
immigrants:
  Austrian, 60
  in Chicago, 59–63
  in economic growth, xvii
  federal assistance and, 63
  German, 59–62
  Irish, 59
  private recruitment of, 62, 63, 64
  Prussian, 60
  Scandinavian, 96
  as skilled workers, 60–62
  in southern states, 105–108
imports, 70
income, national:
  in early new economic history studies, 7–8
  economic growth and, 9
  economics of scale in, 14
income, per capita:
  demographic changes and, xx–xxi
  family income and, xx–xxi
  growth in, 290
  Lebergott on, xx–xxi
income, real, 280
Income and Wealth conferences (1957, 1963), 4, 6, 7–8
income levels:
  capital accumulation and, 279–280
  growth rates for, 259
  regional variations in, 267–268
  savings rate and, 257, 265
Indians, American, 22
  agriculturalists' displacement of, xxi
  in census data, 284
  Lebergott's study on, xxi–xxii
Industrial Commission, 104, 109
industrialization:
  "challenge" in, 40
  culture and, 40
  gender gap ratio in, 146–147
  import substitution in, 70
  interlinked industries in, 39
  internal improvement projects and, 72, 74
  labor market adjustments in, 48
  mercantile capitalism and, 26–27
  in recently settled regions, 53–54
  regional, 40–41
  skilled labor and, 48–49
  skill margins and, 49–50
  western, 70
  world, 41
industrialization, Middle West:
  mechanical industries in, 40–41
  New England migration and, 38
industrialization, New England, 26–47
  British competition and, 34–35
  British Industrial Revolution and, 39
  clock making in, 31–33, 34
  gun manufacturing in, 33, 34
  machine-tool industry in, 34–38
  Marxist view of, 39
  mass production and, 41
  mercantile activity and, 26–27
  metals and machinery in, 27, 30–33
  Puritan "pre-industrial" society and, 24–27
  Southern plantation development and, 35
  textiles in, 27–30
  westward migrations and, 35–36
industry supply curve, 289
inequality, 257, 264–265, 279, 282
institutions, concept of, 115
insurance companies, 265
interchangeable parts, 41
interest rates:
  census capital stock values and, 224–226
  investment goods prices and, 258–261
  market value and, 225–226
  nominal vs. real declines in, 224–226
  patterns of change in, 224–225
  price expectations and, 258–259
  real 1840–1900 decline in, 258–259
  real savings rate and, 255, 261–262
internal improvements:
  Civil War destruction of, 234, 236
  in perpetual inventory estimates, 231, 232–233
  service lives of, 228
Internal Revenue Service, xvi
International Economic History Association, 336
International Labour Office, xv–xvi
International Typographical Union, 91
inventions:
  labor-augmenting, 37
  in machine tool industry, 37–38
  in textile industry, 28
investment goods prices:
  capital accumulation and, 256, 268
  capital losses and, 262–263
  equation for, 260
  interest rates and, 258–261

Index

for machines and tools, 259–260
relative decline in, 259–260
value of wealth and, 261–262
investments:
  Berry-Gallman figures on, 217–218
  capital stock valuation and, 226
  children as, 256, 264, 267
  interregional, 271
  NNP and, 255
  perpetual inventory estimates and, 214, 215, 217
  savings sources and, 257
Ireland, 37, 73
Irish immigrants, 59–63
iron industry, 294, 297
  large vs. small plant efficiency in, 302
  mechanization in, 307–308
ironworkers, 89, 95, 97, 124, 130
Italy, 27, 41

James, John A., 12, 255–285, 289, 314
Jerome, Chauncey, 32
Jim Crow laws, 101, 122–123
job experience:
  in gender gap ratio, 158
  for women, 137, 158
job training:
  expected life cycle and, 140
  for women, 135, 137, 143
Jones and Lamson, 37
Journal of Economic History, xviii, 6
Journal of Political Economy, xxii

Kennedy, Louis Venable, 99
Keynesian categories, 8
Kuznets, Simon, 228, 237–240, 284, 336
Kuznets cycles, 214, 242

labor, semiskilled, 69
labor, skilled:
  in Chicago, 68–70, 91–93
  during Civil War, 68–70, 91–93
  in construction, 71–72
  in 1870 Census, 61
  German immigrants and, 61
  hiring standards for, 65–67
  immigrants and, 60–62
  industrialization and, 49–50
  in machine tool industry, 37
  migrations and, 57–59
  private recruitment of, 63–65
  selective import substitution and, 71
  semiskilled labor and, 69
labor, unskilled, 62–70
  agricultural employment opportunities and, 76–77
  census data on, 272
  during Civil War, 91
  in construction, 71–72

convict-lease system and, 126
depression-formed reserve of, 78–79
hiring standards and, 72
in lumber industry, 80
in manufacturing, 74
mobility of, 79
in railroad construction, 72–74
recruitment of, 73
shortages of, 49, 72
wages for, 72–73, 75, 78–80
labor-capital ratios, 377–381
  in boot and shoe industry, 308
  productivity and, 299–301, 308
laborers, 116
labor force participation rates:
  by age groups, 172–174
  demographic group variations in, 172–174
  Lebergott's work on, 12
  by locations, 185–188
  in population measurements, 171–172
  regional variations in, 171–172
  urban vs. rural, 171–173
labor force participation rates, female:
  by age, 173
  clerical sector expansion and, 138
  cohort underestimation of, 141, 165–166
  compilation of data on, 140–141
  definitions of, 140–141
  earnings and, 140
  gender gap ratio and, 136, 141–143, 164–166
  increase in, 140–141
  life-cycle labor market experience and, 140, 141
  by marital status, 142
  by nativity, 142
  professional jobs and, 137–138
  by race, 142
  by race and marital status, 141
  by region, 173
  succession of generations in, 138
  in urban areas, 182–185
  urban-rural variations in, 172–174
labor force population:
  occupational distribution of, 150
  segmented, 13
  for textile industry, 29
  westward migration and, 35–36
labor force population, urban:
  arithmetical discrepancies in reports on, 204
  definitions of "urban" in, 189–190
  estimated omitted details in, 191–204
  free blacks in, 182–183, 207–209
  free whites in, 182–185, 206–207
  measurement of, 171–172
  sex composition of, 182–185
  slaves in, 146, 148, 176–182, 207, 209–210
  workforce composition in, 188–189

labor markets:
  during Civil War, 91–92
  during demobilization, 93–94
  equality of treatment in, 139
  female participation rates and, 138
  hiring standards in, 49–50, 64–65
  import substitution and, 70
  labor service heterogeneity and, 51
  local living costs and, 51–52
  in Midwest vs. Northeast, 49
  safety-valve hypothesis in, 75
  short vs. long-run conditions in, 51–52
  for skilled vs. unskilled workers, 49–50
  skill margin compression in, 85–87, 91
  in small vs. large cities, 51–53
  supply and demand factors in, 49, 64–65
labor markets, Southern, 85–108
  agricultural ladder and, 116–122
  as anomalies, 98–99
  assumptions about, 100
  convict-lease system and, 126
  "enforced immobility" in, 98–100
  equilibrating pressures in, 125–126
  established vs. mobile categories in, 111
  historical origins of, 106–107
  industrial, 122–130
  industrial vs. farm wages in, 125
  isolation of, 106–108
  national markets vs., 101–103, 106–108
  on-the-job experience and, 130–134
  racial equity in, 130–134
  racial segregation and, 101
  recruitment in, 107
  repressive institutions and, 101
  for sharecroppers vs. wage laborers, 108–116
  wage equality in, 104–105
labor mobility:
  agricultural ladder and, 116–118
  of blacks, 109, 128
  creditworthiness and, 110–111
  debt and, 119–121
  economic opportunity and, 116–118
  nineteenth century views on, 108–109
  public works and, 116
  of unmarried farm workers, 114–115
  for unskilled workers, 79
labor requirements:
  in agriculture, 17–21
  in construction industry, 71–72
labor shortages:
  in agriculture, 77–79
  in Chicago manufacturing sector, 55–57
  David's study on, 12–13
  hiring standards in, 65–67
  migration and, 35–36
  skill upgrading in, 64–66
  of unskilled workers, 49, 72
Labor Statistics, North Carolina Bureau of, 104

Labor Statistics, U.S. Bureau of, xv
labor supervision, 307
labor supply:
  land abundance and, 54
  for manufacturing, 48–49
  rural reservoirs in, 47–48
  in small economies, 51
  westward migration and, 74–75
labor turnover, 102, 103, 362
labor-tying contracts, 111
land erosion, 336–337
land improvement, 255–256
  savings measurements and, 269
land ownership, 110
land requirements, 20
land speculation, 54–56
Laurie, Bruce, 289
Law, John, 57
leather industry, 95, 297
Leatherwood and Camp Branch Plantation, Va., 102
Lebergott, Isaac, xv
Lebergott, Lillian, xv
Lebergott, Stanley, xv–xxiii, 3, 8, 11, 12, 47, 49, 98, 99, 100, 171, 257
  background of, xv–xvi
  Civil War study by, xxii–xxiii
  economic growth factors defined by, xvii–xviii
  as empiricist, xviii–xix
  gender gap study by, 139–141
  household consumption study by, xviii
  Indian displacement study by, xxi–xxii
  labor force movement research by, 8, 11–12, 13
  Manpower in Economic Growth by, xvi–xviii
  "new economic history" and, xvii, 3
  at OSS, xvi
  on per capita income, xx–xxi
  poverty studied by, xxii
  publications of, 399–405
  public housing proposal by, xxii
  quiddity in work of, xix, xxi
  on southern labor markets, 98, 99, 100
Lee, Roswell, 33
Leontief input-output matrix, 8
Lesquereux, Leo, xvii
Lewis, Frank, 233–234, 236, 263–264, 265, 282
Lewis plantations, 119–121
life-cycle consumption, 274
life-cycle labor market experience, 140, 141, 164–165
Lincoln, Abraham, 63
liquor industry, 295, 297
Long, Clarence D., 148
Lowell, Francis Cabot, 28
lumber industry, 294, 297
  expansion of small mills in, 321–322

Index

labor recruitment in, 83–85
productivity in, 80–82
relative efficiency in, 302–304
value-added in, 312–313
wages in, 83–85

McCloskey, Donald, 6
McCormick Works, 93
machine tool industry:
  in New England industrialization, 34–38
  skilled labor and, 37
  technological convergence in, 34
  textile industry and, 29, 34
  westward migration and, 36
machinists, 89, 94, 97
McLane Report (1832), 146, 147
Mandle, Jay, 98, 100, 106
Mangum, Charles S., 123
Manpower in Economic Growth: The American Record Since 1800 (Lebergott), xvi–xviii
manufactories:
  factories distinguished from, 286–287
  industry value-added in, 299
  labor supervision in, 307
  relative efficiency of, 306–307
manufactured producers' durables, 216
  acquisition-cost value of, 222
  Civil War destruction and, 234–236
  declining-balance series on, 242
  in perpetual inventory estimates, 230–231, 246–253
  in post-Civil War years, 237
  price indexes for, 223
  service life for, 228
manufacturing:
  agricultural workforce and, 74–77
  in artisan shops, 286
  in Chicago, 55–57
  child workers in, 145–146
  craft positions in, 162
  employment discrimination in, 122–130
  gender gap ratio in, 143–149
  home production and, 136
  in households, 286
  labor shortages in, 55–57
  labor supply for, 48–49
  lower-quality product lines in, 69–70
  mechanization in, 137
  piece-rate wages in, 163–164
  rural manpower in, 48–50
  selective import substitution in, 70
  sexual segregation in, 148–149
  skill requirements in, 50
  unskilled labor and, 74
  value-added in, 69
  women workers in, 67, 148–149
Margo, Robert A., 121
market:
  active vs. passive elements in, 5

gender distinction in, 13–14
in Lebergott's study, xviii
in new economic history studies, 5
market forces, concept of, 115
market value, 219–220
  census data and, 219–220
  as discounted anticipated income, 224
  as forward looking, 219
  interest rates and, 225–226
  new investment and, 226
  reproduction cost and, 223, 225, 226, 227
Martineau, Harriet, 54–56
Marxist interpretations, 39, 288
Mason, Roswell B., 73
mass production, 40–41
Mears, Bates & Co., 68–69, 83–85
Mears, Charles, 83–85, 94
meat packing industry, 51, 94, 295, 297
  mechanization in, 305
mechanization:
  in boot and shoe industry, 297–298, 305
  capital quality and, 308
  in factory production, 137
  factory vs. mill production and, 297–298
  in iron industry, 307–308
  labor discipline and, 307
  managerial shortcomings and, 307
  of office work, 137
  productivity and, 299–301, 308–310
  relative efficiency and, 305–306
  in tanneries, 305
  women workers and, 137
  work force skill levels lowered by, 300–301
Melville, Herman, 339, 362, 363
mercantile capitalism, 26–27
merchant marine, U.S., 386
metals-smelting industry:
  distribution in, 30
  in New England industrialization, 27, 30–31
Meyer, John, 4
Michigan Central Railroad, 72
Middle colonies, 17–19, 21
Middle West, 38, 40–41
migration:
  capital transferred in, 271
  as destructive to home life, 109
  from Dust Bowl states, 336–337
  frequent, 109
  labor shortages created by, 35–36
  long vs. short distance, 106
  Middle West industrialization and, 38
  to Old Northwest Territory, 57–58
  safety-valve concept and, 75
  southern interstate, 105–106
  from upland South, 57–58
  westward, 35–36
Mill, J. S., 138
milling machines, 38

mills:
    factories distinguished from, 288
    industries dominated by, 297, 299
    labor supervision in, 307
    relative efficiency of, 303–307
    in tobacco industry, 298–299
minimum efficient scale (MES), 318–319
monopoly, natural, 311
Morgan Iron Works, 70
mortgage brokers, 265
mortgage market, 271
mutual savings banks, 265

Nantucket, Mass., 366
Napoleonic Wars, 35
National Accounts Review Committee, xvi
national income and product accounts, xvi
National Longitudinal Survey, 166
National Typographical Union Local No.
    16, 67
Natural Industrial Conference Board, 148
needle trades, 307
net national product (NNP), 255
New Bedford, Mass., *see* whaling fleet, New
    Bedford
New England:
    as adverse environment, 22
    agricultural decline in, 38
    economic development of, 22
    literacy in, 59
    subsistence agriculture in, 19–21
    see also industrialization, New England;
        Puritans
New London, Conn., 366
New York Association for Improving the Condition of the Poor, 79
Nicholls, W. H., 100
Niles Register, 291
North, Douglass, 8, 14
Northrup, Herbert, 122–123

Oaksville Manufactory of Cotton Goods, 147
occupations:
    in capital accumulation equation, 274–276,
        279–281, 282
    census data on, 272
    gender gap ratios and, 153–158
    with greater propensity to save, 258
    inequality and, 279, 282
    mobility in, 128
    sex segregation in, 148–149
Ocerby, Thomas, 83–84
Office of Statistical Standards (OSS), xvi
Ogden, William B., 82
Old Northwest Territory, 57–58
O'Neill, Jane, 149
Ozanne, Robert, 93

Parker, William N., 3–46
patronage capitalism, 110

peddlers, 30, 31
personal and professional services (PERS),
    272, 275–276, 277, 279
personal property, 269
    "true valuation" of, 283–285
Philadelphia and its Manufactures
    (Freedley), 53
piece-rate wages, 163–164
Piore, Michael, 106
plantations:
    agricultural workers on, 111–112
    labor mobility and, 98–103
    labor turnover on, 102, 103
    New England industrialization and, 34, 35
    sharecropping vs. wage labor on, 111–112
    Whatley model of, 111, 112
population:
    age composition of, 256, 263
    textile expansion and growth in, 290
    see also census data; labor force population;
        urban populations
poverty, xxii
power sources:
    census data on, 293–295, 297–299
    in factory production, 287
    inanimate, 293
    "pre-industrial" societies, 24–25
price expectations, 258–259
price indexes, 309
printing industry, 67, 122
production-function estimates, 332–335
production techniques, 336
productivity:
    in agricultural economy, 136
    in boot and shoe industry, 308
    capital stock and, 336, 337, 396
    of child workers, 48, 146
    during Civil War, 68–69
    in factory production, 286–287, 292
    frequent migration and, 109
    hand vs. machine production in, 300
    hiring standards and, 69
    labor-capital ratios and, 299–301, 308
    labor vs. total factor, 300–301
    in lumber industry, 80–82
    mechanization and, 299–301, 308–310
    scale economies and, 14
    in whaling industry, 346, 351–353, 374,
        383–386
    of women workers, 48, 146
productivity, total factor:
    equation for, 300–301
    labor productivity compared with, 300–301
    output elasticities in, 310
    plant organization and, 302–304
    scale economies and, 302–304, 310
professional jobs, 137–138, 156
profits, 122, 389–395
Provincetown, 366
Prussian immigrants, 60

Index

Public Use Sample, 172
public works, 116
Purdue seminars, 6
puritan, concept of, 15
"Puritan mentality," 22
Puritans:
  as artisans, 25, 27
  corporate organization and, 41
  as farmers, 24–27
  land system of, 26
  mercantile capitalism and, 26–27
  in New England economic development, 22
  in "pre-industrial" stage, 24–25
putting-out system, 302

quiddity, xix, xxi

Race Distribution in American Law (Stephenson), 123
racial discrimination:
  established local patterns in, 133–134
  in industry, 122–130
  market pressures and, 101
  occupational mobility and, 128
  on-the-job experience and, 130–134
  profit seeking behavior and, 122
  segregation laws and, 122–124
  wage discrimination and, 124–130
  workplace regulation and, 123
railroad construction:
  in capital stock estimates, 240
  employment levels in, 73–75
  unskilled labor shortages in, 71–75
railroad industry, 66
  service life calculations for, 227–228
  textile industry expansion and, 289–290
rates of return, 389–395
real estate, 269
real income, 280
real property, 283–285
Reinterpretation of American Economic History, The (Fogel and Engerman), 9–11
"Report on Manufactures" (Hamilton), 325
reproduction-cost value, 218–219
  acquisition-cost value and, 220–223, 227
  of capital stock, 219–222, 226–227
  market value and, 223, 225, 226, 227
  new investment and, 226
  perpetual inventory series and, 219, 226
  production relationships and, 218–219
retirement savings, 256, 264
Roback, Jennifer, 101, 116
Rome, 27
Rosenberg, Nathan, 34
"Rosenberg effect," 39
Roys, Captain, 363
rubber industry, 31

saddlery industry, 294–295, 299
safety-valve hypothesis, 75
Salter, W. E. G., 336–338
savings and loan associations, 265
savings behavior:
  acquisition-cost value and, 218
  capital stock in, 218
  "life cycle" model of, 12
savings rate:
  consumption plans and, 266
  conventional vs. unconventional forms in, 269
  dramatic rise in, 255–257
  family size and, 256, 264, 267
  financial intermediation and, 257, 265
  income levels and, 257, 265
  inequality and, 257, 264–265
  interest rates and, 255, 261–262
  investment goods prices and, 259–260
  land improvement and, 255–256
  life-cycle model for, 263–264
  national income and, 255
  population age composition and, 256, 263
  price-inelastic schedules for, 262–263
  traditional measurements of, 255–256
  wealth-income ratio and, 258
  year-specific variables in, 276–277
scale economies:
  actual output and, 311–312
  cost functions and, 311
  factory expansion and, 287
  falling unit costs in, 287
  in iron and sheet metal manufacturers, 302–303
  in machine tool industry, 34
  mechanized production and, 286, 309–310
  output elasticities and, 310
  productivity growth and, 14
  quantitative studies of, 289
  total factor productivity and, 302–304, 310
Scammon, Charles, 345
Scandinavian immigrants, 96
Scheiber, Harry, 4
Schmitz, Mark, 289
schooners, 347, 351, 360–361, 370, 380, 381, 383
securities, industrial, 271
securities markets, 265
segmented labor force, 13
semiskilled labor, 69
service life estimates, 227–230
  in census data, 227–228
  Civil War destruction and, 234, 236
  variations in, 227–228
sewing machine industry, 320
sharecroppers:
  agricultural ladder and, 116–119
  blacks as, 114–115, 117
  creditworthiness and, 110–111
  as distinct legal category, 115–116
  earnings of, 99
  in family-based system, 112, 114

sharecroppers (*continued*)
  in labor-tying contract, 111
  in local moves, 114
  mobility of, 100, 101–104
  term of occupancy for, 104
  white, 117
  see also tenant farmers
sheet metal industry, 302, 319
Shepherd, James, 285
Shifflett, Crandall, 110–111
shipbuilding, 89, 94, 355–358
shipping industry, 234
ships, 347–348, 350–351, 353–354, 355, 360–361, 370, 374
  barks vs., 370, 396–397
  labor-capital ratios on, 377–380
  loss rate for, 381–383
  profitability of, 394–395
  revenue-generating ability of, 386–388
  size of, 396–397
Shlomowitz, Ralph, 99
silverware industry, 31
Simon, Matthew, 270
skill margins, 88
Skinner, Jonathan S., 12, 255–285
Slater, Samuel, 28
slavery:
  in capital accumulation equation, 268–269
  in census labor force statistics, 174
  in city population figures, 177
  during Civil War, xxiii
  in cotton mills, 130–131
  Fogel and Engerman on, 10
  portability in, 110
  in tobacco industry, 132
slavery, urban:
  in border states, 178–179
  city size and, 178
  estimates of labor force in, 176–182, 207, 209–210
  male to female ratios in, 180–182
  variation in importance of, 176–179
sloops, 347, 351, 360–361, 370, 380, 381, 383
Smith, Adam, 7, 139
Smith, Merrit Roe, 33
Social Science History Association, 6, 11
Sokoloff, Kenneth, 13, 48, 286–287, 289, 291, 302, 304, 307, 327
Soltow, Lee, 275
Spain, 27
sperm oil production, 338–341, 366–367
  British market for, 341
  as proportion of catch, 366–367
  whale depletion and, 366–367
  whale oil production vs., 338–339
Stampp, Kenneth, 10
Starbuck, Alexander, 350, 384
steamboats, 290

steam power, 287, 294–295, 297, 298, 299
steam winches, 346
steersmen, 345
Stephenson, G. T., 123
stock market, 271
Stone Fleet, 359
Study of History (Toynbee), 21–22
sugar industries, 320
supply-side shocks, 337–338
sweatshops:
  in boot and shoe industry, 305
  factories distinguished from, 287
  industry value-added in, 299
  in needle trades, 307
  relative efficiency of, 304–307
Sweden, 41
Switzerland, 41

Taiwan, 29
tanning industry, 295
  mechanization in, 305
tariffs, 35
Taylor, Jeremy, xx
Taylorism, 41
technological convergence, 34
technology:
  gender gap ratio and, 135–137
  in industry cost curves, 289
  in whaling industry, 344–346
Temin, Peter, 260, 307–308
tenant farmers:
  asset ownership by, 119–120
  in caste system, 118–119
  creditworthiness of, 110–111, 118
  in debt, 119–121
  on fixed rents, 118–119
  landlords' perspective on, 110–111
  mobility of, 116–118, 122
  see also sharecroppers
Terry, Eli, 31
textile industry:
  black mills in, 132
  demand changes in, 289–290
  entrepreneurs in, 29
  expansion of, 29–30
  factory production in, 298
  labor force for, 29
  machine-tool manufacturing and, 29, 34
  mechanization of, 286
  mill-village system in, 131
  in New England industrialization, 28–30
  population growth and, 289–290
  power sources in, 293, 294, 295
  putting-out system in, 28
  regional repetition in, 40–41
  slaves in, 130–131
  in southern Piedmont, 40
  stolen machinery in, 28
  unskilled labor in, 36–37

Index

wages in, 125
white employment in, 122, 123, 125, 130–131
Thomas, Seth, 32
Tilley, Nannie, 133
Time on the Cross (Fogel and Engerman), 10
tobacco industry, 124, 125, 126
   black workers in, 132–133
   as capitalistic operation, 21
   expansion of small plants in, 322
   factory production in, 298–299
   mechanization of, 298–299
   value-added in, 298–299
Tocqueville, Alexis de, 41
Tolstoy, Leo, 39
town histories, 39–41
Toynbee, Arnold, 23, 40
trade, 24, 25, 26–27
trade occupations, 279
transportation system, 286
Treaty of Ghent, 362
trust companies, 271
Tufts, Aaron, 147
Turner, Frederick J., 41, 75
typewriter keyboard training, 131

Ulmer, Melville, 239
unemployment, 337
urban populations:
   black, 196–201
   estimated share of, 202
   male shares of, 181, 183, 186, 192–193
   slaves in, 177, 180
   by state, 192–201
   by state and region, 175–176
   white females in, 194–195
   white males in, 192–193
   see also labor force population, urban

value-added:
   in artisan shops, 299
   in boot and shoe industry, 305, 309
   during Civil War years, 69
   in different methods of production, 296
   entrepreneurial labor in, 309
   in factory production industries, 293–294, 297–299
   in flour mills, 313–314, 316–317, 320
   in iron plants, 302
   in lumber industry, 312–313
   in manufactories, 299
   regional variations in, 309
   in small firms, 321
   in sweatshops, 299
   in tobacco industry, 298–299
   in Western plants, 309
Van Buren, Martin, 57
Veblen, Thorstein, 7, 8

Wabash and Erie Canal, 72
wage differentials, gender-based, *see* gender gap ratio
wage differentials, North/South, 106, 107
wage differentials, race-based, 104–105, 124–130
   job segregation and, 124–125
   in largest industries, 125–130
wage differentials, skill-associated, 47, 49–50
   during Civil War, 92–93
   in early stage of industrialization, 47, 50–51
   after 1870, 96–97
   industry growth rates and, 95–96
   instability in, 87–89
   during late 1850s, 93
   low margins in, 90–91
   during rapid industrial growth, 50
   in regions of recent settlement, 50–51
   savings levels and, 264–265
   secular widening of, 97
   in various occupations, 89–90
wages:
   for agricultural workers, 75–76, 198, 127
   in clerical trades, 161–162
   convict-lease system and, 126
   during depression periods, 78–79
   industrial vs. agricultural, 125
   labor force participation rates and, 140
   local living costs and, 52
   in lumber industry, 83–85
   of male and female labor forces, 150
   in midwestern agriculture, 75–76
   by occupation, 90, 150
   piece-rate, 163
   racial discrimination and, 124–130
   during rapid industrial growth, 50
   on southern farms, 98, 104–105, 107
   for unskilled workers, 72–73, 75, 78–80
   westward migration and, 75
wagon making industry, 295, 299
Walters, A. A., 311
War of 1812, 35
Warren-Pearson wholesale price index, 269
water power, 287, 293, 295, 297
   changing use of, 294
Waverly plantation, 103, 111–112, 120
Wayne, Michael, 99, 102
wealth:
   census estimates of, 270
   component factors of change in, 278
   growth rates for, 259
   nonlinear and linear regressions on, 273
   see also capital accumulation
wealth-income ratio, 258
Webb, Beatrice, 138
Webb, Sidney, 135, 138, 155
Weber, Adna, 175, 182, 185
Weeks Report, 285

Weiss, Leonard, 291
Weiss, Thomas, 12, 171–212, 291
Wellington, Ruth, xv, xvi
Wells, David A., 68–69, 283
whale bone, 338–339
  price of, 342, 346
  uses of, 339
  whale oil production and, 341–342
  whalers' income from, 366
whale oil production:
  decline in market for, 397
  prices in, 338, 341–342, 346
  whale bone production and, 341–342
  see also sperm oil production
whales:
  hunting grounds for, 343–344, 346–347
  overhunting of, 339, 346, 384–385
  supply of, 384
  types of, 339
whaling fleet, New Bedford:
  active vessel years in, 378
  in Arctic, 363, 366
  in Atlantic grounds, 362–363, 367
  average age of, 354–355, 374–377
  average catch of, 340
  baleen vs. sperm hunting in, 367
  barks in, 347–348, 351, 353–354, 355, 359, 360–361
  brigs in, 347, 351, 360–361, 370, 380, 381, 383
  carrying capacity of, 371–374
  catch per vessel ton in, 390–391
  catch specialization in, 365
  class size in, 351–354, 370–371, 396–397
  composition of, 348, 351, 368–369
  condemned vessels in, 359, 360
  crew quality in, 361–362
  crew size in, 349, 360–361
  decline of, 386
  destinations of, 353–354, 370–371, 377
  as dumping ground, 359
  earnings per ton in, 387
  entrants in, 355, 360, 374, 395, 397
  environmental vs. technical factors in, 374
  exits from, 355, 359, 360
  first voyage figures for, 397–398
  growth of, 350–351
  in Indian Ocean, 367
  labor-capital ratios in, 377–381
  labor intensity in, 361
  loss rate in, 360, 381–383
  marginals in, 371
  men per ton in, 379
  merchant ships in, 358
  new barks in, 355–359, 397
  in Pacific grounds, 363
  productivity in, 351–353, 374, 383–386
  rates of return in, 389–395
  relocation of, 388–389
  reriggings in, 353, 355–358, 359, 360, 397
  sailing vessels in, 347, 350–351
  schooners in, 347, 351, 360–361, 370, 380, 381, 383
  ships in, 347–348, 350–351, 353–354, 355, 360–361, 370, 374, 380
  skill levels in, 362
  sloops in, 347, 351, 360–361, 370, 380, 381, 383
  sources of data on, 349
  specialization in, 366
  sperm whale catches by, 389
  technical characteristics of, 352, 356–357
  technology in, 351–353, 374, 397–398
  tonnage in, 343, 372–373
  tonnage relatives in, 354
  transfers to, 355, 360
  vessel age in, 376
  voyage length for, 375, 397
  voyages by vessels in, 364
  whale bone revenues in, 366
whaling fleets:
  crews on, 344–345, 346
  ports for, 342–343
  total capacity of, 342
  world dominance of, 337–338
whaling industry:
  capital investment in, 360
  capital stock shifts in, 360
  captains in, 344–345
  catch composition in, 363–366
  environmental changes and, 346–347
  fleet size in, 344
  historiography of, 24
  intense hunting in, 384–385
  labor turnover in, 362
  market shifts in, 346
  in Nantucket, 366
  organization of, 344–345
  profit rates in, 360
  revenues in, 386–389
  rise and fall of, 337–338, 347
  risks in, 360
  search and capture in, 344–345
  in separate areas, 363
  supply-side shocks in, 337–338
  technology in, 344–346
  tonnage in, 343
  total output of, 343
  voyages as "firms" in, 344
  voyage specialization in, 363–366
Whatley, Warren, 111
Whitin works, 37
Whitney, Eli, 30, 31–32, 38
Wiebe, Robert, 110
Wiener, Jonathan, 98
Williamson, Jeffrey G., 257, 260, 264–265, 279

women workers, 48
  in agriculture, 136
  in clerical jobs, 136, 137–138, 156
  comparable worth and, 138, 139
  in craft positions, 162
  hiring standards and, 66–67
  increased demand for, 138
  job experience for, 137, 158
  job training for, 135, 137, 143
  in manufacturing, 67, 148–149
  mechanization and, 137
  occupational distribution of, 150
  productivity of, 48, 146
  in professions, 137–138, 156
  in segregated occupations, 148–149
  in urban labor force, 182, 194–195
  see also gender gap ratio
Woodward, C. Vann, 122
woodworking, 32
Woofter, T. J., 115, 118
woolen textiles, 298, 299, 303
World War I, 139
Wright, Carroll D., 139, 287, 299, 301, 308
Wright, Gavin, 13, 98–134

Zevin, Robert, 289

About the Book

Quantity & Quiddity was composed in Times Roman by G&S Typesetters, Inc. of Austin, Texas. It was printed on 60-pound Miami Book Vellum paper and bound by Arcata Graphics/Kingsport, Inc. of Kingsport, Tennessee. Design by Joyce Kachergis Book Design and Production, Inc. of Bynum, North Carolina.

Wesleyan University Press, 1987